Interspecific Cor

Oxford Avian Biology Series
Series Editor: Professor T.R. Birkhead FRS (University of Sheffield, UK)

A new series of exciting, innovative contributions from the top names in avian biology. Topics have been carefully selected for their wider relevance to both students and researchers in the fields of ecology and evolution.

Volume 1: Avian Invasions
Tim M. Blackburn, Julie L. Lockwood, Phillip Cassey

Volume 2: Interspecific Competition in Birds
André A. Dhondt

Interspecific Competition in Birds

André A. Dhondt
Edwin H. Morgens Professor of Ornithology, Cornell University

OXFORD
UNIVERSITY PRESS

OXFORD
UNIVERSITY PRESS

Great Clarendon Street, Oxford ox2 6DP

Oxford University Press is a department of the University of Oxford.
It furthers the University's objective of excellence in research, scholarship,
and education by publishing worldwide in

Oxford New York

Auckland Cape Town Dar es Salaam Hong Kong Karachi
Kuala Lumpur Madrid Melbourne Mexico City Nairobi
New Delhi Shanghai Taipei Toronto

With offices in

Argentina Austria Brazil Chile Czech Republic France Greece
Guatemala Hungary Italy Japan Poland Portugal Singapore
South Korea Switzerland Thailand Turkey Ukraine Vietnam

Oxford is a registered trade mark of Oxford University Press
in the UK and in certain other countries

Published in the United States
by Oxford University Press Inc., New York

© André A. Dhondt 2012

The moral rights of the author have been asserted
Database right Oxford University Press (maker)

First published 2012

All rights reserved. No part of this publication may be reproduced,
stored in a retrieval system, or transmitted, in any form or by any means,
without the prior permission in writing of Oxford University Press,
or as expressly permitted by law, or under terms agreed with the appropriate
reprographics rights organization. Enquiries concerning reproduction
outside the scope of the above should be sent to the Rights Department,
Oxford University Press, at the address above

You must not circulate this book in any other binding or cover
and you must impose the same condition on any acquirer

British Library Cataloguing in Publication Data
Data available

Library of Congress Cataloging in Publication Data
Library of Congress Control Number: 2011933729

Typeset by SPI Publisher Services, Pondicherry, India
Printed in Great Britain
on acid-free paper by
CPI Group (UK) Ltd, Croydon, CR0 4YY

ISBN 978–0–19–958901–2 (Hbk.)
      978–0–19–958902–9 (Pbk.)

1 3 5 7 9 10 8 6 4 2

To the memory of my parents

Jan Dhondt and Lea Sevens

Preface

Depending on who you believe, interspecific competition is so important in nature that it only occurs transiently and cannot easily be observed—only its effects can—or interspecific competition is ongoing in nature, can be observed, and its existence and importance can therefore be tested experimentally. Since I discovered evidence in the mid-1970s that interspecific competition occurs in tits, I have used long-term experiments in two Belgian locations (Ghent and Antwerp) to explore the existence and effects of interspecific competition in great and blue tits. After my plenary talk at the International Ornithological Congress in Durban in 1998, John Krebs asked me if I had ever considered writing a book about the subject. I answered honestly that I had not, but that if I ever were to write a book, it would be about interspecific competition. During two sabbaticals that I was granted by Cornell University I spent one semester in Fort Collins in 2002, hosted by John Wiens, and a full year in Montpellier, France, hosted by Jean-Dominique Lebreton, Thierry Boulinier, Jacques Blondel, and colleagues at the CEFE-CNRS in 2008–09. This gave me the chance to conceive the book, and during the second sabbatical to write most of it. What lies in front of you is the result of these efforts in which I have tried to bring together most of the work on interspecific competition in birds and summarize my own.

I am very grateful to a number of colleagues who were willing to read parts of earlier drafts and comment on them critically and constructively. I am particularly grateful to Ian Newton who read more than 90% of the book manuscript and gave very helpful comments indeed, sometimes even taking my text with him on his travels and returning his comments from exotic places. Ian's comments and insight were extremely helpful. Andy Gosler, Ken Otter, Erik Matthysen, Stefan Hames, Christine Cockle, Peter Grant, and Nick Gotelli commented on one or several chapters of the manuscript. Wesley Hochachka and Richard T. Holmes commented on the section in Chapter 6 that attempted to summarize some of their work. I am very grateful to all for their willingness to read and comment. Their comments have certainly improved the book.

Diane Tessaglia-Hymes and Susan Spear from the Cornell Laboratory of Ornithology redrew Fig. 7.2 so that I could include it. Christine Cockle, Katie Martin, and Katie Aitken generously provided me with unpublished manuscripts, making it possible to include their important work in my book. Many colleagues provided me with electronic versions of their graphs so that I could alter them to make the illustrations more homogeneous. I have always admired the drawings of Robert Gilmore. I am, therefore, very grateful to him that he allowed me to reprint one of his drawings from David Lack's 1971 book here as Fig. 1.1.

I am grateful to Helen Eaton at Oxford University Press for her patience and for always answering my queries immediately; to Cornell University for the sabbatical leaves that were essential in giving me the time off to structure my thoughts and write without too many interruptions. Finally I am very grateful to my wife Keila for believing that I would be able to finish this effort and encouraging me.

Ithaca, April 2011.

Contents

Introduction 1

1.1 The study of biotic interactions in nature 1
1.2 Criticism as to the importance of interspecific competition 2
1.3 Tits to the rescue 4
1.4 The paradox of competition as illustrated by Kluijver and Lack 8
1.5 The conflict on the importance of interspecific competition in North America 11
1.6 Conclusions 11

2 Definitions, models, and how to measure the existence of interspecific competition 13

2.1 Definitions: effects on individuals or populations? 13
2.2 Models and equations: logistic, theta logistic, and Lotka–Volterra 16
2.3 Conclusions 22
2.4 The structure of the rest of the book 24

3 Space as a limiting resource 25

3.1 Introduction 25
3.2 The Buffer Hypothesis was developed from studies of tit populations and is probably generally important 25
3.3 Winter social organization determines when space is limiting 28
3.4 Interspecific territoriality 34
3.5 Conclusions 37

4 Food as a limiting resource 39

4.1 Introduction 39
4.2 The classical case of beech mast: correlation is not causation 40
4.3 Experimental evidence that food does actually influence winter survival or the size of the following breeding population 42

4.4	Behavioural responses to winter cold and predation risk: costs and benefits of flocking	49
4.5	Individual responses to managing body fat reserves in the context of food availability and predator presence	50
4.6	Pre-breeding food supplementation effects on reproduction	55
4.7	Food manipulations during the breeding season	57
4.8	Predation by birds and other taxa can reduce food availability and thus have indirect effects	63
4.9	Food supplementation experiments as a conservation tool	65
4.10	Conclusions	66

5 Nest sites as a limiting resource 69

5.1	Are nest sites limiting in cup-nesting species?	69
5.2	Are cavities limiting for cavity nesters?	72
5.3	Are cavities in natural forests superabundant?	73
5.4	Studies of nest web communities	77
5.5	Conclusions	80

6 The effect of intraspecific competition on population processes 83

6.1	Intraspecific competition seems to be generally important in birds	83
6.2	Case studies show variation in what processes are affected by density-dependence	84
6.3	Density-dependence in introduced populations	91
6.4	Mechanisms resulting in density-dependence: the importance of habitat heterogeneity	92
6.5	Density-dependence in titmice	94
6.6	Conclusion	101

7 Studies of foraging niches and food 103

7.1	The early studies of foraging behaviour emphasized differences between species	104
7.2	In the 1970s observational arguments were used to document the existence of interspecific competition. These arguments only convinced the believers	105
7.3	Field and cage experiments provided conclusive evidence as to the effect of interspecific interactions on the foraging niches used	108

7.4	Measures of fitness-related traits are needed, however, to prove the existence of interspecific competition	109
7.5	The story of the coal tit on Gotland: alternative explanations can be right	111
7.6	Altitudinal replacement of closely related species	113
7.7	Seasonal variation in niche overlap	114
7.8	Effects of migrants on residents	115
7.9	Conclusions	116

8 Field experiments to test the existence and effects of interspecific competition — 117

8.1	Effect of manipulation of cavities available on reproductive or foraging success of presumed competitors (Table 8.1)	119
8.2	Effect of resource manipulation on population size of presumed competitors: effects on single species (Table 8.2)	125
8.3	Studies of communities of cavity nesters: experiments in which natural cavities were blocked or nest-boxes added generated a diversity of results (Table 8.3)	128
8.4	Interactions between cavity and open nesters: does adding nest-boxes influence the density of open-nesting species? (Table 8.4)	135
8.5	Effects of direct removals on habitat use and population size of subordinate species (Table 8.5)	139
8.6	Competitive interactions between birds and species of a different class	145
8.7	Competition between burrow-nesting seabirds can have a severe impact on numbers: application of our understanding of interspecific competition for conservation (Table 8.9)	156
8.8	Heterospecific aggression and interspecific territories	158
8.9	Heterospecific attraction	160
8.10	Conclusions	168

9 Long-term experiments on competition between great and blue tit — 171

9.1	Interspecific competition in tits: the origin of the idea	172
9.2	Is winter competition between great and blue tit for roosting sites only, for food only, or for both resources?	176
9.3	Experimental manipulations to vary the intensity of intra- and of interspecific competition	179
9.4	Effects of intra- and interspecific competition on blue tit density and demographic variables	182

9.5	Effect of intra- and interspecific competition on great tit density and demographic variables	190
9.6	How similar are the results of experimental and correlational studies?	192
9.7	Density and dispersal	193
9.8	What have we learned about competition between blue and great tit?	196
9.9	Concluding comments	200

10 Evolutionary effects of interspecific competition — 203

10.1	Ecological character release and the Niche Variation Hypothesis	203
10.2	Testing the criteria for ecological character release	206
10.3	How rapidly can interspecific competition cause evolutionary changes in morphology? Observational data	211
10.4	How rapidly can interspecific competition cause evolutionary changes in morphology? Experimental data on selection pressures and evolutionary change	213
10.5	Community composition and interspecific competition	216
10.6	Interspecific competition and life-history traits	217
10.7	Conclusions	221

11 Concluding thoughts — 225

Appendix 1—Common and scientific names of bird species mentioned in the text	229
Appendix 2—Common and scientific names of other species mentioned in the text	233
Appendix 3—Detailed results of analyses summarized in Chapter 9. All pertain to the Ghent and Antwerp study sites in Belgium	234
References	245
Index	275

1
Introduction

> It is difficult to perceive a real sense of progress in our understanding of the role of competition in nature. We cannot point to an accretion of knowledge, to empirical questions clearly framed and unequivocally answered; rather we are left with an impression of a subject moving from one world view to another depending on the influence of the prevailing protagonists
> (Law and Watkinson (1989)).

1.1 The study of biotic interactions in nature

One of the challenges of the future is to maintain biodiversity. A very large number of factors influence global biodiversity. At a more local scale, biodiversity can be maintained by managing communities in ecosystems. To do that effectively one needs to understand how populations, communities, and ecosystems function. Organisms interact with their biotic and abiotic environment. The outcome of these interactions generates a dynamic equilibrium that determines community structure and species richness. Three types of between-species interaction contribute to this dynamic flux: competition, predation (including parasitism; see Raffel *et al.* 2008), and mutualism. All exert powerful selection pressures on organisms and all shape communities. Although I am not sure I would go as far, some authors believe that of these interactions, competition is the most important; or as Paul Keddy phrased it *'As one of the three major kinds of biotic interactions we may anticipate that it could be a force as fundamental to ecosystems as gravity is to planetary systems'* (Keddy 1989). He defined competition as *'the negative effects which one organism has upon another by consuming, or controlling access to, a resource that is limited in availability'* (Keddy 1989). The word 'consuming' refers to exploitation competition, while the words 'limiting access to' refer to competition by interference. I will discuss definitions of competition in more detail in Chapter 2.

There are a couple of important differences, though, between studies of competition, predation and parasitism, and mutualism. When a predator takes prey and eats it, when a flea sucks blood from its host, or when a zebra eats grass we do not really need to prove that the interaction took place: we can directly observe it. Predator–prey studies therefore concentrate on the effects of the interactions: are prey numbers reduced by predation? How do predators respond to changes in prey abundance? What responses do hosts develop to infection? How do pathogens overcome host defences? Can plants profit from being grazed upon? In contrast, a major part of the research on mutualism and on competition in animals simply tries to answer the question: can we prove that the interaction takes place at all, and can interactions between species vary in type and outcome (Bronstein 1994)? For intraspecific competition we often ask: 'how common is it?' (Sinclair 1989), and why is it detected in some cases but not in others (Dhondt *et al.* 1992)? As regards interspecific competition, reviews do not always agree on the interpretation of the evidence, as illustrated by the notorious discussion between Schoener (1983) and Connell (1983). For logistical reasons many studies of interspecific competition even stop when an effect on only one of the competing species has been demonstrated, whereas strictly speaking individuals from both species should be adversely affected by the interaction (see Chapter 2). Furthermore, the quality of many of the older field experiments is dubious as only a small percentage of field experiments addressing interspecific competition were carried out with adequate replication and controls and lasted more than a few weeks (Hairston 1989; Wiens 1989).

Another important difference between predation and competition is that experiments on predator–prey, host–parasite, or herbivore–food interactions usually study mechanisms, fitness effects, and evolutionary responses, while this is rarely the case for interspecific competition. Furthermore, when we look at field experiments testing the existence and effects of competition, not only do we only rarely look comprehensively at how the populations involved are affected, but, with the exception of Dolph Schluter's work on sticklebacks,[1] the question of possible evolutionary effects in ecological time is not asked (Schluter 1994).

1.2 Criticism as to the importance of interspecific competition

Whereas some authors are convinced that interspecific competition is a very powerful interaction in nature (see Keddy's citation above) many others have severely criticized the uncritical approach taken in accepting interspecific competition as a major and pervasive force structuring communities. They reject the poor quality of field experiments and their lack of integration in an overall framework. This leaves the question of how important interspecific competition in nature really is unanswered. Connell (1980) explicitly concluded: '. . . *until some*

---

[1] Scientific names of species mentioned in the text are given in Appendix 2 (birds), and in Appendix 3 (other species).

strong evidence is obtained from field experiments along the lines suggested above, I will no longer be persuaded by such invoking of 'the Ghost of Competition Past'.

The critics' arguments are:

1. **Circumstantial evidence from descriptive data cannot be used to exclude alternative hypotheses**. Circumstantial evidence, therefore, does not provide compelling evidence that competition is truly important and that interspecific competition is more likely than some alternative explanation. Thus, the observation that different closely related bird species use different substrates for foraging can reflect the existence of interspecific competition in the past, the current existence of interspecific competition, or simply the fact that they are different species. If the observational data do not fit predictions from interspecific competition they can be used to reject the existence of interspecific competition (Wiens 1977). If they conform with the predictions of interspecific competition they can, at best, be used to generate hypotheses that need to be tested experimentally (Underwood 1986).

2. **Interspecific competition may not occur continuously**. Although interspecific competition may have helped to shape community structure in the past, it may not be possible to detect it as a current process. Alternatively, current interspecific competition may only occur occasionally, and there is no need to assume that it plays an important role in current communities. Wiens (1977) believed that, because interspecific competition will only take place when resources become limiting, interspecific competition may be rare, and hence be relatively unimportant. David Lack used a similar argument as concerns the importance of spring territorial behaviour in limiting great tit numbers in Wytham Woods, Oxford. He argued that great tits were not limited by spring territorial behaviour, because the size of the breeding population in 1961 was very much larger than in any of the previous 15 years. Therefore, if territorial behaviour limited numbers, that could only have been the case in 1961. He then concluded that there was no reason to assume that even in 1961 spring territorial behaviour had limited numbers ((Lack 1966) see also (Chitty 1967)). He was later proved wrong (see Chapter 3).

3. **The quality of field experiments testing for the existence of interspecific competition is often poor**. Many experimental tests of interspecific competition in the field are unreplicated, lack adequate controls, are poorly executed, and are carried out in non-natural conditions. They cannot, therefore, be used at all, or cannot be used to draw inferences as to the generality and/or importance of interspecific competition (Hairston 1989; Underwood 1986; Wiens 1989).

4. **Field experiments testing for the existence of interspecific competition lack a comprehensive approach**. Very few experiments test for the existence

of intraspecific competition along with interspecific competition, or document which resources are limiting. Few also test alternative hypotheses (Underwood 1986; Wiens 1989), or reciprocal effects between species (Keddy 2001).

5. **The species used for testing the existence of interspecific competition are a biased sample.** The species chosen to test experimentally the existence of interspecific competition are biased in that experiments tend to be carried out using species for which there is a priori evidence that interspecific competition would be very likely (Connell 1983; Keddy 2001; Schoener 1983; Underwood 1986). Again, while the results may well demonstrate interspecific competition in the species concerned, they cannot be used to draw conclusions as to how general the role of interspecific competition is in nature.

The critics therefore claim that it is difficult to draw any conclusions as to the frequency at which interspecific competition occurs in nature, but especially as to its importance in structuring communities and traits of individual species. Wiens (1989: p. 41–42) explained:

'Good experiments are founded on good natural history. An experiment designed to explore the possibility of competitive interactions among species, therefore, should be conducted in situations in which previous observational studies have provided at least suggestive evidence of competition. The extent to which experiments are designed and conducted in this manner, the literature will contain an intrinsic bias favouring competition.... Attempts to discern the frequency of competition in nature from the frequency with which it is reported in experimental studies are of doubtful value.'

The same holds for experiments on any other potentially limiting factor in which the study species were chosen on the basis of prior experience (Newton 1998).

1.3 Tits to the rescue

Between about 1950 and 1980, tits, species in the family Paridae, have played a major role in the study of density dependence and population regulation which is, after all, the study of the impact of intraspecific competition on population processes and population size. Three of David Lack's ideas that were clearly formulated in his ground-breaking 1954 book (Lack 1954), based in part on his research on tits, were central to this debate. The first idea is that through natural selection birds have been selected to lay an optimal clutch size, which is the number of eggs that will result in the largest number of surviving offspring. The second is that populations are limited through food shortage during winter. The third was Lack's conviction that breeding population size is not limited through spring territorial behaviour. Since about the mid-1970s, tits have also played a major role in the debate concerning interspecific competition. A lot of this latter work has been experimental. Tits are, therefore, a useful group to comprehensively evaluate the role interspecific competition plays in nature (criticism 4 above), even though, in the absence of appropriate studies, it may not be safe to extrapolate the conclusions to other species.

Until about the mid-1970s most studies on tits emphasized the absence of current interspecific competition, emphasizing mechanisms that allowed coexistence by its avoidance. This was largely the result of David Lack's view that closely related species could coexist because they are ecologically segregated and therefore do not compete with one another. This view was influenced by Gause's book (Gause 1934) and by laboratory experiments that tested the mathematical models of Lotka and Volterra (Lotka 1925, 1932; Volterra 1926) on interspecific competition. The importance of the Lotka–Volterra models for interspecific competition was that they predicted that in certain conditions—when intraspecific competition was stronger than interspecific competition—two competing species would be able to coexist, even when conditions remained constant. Gause showed that in certain experiments one species would out-compete the other, whereas in other experiments two competing species could coexist, as mathematical theory predicted. He concluded that closely related coexisting species usually differed in niche (1934, p. 19):

'It is admitted that as a result of competition two similar species scarcely ever occupy similar niches, but displace each other in such a manner that each takes possession of certain peculiar kinds of food and modes of life in which it has an advantage over its competitor. Curious examples of the existence of different niches in nearly related species have recently been obtained by A. N. Formosov ('34). He investigated the ecology of nearly related species of terns, living together in a definite region, and it appeared that their interests do not clash at all, as each species hunts in perfectly determined conditions differing from those of another. This once more confirms the thought mentioned earlier, that the intensity of competition is determined not by the systematic likeness, but by the similarity of the demands of the competitors upon the environment.'

In the middle of the debate about the validity of Gause's views in the 1940s, Lack reviewed factors allowing closely related species to coexist (Lack 1944) and started his research on tits in England as part of his campaign to try and convince the world that closely related species could coexist in the same habitat *because they avoided* interspecific competition by using different resources, or by using the same resources in different ways. In Lack's view, interspecific competition may have been transiently important at some time in the evolutionary past, but competing species either excluded one another or rapidly evolved traits that eliminated interspecific competition (see below). This idea remained an important theme throughout his career, culminating in his 1971 book in which, for a suite of bird families, he explained how species avoided interspecific competition by ecological isolation (Lack 1971).

Fig. 1.1 In Wytham Woods near Oxford (UK), multiple tit species coexist. David Lack believed they could coexist because they used different resources or the same resources in different ways. This is illustrated in this figure (drawn by Robert Gillmor), in which it is illustrated that great, marsh, and blue tits preferentially forage at different heights, and that blue tits often forage hanging upside down from twigs, while other species do this less frequently. From Lack (1971, p.23) with permission from Robert Gillmor.

Thus, the first studies discussing interspecific competition in tits were carried out to document that different species of coexisting tits *did not compete*. In 1979, Christopher Perrins, one of David Lack's students and a leader in the study of tit ecology, concluded his book chapter on interspecific competition among tits (p. 98) by re-emphasizing ecological segregation among the species that coexist: *'Over long periods of time, natural selection has favoured those individuals of a species which have evolved ways of life which differ from those of other species; in this way interspecific competition has been avoided'*. He pointed out, however, that both within and between tit species there continue to be fights over food and over nest sites, in which the larger species normally have the upper hand (Perrins 1979).

Although a few of the earlier papers mentioned the probable existence of interspecific competition amongst tit species (between coal tit and blue tit in pine forest (Gibb 1960); between great and blue tit (Kluyver 1966)), my 1977 paper (Dhondt 1977) was the first to provide non-experimental results showing significant inverse correlations between blue tit density and various components of great tit reproductive success, and weaker inverse correlations between great tit density and some aspects of blue tit reproductive success. After the late 1970s, the number and diversity of studies of interspecific competition in tits increased dramatically, especially in Europe, and many field experiments tested predictions of interspecific competition.

Tits are a good model group to evaluate the importance of interspecific competition in a comprehensive way, because studies on interspecific competition in tits were started even though it had been assumed for 25 years or more that tits coexisted in the same habitat precisely because they avoided interspecific competition. Studies in this group, therefore, are not biased in favour of discovering evidence for the existence of interspecific competition.

There are now a great number of observational and experimental studies carried out by many different scientists in many different geographic areas using a fair number of different, but closely related, species that allow a critical evaluation of the role of interspecific competition in tits. By using both experimental and observational data, my goal in this book is to evaluate the combined body of evidence available for tits as a group, and to determine the intensity and importance of interspecific competition in this well-studied group. I will, however, exhaustively refer to relevant work in other bird species, to explore the generality of conclusions based on tit studies.

So who is right? Is David Lack correct in claiming that tits can coexist because they mostly avoid interspecific competition; is Rauno Alatalo (1982, p. 315) right in his conclusion *'that the present evidence for interspecific competition between tits is quite strong'* (Alatalo 1982), or is Underwood right in his conclusion that because none of the experiments involving tits has been performed adequately we cannot conclude anything about the importance of interspecific competition in tits (Underwood 1986)?

My hope is that this book will clarify these issues. Although I initially intended to cover interspecific competition among organisms in general, I decided to limit myself primarily to competition among birds, often using the well-studied tits as examples. Like David Lack, I believe it is wise to limit oneself to a group with which one has personal experience, making it easier to critically evaluate the evidence (Klomp 1967).

1.4 The paradox of competition as illustrated by Kluijver and Lack

The controversy about the importance of interspecific competition has an interesting history. On 21 March 1944 the British Ecological Society held a meeting to discuss to what extent '*Gause's contention that two species with similar ecology cannot live together in the same place...*' (Anonymous 1944)[2] was valid and to what extent, therefore, interspecific competition was an important force in nature. '*A distinct cleavage of opinion revealed itself*'. The majority opinion was that there existed no real evidence that interspecific competition was really important, and that closely related species could live together without competing, although a small group (Lack, Elton, Varley) provided arguments and examples supporting the so-called 'Gaussian concept'. David Lack argued that interspecific competition was so important that we could only rarely observe it, because competition was transient. He argued that when two geographic races came together after having been isolated this would lead to one of four outcomes: (1) elimination of one by the other; (2) elimination of one in the area of overlap, but its survival in parts of the range where it was better adapted, leading to geographical replacement; (3) habitat separation due to adaptation to different zones of the ancestral habitat; (4) cosurvival resulting from divergences in food preferences (Anonymous 1944); see also (Lack 1944). Thus, although we could not often observe competition in action, we could observe its effects. And this was what the Principle of Gause was all about: species living in the same habitat and using similar resources to coexist needed to use different resources or the same resources in different ways.

For a number of years Lack devoted his considerable talents to demonstrating that this was true in cases in which the less careful observer would assume that the coexisting species did actually use the same resources: Darwin's finches on the Galápagos islands differed in beak size, and hence used seeds of different sizes (Lack 1947); cormorants and shags, although fishing in the same sea, dove to different depths and hence avoided competition by eating different fish and invertebrate species (Lack 1945). (Recent year-round work confirms that during

---

[2] The Anonymous reporter of the meeting was David Lack himself (C.M. Perrins, pers. comm.).

the breeding season, when the distribution of the two species overlaps most, the diet similarity is smallest (Lilliendahl and Solmundsson 2006). The jewel on the crown became Lack's tit studies. In Wytham Woods the five tit species were able to coexist by using different foraging niches, thereby avoiding competition. He further elaborated on his ideas in his 1971 book *Ecological Isolation in Birds* (Lack 1971). In this book he explained in detail for family after family of birds what mechanisms they used to *avoid* competition.

David Lack's idea—that interspecific competition is so important that it can only rarely be observed, because when it occurs it will be transient—is what the paradox of competition is all about. This idea was predominant in the 1950s as illustrated by a quote from Brown and Wilson (1956) who wrote, when reviewing character displacement: '*However, interspecific competition of the direct conspicuous, unequivocal kind is apparently a relatively evanescent stage in the relationship of animal individuals or species, and therefore it is difficult to catch and record*' (p. 60). '*What we usually see is the result of an actually or potentially competitive contact, in which one competitor has been suppressed or is being forced by some form of aggressive behaviour to take second choice, or in which an equilibrium has been established when the potential competitors are specialized to split up the exploitable requisites of the environment.*'

According to this view, proving directly or experimentally that interspecific competition occurs would therefore be nearly impossible and the only arguments that can be invoked to support the tenet that interspecific competition is an important force in nature are indirect ones (Dhondt 1989b). I will address these indirect arguments in Chapter 7, elaborating on the idea that we need to invoke the powerful force of interspecific competition in order to understand and explain large-scale patterns of community structure (see also Section 10.5).

In the Netherlands, Huib Kluijver's approach, initiated in the late 1930s (and building on Wolda's nest-box studies begun in 1912), was to actually try to measure possible effects of intraspecific (and interspecific) competition on life-history traits. When Kluijver published his series of papers on great tit population ecology and behaviour ((Kluijver 1950; Kluijver 1951; Kluyver 1952)[3] together close to 200 pages), there was no evidence from any field study that density dependence could regulate numbers. In the great debate between Andrewartha and Birch, on the one hand, and Nicholson and Lack on the other, experimental data were lacking (Andrewartha and Birch 1954; Lack 1954; Nicholson 1933). Density dependence was a necessary condition for population regulation to occur in the Nicholsonian sense, but other than verbal arguments about the logical necessity of density dependence there was little data to support this idea, and certainly no compelling field data. In his 1951 paper Kluijver, not surprisingly,

---

[3] The spelling of the name Kluijver or Kluyver varies between publications. I use the spelling as in the paper cited.

concentrated on documenting that intraspecific competition could be shown by establishing the existence of density-dependent reproduction. His figure 7 (here reproduced as Figure 1.2) shows a clear inverse relation between breeding density and fecundity. (Kluijver points out that the relationship is not linear but hyperbolic: 1951, p. 79). As concerns the existence of interspecific competition, Kluyver (1966, p. 390) wrote '*In the Great Tit it is not only the numbers of Great Tit which play a role in its population dynamics, but also those of the related Blue Tit* Parus caeruleus, *which competes to a certain extent in food and nesting sites. I will restrict, however, my further considerations to the intraspecific aspect of the problem, as I consider it to be the most important aspect*'. Knowing how careful Kluyver was in his work, he must have had evidence that blue tit density also influenced great tit reproduction, but he never came back to this. When I visited him after his retirement, in the late 1960s, he showed me some of his impressive unpublished data on dispersal on Vlieland (as I was then very interested in dispersal; it would take me another eight years to become interested in interspecific competition). When I asked him why he did not publish these results he said 'I only write up things when I am asked to give a talk'. (I wish more people had invited him to give talks!) Kluijver's 1951 paper (using data collected 1930–34), therefore, documented for the first time in any field study that reproductive output per pair per season declined as density increased. Although both Lack and Kluijver agreed

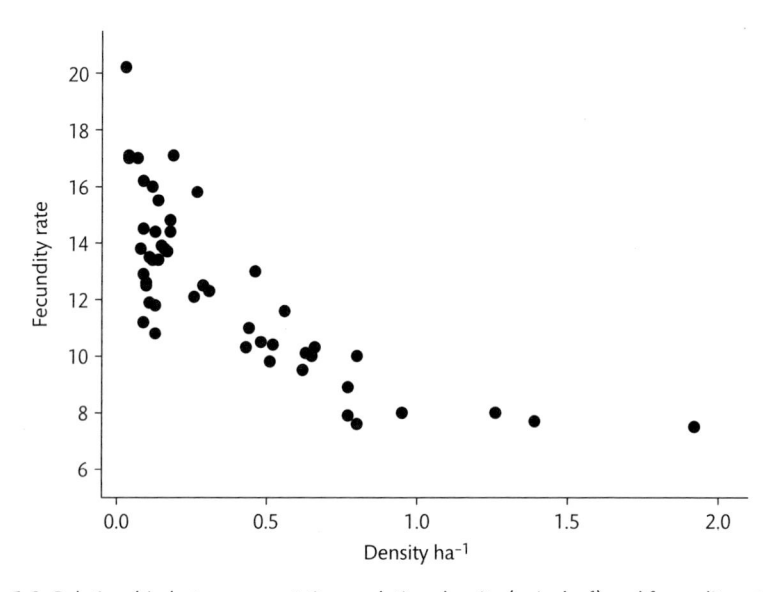

Fig. 1.2 Relationship between great tit population density (pairs ha^{-1}) and fecundity rate (mean number of eggs laid per pair and per season) based on data from 16 areas in the period 1930-34. Redrawn using data in Table 27 from Kluijver (1951).

that the different tit species avoided competition during winter by using different foraging niches, Kluijver (1951, p. 109–110) concluded that food competition between the different species existed to a limited extent in winter. Likewise, both Lack and Kluijver agreed that during the breeding season this '*ecological differentiation disappears, and the nestlings of all four species are fed on the same insect species*' (Kluijver 1951, p. 110), but their interpretation of the same observation differed strongly. Kluijver (1951, p.110) wrote: '*Hence there might be competition for food in spring*', whereas Lack (1966) argued that food in spring was superabundant and that, therefore, interspecific competition did not take place. This disagreement underlines the necessity to test hypotheses experimentally.

1.5 The conflict on the importance of interspecific competition in North America

In the late 1960s and 1970s discussions about interspecific competition in the North American literature became very closely linked to discussions about community structure. The 'clash-of-the-giants' interspecific competition debate focused on whether or not interspecific competition was the main force structuring communities, and about what type of evidence was sufficient to prove it. On one side of the debate were MacArthur, Cody and Diamond, and on the other Wiens, Simberloff and later Hairston, joined by Underwood from Australia. The MacArthur camp (like Lack) championed indirect arguments and concluded that, given a sufficient number of examples of communities that seemed as if they were structured through interspecific competition (occurring in the past), interspecific competition was the main force structuring communities. The other camp argued that unless sufficient, high quality, experimental evidence was available, that conclusion was not supported. Furthermore, Wiens (1977) emphasized that although interspecific competition might occur, it was unlikely to occur continuously, and hence could not structure communities as proposed by the MacArthur camp. In contrast, since predator–prey interactions were continuous, that biotic interaction was more likely to play that role. It is only relatively recently that evidence that interspecific competition can cause rapid evolutionary change has become available (see Chapter 10), which would have provided a mechanism needed to support the hypothesis that interspecific competition can impact community structure.

1.6 Conclusions

There is a clear disagreement in the literature concerning the possible existence of current and ongoing interspecific competition in natural populations. If the

'Ghost of Competition Past' argument is correct, that is that strong competition in the past *'has acted as a sieve, leaving behind species with different niches'* (Law and Watkinson 1989) then interspecific competition in current populations will be at best weak and intermittent. On the other hand, if interspecific competition regularly occurs in extant populations, manipulative field experiments will confirm its existence. Thus if one observes that coexisting species use niches with no or limited overlap, as illustrated in Fig. 1.1, and removal of one of these species does not result in niche expansion of the remaining one(s), this will reinforce the idea that current resource partitioning reflects the ghost of past competition. If, on the other hand, removal of the dominant species (manipulative experimentation) results in the subordinate species shifting to occupy the vacated niche, this would suggest that competition is current and ongoing and reinforces resource partitioning. In the former case field experiments will not help to detect the existence of interspecific competition, while in the latter case manipulative field experiments will prove its existence (Dhondt 1989b). Field experiments, therefore, can distinguish between the alternate hypotheses that current interspecific competition is or is not an important force in nature. As I develop my argument in this book I will therefore rely heavily on results from field experiments that have tested the hypothesis that interspecific competition occurs. In Chapters 8 and 9 I review these experiments.

2
Definitions, models, and how to measure the existence of interspecific competition

2.1 Definitions: effects on individuals or populations?

It is fascinating to go through ecology textbooks and compare definitions of interspecific competition. Their differences underline why I believe that a set of equations is more useful than verbal models as a starting point for definitions and for identifying ways to measure the existence and effects of interspecific competition. *Intraspecific competition* differs from *interspecific competition* in that in the former the competing individuals belong to the same species while in the latter competing individuals belong to different species.

In both cases interaction can be asymmetric when one individual is dominant over another. Complete asymmetry of the effects on individuals of different species is called *amensalism* (also called asymmetrical competition by Connell (1983)): fitness of one of the competing individuals is reduced, while fitness of the other is not affected (a '0,-' interaction). When competitive interactions involve individuals belonging to many different species, the cumulative effect of many species on one is called *diffuse competition*.

In the second volume of his important book on bird communities, John Wiens discusses a selection of definitions of interspecific competition in great detail (1989). Some definitions propose that interspecific competition operates at the individual level, whereas others define interspecific competition as operating at the population level. Very few claim it operates at both levels. Most definitions state or imply that there are two types of competition (exploitation or scramble, interference or contest) that differ both in their mechanisms and in their effects. In exploitation competition, individuals that use the same resources do so without interfering with one another, but reduce resource availability for other individuals simply by exploiting them. Thus when in Madagascar Verreaux's sifaka feed on fruits of kapok trees in daytime and Madagascan flying foxes feed on them at night they compete for the same food, although never encountering one another. This is exploitation competition. In interference competition there is usually a behavioural component. Competition for space through territorial behaviour is an example of interference competition.

Competition is defined through its effects (on population size, on population growth rates, on individual fitness) or by pre-existing conditions (limiting resources). Finally, Wiens concludes that amensalism (a '0,-' interaction) and competition (a '-,-' interaction) should be included in the same definition. I believe they should be treated separately, if possible, because the evolutionary effects of competition (both species will evolve to avoid it) and of amensalism (only the species suffering a fitness effect will evolve to avoid it) are different.

Wiens summarizes that for interspecific competition to exist the competing species must share resources and the joint exploitation of those resources and/or interference related to the resources must negatively affect the performance of individuals of either one or both species; often these effects have population consequences as well (p. 7). Rephrasing these conditions into a definition, Wiens (1989) writes: '*Interspecific competition is an interaction between members of two or more species that, as a consequence either of exploitation of a shared resource or of interference related to that resource, has a negative effect on fitness-related characteristics of at least one of the species.*'

Explicitly expanding Wiens' definition discussed above, which emphasizes that interspecific competition should primarily be studied through its effects on individuals, Ian Newton (1998, p. 320) writes:

'At the population level competition can thus be defined as a reduction in the distribution or numbers of one or more species that results from their shared use of the same resources. It may involve depletion, where individuals of one species reduce the amount available to individuals of another species. Or it may involve interference, where individuals of one species, by aggressive or other means, reduce access to a resource by individuals of another species.'

His approach was to see whether individual fitness effects would translate into effects on population size or distribution of breeding populations. This is not necessarily the case even when interspecific interactions result in a reduction in reproduction of survival rate in a local population, if compensatory mortality or increased immigration result in there being no overall change in population size. He writes (p. 327):

'As with any other adverse factor, declines in individual performance are not necessarily translated into declines in subsequent breeding numbers.... In the rest of this chapter, therefore, I shall concentrate on those forms of evidence that implicate interspecific competition in limiting the distribution or breeding densities of birds.'

The point Newton makes is an important one and one can ask to what extent Newton's definition and more traditional definitions, such as that formulated by Wiens (see above), are different. I will return to this in Section 2.2 when I discuss models.

Using Newton's expanded definition, only a few experimental tests demonstrate the existence of interspecific competition, since very few experiments have lasted long enough to measure effects on population size. Most experiments to test for the existence of interspecific competition measured whether, with increasing density, some population process (usually related to reproduction) was adversely affected. Only a small number of field experiments measured both effects on numbers and effects on population processes that would explain how the change in numbers could have come about. On the other hand a lot of the data that are traditionally used as indirect evidence for the existence of interspecific competition—allopatric distributions of closely related species or inverse fluctuations of abundance of species that are suspected to compete—are rejected as unconvincing evidence by the critics of interspecific competition (see Chapter 1).

There are situations, however, when non-experimental data can be used to document effects of interspecific competition on distribution or abundance (Underwood 2009). Inverse changes in abundance between two species observed across large scales can form strong evidence for the importance of interspecific competition, especially if patterns of change vary over time. One example would be as follows: as one species expands its range or increases in abundance (either naturally or through human intervention), another species declines either simultaneously or just after, while the opposite happens at a later time. House sparrow numbers in Eastern North America declined as house finch abundance gradually

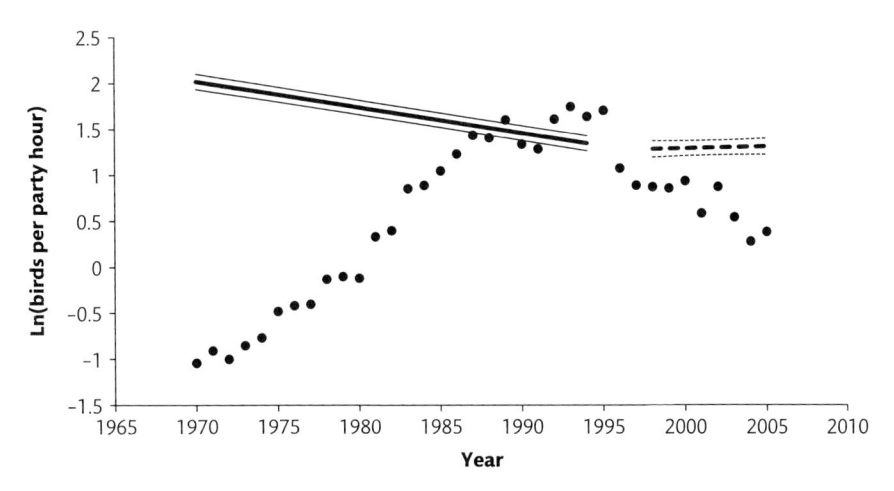

Fig. 2.1 Solid circles represent the mean annual abundances of house finches based on data from the Christmas Bird Count in nine north-eastern US states. After their introduction house finches initially increased (1970–95), but later decreased because of a disease epidemic. House sparrow abundance decreased during the period of house finch increase (the solid thick lines; thinner lines are the 95% confidence limits), but started to increase when house finches declined (dashed thick line; thinner dashed lines are the 95% confidence limits) suggesting the existence of interspecific competition between these two introduced seed eating species. After Cooper et al. (2007), with permission from the Ecological Society of America.

increased across that region. Some authors argued that this decline was the result of interspecific competition (Bennett 1990; Wootton 1987), while others questioned this (Kricher 1983). When house finch numbers declined by half because of the epidemic of the bacterial pathogen *Mycoplasma gallisepticum* (Hochachka and Dhondt 2000), the further decline in house sparrow numbers stopped, supporting the idea that the two species do compete with each other (Cooper *et al.* 2007). Note that these authors tested both for intra- and for interspecific competition: they found that change in house sparrow numbers was explained both by house finch and by house sparrow numbers. Some of the results of Cooper *et al.* 2007 are illustrated in Fig. 2.1.

2.2 Models and equations: logistic, theta logistic, and Lotka–Volterra

In this section I discuss various mathematical models related to intra- and interspecific competition. If the reader is worried about equations, it is possible to skip these, and simply study their graphical representations in Figs 2.2 to 2.5 and then proceed to Section 2.3.

2.2.1 Linear models

The basic model describing intraspecific competition is the logistic regression model first formulated by Verhulst (Verhulst 1838, 1845, 1847) and later reinvented by Pearl and Reed (1920).

$$\frac{dN_1}{dt} = r_{m1}.N_1[1-\frac{N_1}{K_1}] \qquad (2.1)$$

The idea is simply that as the population size of species 1 (N_1) increases, per-capita growth rate of species 1 $\left(\frac{dN_1}{N_1.dt}\right)$ decreases at a constant rate because of an ever increasing adverse effect of intraspecific competition on per-capita growth rate as the size of the population N_1 approaches its equilibrium value K_1 [1]. When the $N_1 = K_1$ growth rate becomes zero, the population has reached its equilibrium value and numbers no longer increase. r_{m1} is the maximal per-capita growth rate.

There are three ways to plot this equation (Fig. 2.2): population size (N) against time (t) which generates the well-known sigmoid curve (Equation 2.1); population growth rate $\left(\frac{dN}{dt}\right)$ against population size (N), which generates a parabolic curve;

---

[1] K is often incorrectly called 'carrying capacity'. See Dhondt, A. A. (1988). Carrying-capacity - a confusing concept. *Acta Oecologica-Oecologia Generalis* 9, 337–46.

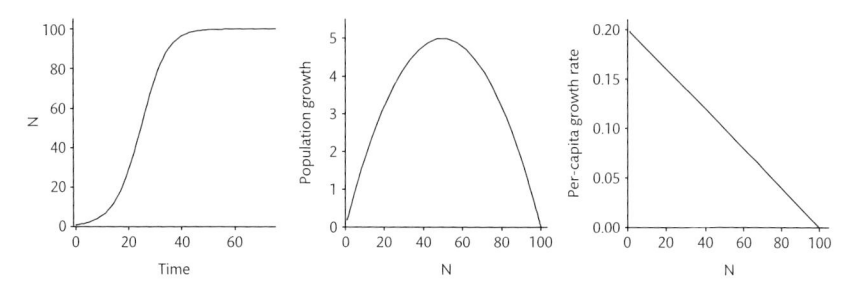

Fig. 2.2 Three representations of the logistic model for population growth. On the left the plot of population size (N) against time (t); in the middle the plot of population growth ($N_{t+1} - N_t$) as a function of population size (N_t); on the right the plot of per-capita growth rate against population size. In this example K=100 and r_m = 0.2; hence the inflection point is at N = 50.

and a plot of per-capita growth rate $\frac{1}{N_1} \cdot \frac{dN_1}{dt}$ against population size (N) (Equation 2.2).

$$\frac{1}{N_1} \cdot \frac{dN_1}{dt} = r_{m1}[1 - \frac{N_1}{K_1}] \qquad (2.2)$$

Because in the logistic model this latter plot is a straight line, I call this a linear model. What it basically means is that whatever the size of the population, each individual that is added to the population decreases the per-capita growth rate $\left(\frac{dN_1}{N_1 \cdot dt}\right)$ by the same amount $\frac{r_m}{K}$, clearly a most unrealistic assumption in most species. The straight line can be represented by the equation $r = r_m - \left(\frac{r_m}{K}\right) \cdot N$, meaning that the maximal per-capita growth rate is r_m (when N = 0), and that for each individual added to the population the per-capita growth rate decreases by a fixed amount $\left(\frac{r_m}{K}\right)$. At N=K the equilibrium value is reached because r = 0.

If we now add interspecific competition to the logistic model (2.1), following Lotka (1932) and Volterra (1926), the equation for species 1 becomes:

$$r_1 = r_{m1}\left(1 - \frac{N^1}{K_1} - \frac{\alpha_{12} N_2}{K_1}\right) \qquad (2.3)$$

The subscripts of r, N, and K indicate the species to which these variables apply; α_{12}: represents the competition coefficient, which simply translates the number of individuals of species 2 into units of species 1.

This equation (2.3) can be rewritten as

$$r_1 = r_{m1}\left(1 - \frac{\alpha_{12}}{K_1} N_2\right) - \frac{r_{m1}}{K_1} \cdot N_1 \qquad (2.4)$$

Equation (2.4) is the equation of a straight line plotting r_1 against N_1, whereby the first term $\left(r_{m1}\left(\frac{1-\alpha_{12}N_2}{K_1}\right)\right)$ is the intercept and $\left(\frac{r_{m1}}{K_1}\right)$ is the slope. By adding interspecific competition to the logistic model the intercept is reduced by $\frac{\alpha_{12}N_2}{K_1}$, but as the coefficient of the term in N_1 does not change the slope of the new line remains unchanged $\left(\frac{r_{m1}}{K_1}\right)$. Since the intercept in Equation 2.4 is smaller than that in Equation 2.3, but the slopes in both equations are the same, the two lines expressing per-capita growth rate versus numbers are parallel and the line including interspecific competition lies below that describing the effect of intraspecific competition only (Fig. 2.3).

We can now use these graphical representations of per-capita growth rate with and without interspecific competition to describe how to measure the effect of interspecific competition on a population and on an individual. At equilibrium

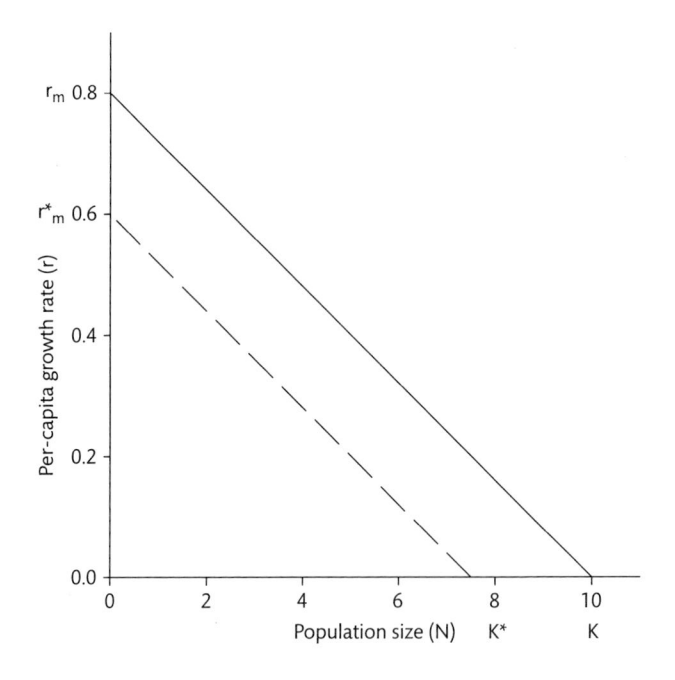

Fig. 2.3 Per-capita growth rate as a function of population size with intraspecific competition only (top line) and with both intra- and interspecific competition present. Note that the two lines are parallel, but that at each population size per-capita growth rate with interspecific competition is smaller by $\alpha_{12}N_2/K_1$ than with intraspecific competition only. The difference between K and K*, or the effect of interspecific competition on population size at equilibrium, is $\alpha_{12}N_2$. The graph thus illustrates both the effect of interspecific competition on individual fitness (by comparing per-capita growth rate at any population size) and its effect on equilibrium population size (by comparing K versus K*).

the K^*_1-value with interspecific competition will be smaller by $\alpha_{12}N_2$ than the K_1-value without interspecific competition. *At any given density* a population subjected to interspecific competition will have a per-capita growth rate that is lower by $\frac{\alpha_{12}N_2}{K_1}$ than when no interspecific competition takes place.

Using this latter conclusion we can revisit the definitions of competition we have discussed earlier. Both the definitions that express interspecific competition as operating on individual fitness and those which express interspecific competition as having an effect on population size are valid, because in the simplified world of models of closed populations both conclusions follow from the model. When comparing the effect of interspecific competition to that of intraspecific competition only at the *individual level*, per-capita growth will be reduced at any density by the same amount. Note, however, that in the real world this reduction may affect individuals unequally. At the *population level* the effect of interspecific competition will reduce K, so that a population subjected to interspecific competition will always reach a lower equilibrium value than one subjected to intraspecific competition only.

2.2.2 Non-linear models

In the late 1960s Francisco Ayala experimentally tested the predictions of the Lotka–Volterra equations and concluded that he had '*invalidated the Principle of Competitive Exclusion*' (Ayala 1969) when keeping two *Drosophila* species together in an environment that was limited for food and space. Initially he defended his position (Ayala 1970) when his conclusions were criticized by Gause (Gause 1970) and many others. Two years later, however, Gilpin and Justice provided a reinterpretation of Ayala's results and showed that intraspecific competition was non-linear (Gilpin and Justice 1972) which resulted in zero-growth isoclines to be non-linear. This made coexistence of two competing species possible. Collaboration between Ayala and Gilpin resulted in the proposition that by adding one additional parameter to the logistic model one could generate non-linear isoclines that described Ayala's experimental results well (Ayala *et al.* 1973; Gilpin *et al.* 1976). They thereby illustrated how in science the wheel is regularly reinvented because the expanded logistic model (later called the theta logistic (Saether *et al.* 2002)) had been originally introduced by Richards, 15 years earlier, to describe plant growth (Richards 1959). This model provides a much more realistic framework to describe variation in population sizes, because its adds flexibly non-linear effects of intraspecific specific competition (Fig. 2.4). Nevertheless it is rarely used to describe variation in the shape of the effects of intraspecific competition on per-capita growth of populations.

The expanded logistic model developed independently by Richards (1959) and by Gilpin *et al.* (1976) adds a third parameter to the logistic model, the exponent θ.

20 | Interspecific Competition in Birds

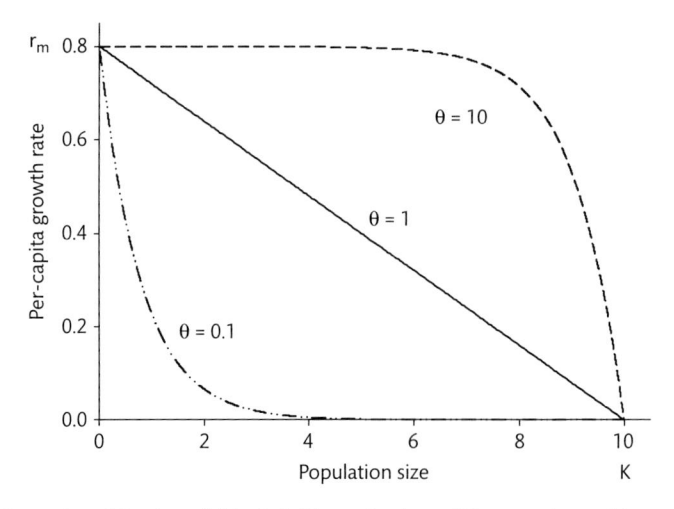

Fig. 2.4 Examples of the theta (θ) logistic illustrating how different values of θ generate concave ($\theta < 1$), convex ($\theta > 1$), or linear ($\theta = 1$) plots for per-capita growth rates against population size.

$$r = r_m \cdot \left[1 - \left(\frac{N_1}{K_1}\right)^\theta\right] \tag{2.5}$$

The logistic model is therefore simply a special case of this expanded model for $\theta = 1$. For values of θ that are different from 1, not only does the graph of per-capita growth rate against numbers become non-linear, but so do the zero-growth isoclines which summarize the effects of interspecific competition between two species (Dhondt 1985a). The equation still produces a graph of per-capita growth rate against numbers that begins and ends at the same points as the logistic (N,0, and 0, r_m), but what makes this model interesting is that the impact of one individual added to the population varies with density (Fig. 2.4). If $\theta < 1$ the curve is concave. The impact of one additional individual on per-capita growth rate at low density (as compared to K) is very large, whereas individuals added at high density have small effects only. If $\theta > 1$ the curve is convex: additional individuals at low density have almost no effect on per-capita growth rate, but when population size approaches K, each additional individual has an ever increasing impact (Fig. 2.4). That such variation does exist has been illustrated by Fowler (1981), for example, for large mammals and by Smith (1963) for *Daphnia*.

Adding interspecific competition to this equation has a very interesting effect (Dhondt 1985b). For species 1 the equation becomes:

$$r_1 = r_{m1} - \frac{r_{m1}\alpha_{12}N_2}{K_1} - r_{m1}\left(\frac{N_1}{K_1}\right)^\theta \tag{2.6}$$

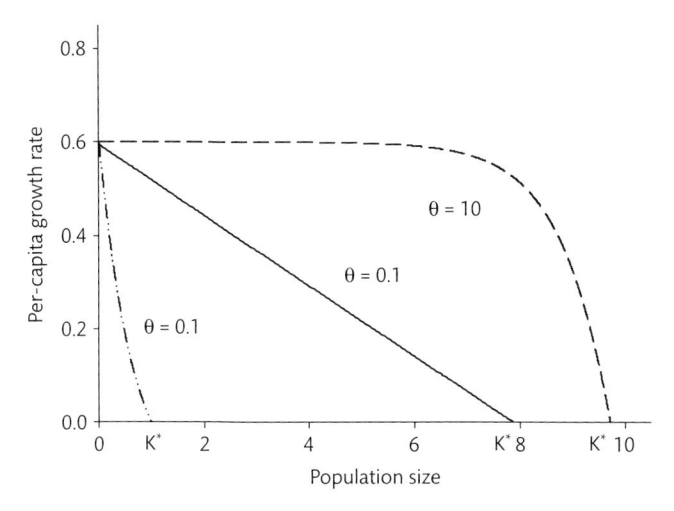

Fig. 2.5 Effect of interspecific competition using the θ-logistic for different values of θ. Note how the impact of interspecific competition for species growing according to a model with large values of θ is very small (K =10 in the absence of interspecific competition), while the impact of interspecific competition on species growing according to a model with small values of θ is very large.

Similarly to the linear model, the term representing interspecific competition simply causes the intercept to change, but the slopes of the curves remain unchanged. The equilibrium values K*, reached when per-capita growth rates become zero, however, depend strongly on the value of θ. This is a simple graphical illustration of Pianka's description of effects of interspecific competition when comparing r- and K- selected species (Pianka 1970). r-selected species (with $\theta < 1$) are very sensitive to competition, whereas K-selected species (with $\theta > 1$) are not. In the graph shown in Fig. 2.5, the K-value in the absence of interspecific competition is arbitrarily set at 10. The new equilibrium values (for r = 0) vary strongly with θ. For $\theta = 10$ the equilibrium value in the presence of interspecific competition is about 9.5. In contrast, for $\theta = 0.1$ the equilibrium value is only about 0.8.

The θ-logistic model thus leads to the conclusion that, since the effect of competition is weak on K-selected species and strong on r-selected species, it should be much easier to document the existence of interspecific competition in field experiments using r-selected species. Note also that these models imply that only a numeric change in population size documents the existence of interspecific competition.

In most textbooks and papers, the isoclines describing the relationship between the densities of two interacting species are linear. These isoclines become non-linear when the per-capita growth rate is described using Equation 2.5. Tom

Schoener explored various models generating such non-linear isoclines, one of which had refugia from which a species could not be expelled by the competing species because the isoclines were concave curves that ended parallel to the axes (Schoener 1973, 1974).

I present these models here in some detail for two reasons:

1. When discussing definitions it was not clear to what extent a numeric response at the population level was required to conclude that interspecific competition exists. If we follow these equations we will accept that interspecific competition exists if a numeric response is observed, but also if we can demonstrate a decrease in the value of r at any given density.
2. Kluijver (1951) has already shown that in great tits, reproductive rate had a non-linear relationship with density (see also Klomp (1980)). I showed experimentally that the isoclines relating great and blue tit numbers were non-linear (Dhondt 1985a). This result implies a non-linear density dependence in at least one rate that influences breeding population size. We can therefore expect non-linear density dependent effects not just in tits, but in most organisms.

2.3 Conclusions

Wiens (1989) listed a set of criteria that could be used to establish the existence of interspecific competition. He listed them in increasing degree of certainty. I will here use these criteria as modified slightly by Newton (1998). I reorganized these criteria and combined them with components of the definition to answer the double question: what conditions need to be met so that the possible existence of interspecific competition can even be considered (necessary conditions), and what constitutes evidence that interspecific competition occurs (sufficient conditions)? For this reason I omit Wiens's first type of 'evidence' for interspecific competition, namely that the observed distribution (geographic, by habitat) must be consistent with expectations of the operation of interspecific competition.

There are three necessary conditions that need to be fulfilled before we can even consider the existence of interspecific competition, and three kinds of evidence that prove its existence with increasing degrees of certainty.

- Necessary conditions: for considering the existence of interspecific competition:
  1. One or more resources must be limiting.
  2. Intraspecific competition must occur (density dependence can be documented).
  3. Resource use between potential competitors must overlap.

- Sufficient conditions: required to prove the existence of interspecific competition:
  4. Resource use of one species affects the resource availability (use) of another.
  5. Fitness of individuals of one species is reduced by the presence of individuals of another species.
  6. Distribution or abundance of one species is reduced by the presence of another.

If we require competition to have an adverse effect on both species, the sufficient conditions can be rephrased to imply this (for example: distribution and abundance of each species is affected by the presence of individuals of the other).

Necessary conditions might not be fulfilled if other factors, such as predation, keep populations well below equilibrium values.

In the above list conditions 1 and 3 follow directly from the definition of competition. Although Martin (1986) claims that condition 2 is not a necessary condition for the possible occurrence of interspecific competition, I agree with Reynoldson and Bellamy (1971), Underwood (1986), and Wiens (1989) that intraspecific competition must exist for interspecific competition to have an effect.

Evidence 1 is what most authors refer to as a niche shift. Evidence 2 and 3 were grouped by Wiens (1989), but split into two categories by Newton (1998). Newton makes the important point that negative effects on individuals do not necessarily translate into negative effects at the level of the breeding population and hence, in the strict mathematical definition of interspecific competition in closed populations, would not qualify as interspecific competition. Condition 2 refers to fitness, but many authors study effects on some component of fitness only. If during the breeding season some component of reproduction declines with increasing con- or heterospecific density because, for example, food is more limiting at higher densities, that would be interpreted, in most studies, as an effect of competition. If there is no compensatory survival this would have an effect on the breeding population size of the following year and would indicate the existence of competition. If there was compensatory immigration, however, there would be no effect on population size, although, in this case, competition would still exist. The point here is that natural populations are open populations, and that, as will be shown later, immigration and emigration can be important processes. The interpretation of field experiments, therefore, needs to be done carefully.

An important comment is that when individuals that differ in heritable traits respond in different ways to interspecific competition, interspecific competition may constitute an important selection force. I will argue, for example, (see Chapter 10) that high densities of blue tits, although having no effect on the size of the breeding population of great tits, might not only result in great tits fledging fewer young, but might actually select for a smaller optimal clutch size of great tits, potentially leading to an evolutionary change in the great tit population, because clutch size is heritable (Van Noordwijk and Scharloo 1981).

2.4 The structure of the rest of the book

This book is structured along the points listed above. In Chapters 3–5 I will discuss the strength of the evidence that resources are limiting. I will explore the importance of competition for space (Chapter 3), for food (Chapter 4), and for nest sites (Chapter 5). In Chapter 6 I will evaluate the evidence that intraspecific competition exists. Here I can take advantage of the extensive literature on population regulation, and explore which population processes are density dependent. I will also use my own extensive data and will explore which population processes are a non-linear function of density. In Chapter 7 I will discuss foraging niches. I combine both the condition that resource use overlaps (condition 3) and that foraging niches and food eaten vary depending on what other species are present (evidence 1) because many of the studies address these questions together. In Chapter 8 I will discuss evidence 2 and 3 by reviewing all field experiments that tested for the existence of interspecific competition. To underline that interactions between species are not always negative I have added a section discussing experiments proving the existence of heterospecific attraction (Section 8.9). I have separated my own work from Chapter 8 and discuss experiments on competition in tits in Chapter 9. Finally, in Chapter 10, I will discuss interspecific competition as a selective force on morphology and life-history traits, and its possible effect on structuring communities, and offer some concluding thoughts in Chapter 11.

3
Space as a limiting resource

3.1 Introduction

For competition to occur, resources must be limiting. For interspecific competition to occur, resource use by individuals of one species must affect resource availability for individuals of the other species. In this and in the two next chapters I will explore the evidence for the existence of resource limitation, often using the work carried out on tits that has historically played a major role in this debate, but including studies on other species when appropriate. I will address three questions: can we document shortage of space? Can we document shortage of food? Can we document shortage of nest sites; for cavity nesters in particular, is there a shortage of cavities? In Chapters 3, 4, and 5 I will address these questions primarily at the level of intraspecific competition. I will explore resource limitation caused by interspecific competition mostly in Chapters 7 and 8.

Like many bird species, titmice are very territorial and can hence suffer from a shortage of high quality territories. During the breeding season all tit species defend Type A pair territories on which they nest and collect the food to raise their nestlings. Detailed work by Raymond Stefanski showed how the actual area used and defended by black-capped chickadees varied by an order of magnitude over the reproductive season (Stefanski 1967). In a few species the adult pair is joined by helpers (non-breeding adults) who also live on these territories and help raise the young (Southern black tit (Tarboton 1981); tufted titmouse (Cimprich and Grubb 1994); bridled titmouse (Christman and Gaulin 1998; Nocedal and Ficken 1998); stripe-breasted tit (Shaw 2003)). During the non-breeding season most species defend group territories against other similar groups. Quality space, therefore, could be limiting both during the breeding season and during the non-breeding season. I will discuss the evidence for this.

3.2 The Buffer Hypothesis was developed from studies of tit populations and is probably generally important

All landscapes are heterogeneous but this heterogeneity can occur at different scales for different organisms. For birds there are two important non-exclusive scales: the scale of a bird's territory within a habitat, and the scale of the habitat within a landscape. At low density birds will only occupy high quality sites in high quality habitats. In one of my study sites, for example, blue tits at low densities only occupied those parts of a mixed deciduous woodland in which mature oaks were abundant (Dhondt *et al.* 1982). When I experimentally increased blue tit density, they occupied the entire study plot, but in the 'second choice' sites birds laid a smaller clutch (Dhondt *et al.* 1982; Dhondt *et al.* 1992). As density increases birds gradually spill over into poorer quality habitat. When studying tits in a heterogeneous landscape Kluijver and Tinbergen (1953) found that between-year density fluctuations were much stronger in poor quality pinewood than in high quality oak wood. Based on these observations they formulated the 'Buffer Hypothesis' in which they proposed that birds prefer to settle in high quality sites, but will move to lower quality sites when forced out because of the territorial behaviour of birds already occupying the best sites. In tits this happens in early spring (Dhondt 1971; Smith 1967). This idea was extended by Brown (1969) who added that when both high and low quality habitats were fully occupied any remaining individuals would become floaters, and be unable to breed. Fretwell and Lucas (1969) translated this idea into a theoretical model. Depending on the importance of territorial behaviour, birds would distribute themselves following an ideal despotic distribution (if territorial behaviour made settlement costly) or following the ideal free distribution (if this was not the case). Under the ideal free distribution in any one year the fitness of all birds in any habitat should be the same. Under the ideal despotic distribution, on the other hand, fitness would be highest in the best habitats, as birds attempting to settle would need to overcome the territorial behaviour of already settled birds, which made the apparent quality of the site lower. Recently it has been suggested that one of the ways birds find out about habitat quality is via 'public information'. Habitat quality can be measured by the breeding success in the area in the year prior to settling (Danchin *et al.* 1998; Doligez *et al.* 2003; Mattsson and Niemi 2008), or by the density of con- or heterospecifics (Forsman *et al.* 2009; Mönkkönen *et al.* 1990).

This brief overview illustrates that intraspecific competition for space can play an important role in population regulation because, as density increases, average reproductive rate often decreases (see Chapter 6). If a habitat is heterogeneous at the level of the territory different individuals will be affected unequally (Dhondt *et al.* 1992; Ferrer *et al.* 2006; Martinez *et al.* 2008). Individuals forced to breed in low quality sites or in low quality habitat will have reduced reproduction, and some can be excluded from reproduction altogether (see Fig. 3.1; Fig. 3.2). This means that, as numbers increase, mean per-capita reproductive output declines in a density dependent manner.

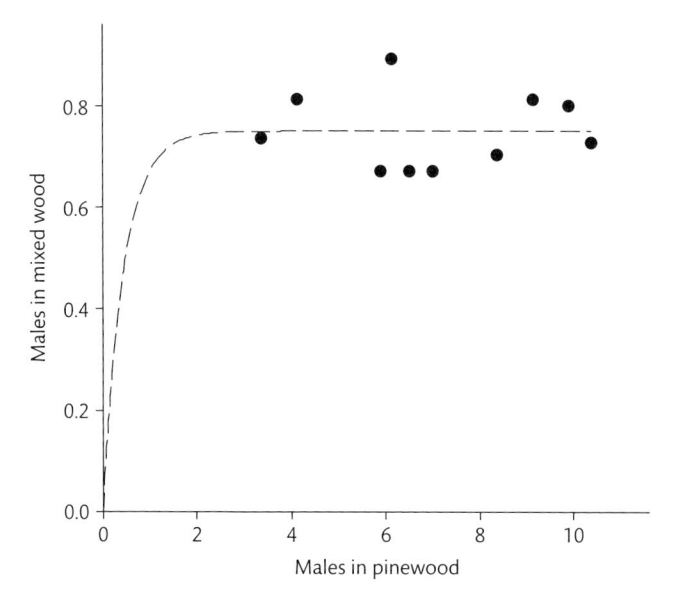

Fig. 3.1 Between-year variation in the number of territorial males in preferred mixed wood and in lower quality pinewood. Note how density in the mixed wood varies much less than that in pinewood, indicating that in each year birds preferentially settle in the mixed wood. Only when most territories in mixed wood are occupied do surplus individuals settle in pinewood, illustrating the Buffer Hypothesis. After Glas (1960).

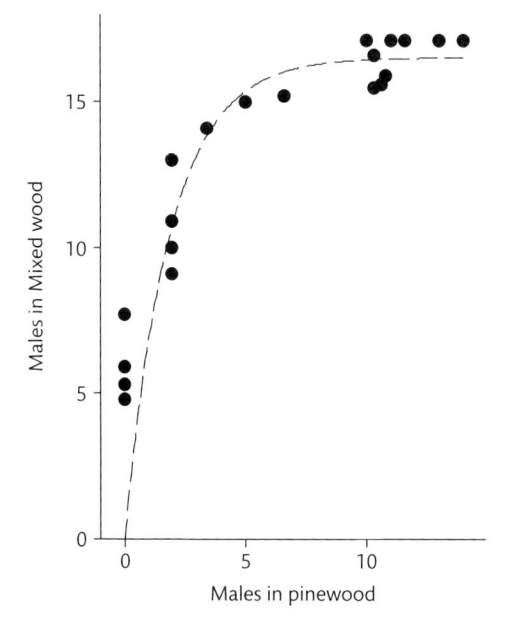

Fig. 3.2 Changes in number of territorial male chaffinches in the course of spring. Note how birds initially settle in mixed wood (preferred habitat) and that numbers only increase rapidly in pinewood (less preferred habitat) once the mixed wood is mostly fully occupied. After Glas (1960).

In the next sections I will summarize removal experiments on tits that vary in winter social organization and document that the period when space limitation occurs differs with the type of winter organization.

3.3 Winter social organization determines when space is limiting

Like many other bird species tits are very territorial during the breeding season and can be extremely aggressive. On one occasion a stuffed decoy great tit that I had left unsupervised for a few minutes was destroyed by the territory owner. Although territory size varies with habitat (Wesołowski *et al.* 1987), age of the male (Dhondt 1966), and with synchronicity of settlement (Knapton and Krebs 1974; Krebs 1971) one can expect territorial behaviour to limit breeding population size, at least in optimal habitats. To test this experimentally one can remove established birds to determine if they will be replaced, that is carry out a *removal experiment*. The results of removal experiments vary between season and species. Removal experiments suggest that the non-breeding social system influences whether tits are primarily winter or summer limited. A population is summer limited if birds that are removed at the onset of the breeding season in spring are replaced, as this implies that there is a surplus of non-breeding birds. A population is winter limited if birds removed in fall or winter are replaced but those removed later are not. Winter limited populations most probably suffer primarily from food shortage during winter, when the floaters especially are affected. If summer limited birds suffer from food shortage, this should be primarily during the breeding season. Limitation outside the breeding season occurs gradually as illustrated by the 13 red grouse removal experiments carried out between August and April by Watson and Jenkins (1968). They concluded that autumn territorial behaviour prevented surplus birds from settling on territories. Their observation that the proportion of removed individuals that were replaced gradually declined from autumn to spring shows that floaters did not survive as well as the territorial birds. None survived until the next breeding season. In August and September the replacement rate was more than 100%, while by February territories of removed birds remained unoccupied (Watson and Jenkins 1968) (see also Fig. 3.3 for an example with tits).

There are two basic types of non-breeding social organization in tits (Ekman 1989; Matthysen 1990). One type is that used by great and blue tits. They live in flocks of varying size and composition during most of the non-breeding season (Saitou 1978). These flocks roam over extended areas, and the individuals are ranked through dominance hierarchies that vary with location (Saitou 1979, 1982). This is caused by the fact that a limited number of individuals (most adult males and some juvenile males) display site-specific dominance, although they do not defend territories from which they exclude conspecifics (Drent 1984). Most other well-studied resident tits are territorial year round. In some species/populations these territories are occupied by a single pair (Ludescher 1973), but in most they are occupied by a group, at least in winter. These groups defend group territories against other similar groups. Groups are formed in late summer, usually by non-kin juveniles joining one (sometimes several) adult pair on their former breeding territory. Individuals that cannot join such a group territory survive less well than established birds (Ekman 1979). Usually these groups have an even sex ratio, and each bird is closely associated with one other bird of the opposite sex. Similarly to winter-flocking species, group members have a strict dominance hierarchy. At the onset of the breeding season the groups break up and breeding territories are established by the surviving pairs. Any extra birds are expelled and become floaters (Smith 1967), although in bridled titmice some philopatric sons defend a joint territory with their father and breed side by side with him (Christman 2001).

Because of these very different social systems, I will review the removal experiments separately for non-territorial winter-flocking species and for the winter-territorial species.

3.3.1 Winter-flocking species are summer limited

In three countries (England, Belgium, and Poland) established great tits of one or both sexes were removed after territories had been established in spring. In all cases all or most of the removed birds were replaced, and the authors concluded that some birds were either excluded from breeding altogether, or at least excluded from optimal breeding habitat (Krebs 1971; Krebs 1977; Lambrechts and Dhondt 1988; Wesołowski *et al.* 1987). The most surprising result here was that in the Polish study, which was carried out in primeval forest in which the breeding density was 5–10 times lower than in the other populations—possibly because of more numerous predators—the authors still found evidence for territorial exclusion. In Belgium, we carried out one more such experiment, in which we removed only already paired females in late March. Here also, removed females were rapidly replaced by new females, mostly of unknown origin (Plompen, Adriaensen, and Dhondt, unpublished). The replacement females, however, were expelled when the original females returned to their territories after release, as happened when Ken Otter and Laurene Ratcliffe performed a similar experiment with black-capped chickadees (Otter and Ratcliffe 1996). In this latter case, however, replacement females were not floaters but neighbouring females that had deserted their partners and that had been subordinate to the new partner in the previous winter, suggesting the absence of surplus floaters. Most great tit populations seem to have surplus birds that are excluded from breeding, at least in optimal habitat. That this surplus can be quite large in great tits was illustrated by the experiment that John Krebs carried out in a small woodlot near Oxford (Krebs 1977). At three-week intervals he twice removed all eight pairs of great tits from a woodlot and both times all the birds were replaced within a few days. Whether they came from hedgerow territories or were floaters was unknown.

Given that there is a surplus of great tits in spring, it is not surprising that the only removal experiment with great tits in the autumn resulted in rapid replacement of the removed site-dominant juveniles (Drent 1984). The removed birds were replaced by a larger number of juveniles that were locally present but not yet established.

As blue tit winter social organization is much like that of great tits, I would predict a surplus of non-breeding floaters in that species too. Although no removal experiments have been carried out with blue tits, some observations suggest the existence of floaters. Dhondt et al.(1990a) documented that in some study plots male blue tits postponed first breeding until 2, 3, and sometimes even 4 years of age, suggesting they were excluded from breeding at an earlier age, as breeding postponement reduces lifetime reproductive success (Dhondt 1989a).

In conclusion, great tits (and probably also blue tits) are summer limited: in most years there are surplus birds, many of which probably do not breed. Some pairs try to intrude into plots in which cavities remain unused, most likely because they do not have a suitable cavity for breeding in their own territory. Initially these intruders do not display territorial behaviour, but behave inconspicuously. These 'intruders' (Dhondt and Schillemans 1983) or 'guest pairs' (Drent 1987) are less successful than birds that are able to breed on their spring territory (Dhondt and Schillemans 1983).

The conclusion that these species are summer limited does not mean, however, that food shortage outside the breeding season or harsh winter conditions may not also influence the population size also (see below).

3.3.2 Winter-territorial species in cold climates are primarily winter limited

With the exception of the great and blue tit discussed in the previous section, most tit species defend group territories during winter. In species with this kind of social organization, individuals that cannot settle on a territory in late summer may have problems surviving winter (Ekman 1979) and below (Fig. 3.3). The first removal experiments with winter-territorial species were carried out by Göran Cederholm and Jan Ekman. When removing established crested and willow tits just before the breeding season no birds came to replace them (Cederholm and Ekman 1976). The authors concluded that there were no surplus birds and wrote (p. 212): '*As our removals were done late in spring they do not, however, exclude the possibility that territorial behaviour limits numbers earlier on. ... the question should be pursued by further removal experiments outside the breeding season*'. Most other removal experiments in this group of species were done in autumn and winter, and nearly all showed that in these winter-territorial tits very few, if any, surplus birds remained alive by the beginning of the breeding season (Table 3.1).

In his long-term study on willow tits in Norway, Olav Hogstad carried out a series of removal experiments between September and February (Hogstad 1989b). Throughout autumn and winter he removed single birds from territorial groups, releasing them back into their flock after a few days. In October and November all removed birds were replaced by non-territorial floaters and the re-released birds expelled the replacement birds. Later in the season removed birds were no longer replaced indicating that no non-territorial birds remained in the population. His data, summarized in Fig. 3.3, show very clearly how the number of floaters available for replacing vacancies declined through winter, so that there remained no non-territorial surplus birds by the time the breeding seasons began. Using ptilochronology, Hogstad confirmed that floaters had problems obtaining sufficient food. Daily growth bars of induced feathers of floaters were about 10% narrower than those of subordinate flock members, which probably contributed

Table 3.1 Extent to which removed individuals are replaced.

	Willow tit	Crested tit	Coal tit	Black-capped/ Carolina chickadee	Tufted titmouse
August	all [a]	none [a]			
September	all [a,b,c]	none [a]		none [d]	some [d]
November	some [b]; none [e]	none [e]	none [e]		
December	none [b]			none [j]	
January	none [a,b,e]	none [a]		some [k]	
February	most [l]			some [f]	
April/May	none [g]	none [g]		some [m]; none [h]	

a: Ekman et al. (1981); b: Hogstad (1989a); c: Hogstad (1990); d: Samson and Lewis (1979); e: Alatalo et al. (1985b); f: Smith (1987): high ranking individuals only are replaced by flock switchers; g: Cederholm and Ekman (1976); h: Otter and Ratcliffe (1996); i: Cimprich and Grubb (1994); j: Dolby and Grubb (1998); k: Groom and Grubb (2006): removed birds were replaced after an extended period by birds from unknown origin, likely subordinates from other flocks; l: Hogstad (1999): removed birds were replaced by subordinates from adjacent flocks; m: Desrochers et al. (1988): some removed black-capped chickadees were replaced.

all: all vacancies filled by floaters; some: some vacancies filled by floaters; none: no vacancies filled by floaters. In some studies removed individuals were replaced by birds settled in adjacent flocks, but not by floaters. In some studies birds were removed over a period of several months. The month listed in the table is the earliest month of the removal period.

to their disappearance when winter conditions became really tough (Hogstad 2003). Hogstad (1999) repeated his removal experiment by removing willow tits from 11 flocks in the period December to February when, according to his earlier results (Hogstad 1989a), there were no longer floaters around. In this experiment removed birds were replaced in 8 of the 11 experiments, but this time by subordinate juvenile pairs from adjacent flocks, not by non-territorial floaters. In some flocks the adult males disappeared from their flocks after a juvenile pair had joined them, while in other flocks the adult males remained in the α position after the juvenile pair settled. Males that remained dominant were 2 years old, while those that lost their top position and eventually disappeared were 3 years or older, implying a decrease of dominance with age. This result, combined with the observation that juvenile pairs were regularly found up to 1,200 m from their own territory, clearly suggests that during winter, subordinate willow tits monitor an area of about 150 ha (5–6 territories) in order to be able to replace disappearing individuals rapidly, thereby increasing their dominance position and hence their chance to breed in the following breeding season. Unlike Cederholm and Ekman (1976) in Sweden, Hogstad did observe spring floaters in some years (Hogstad 1999).

North-American black-capped chickadees have a somewhat different winter social organization than their European counterparts. Although most live in

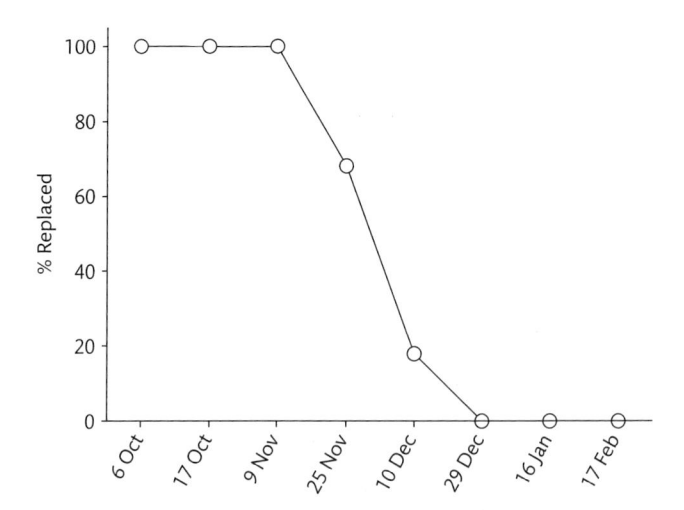

Fig. 3.3 Seasonal variation in the percentage of willow tits that were replaced after removal of single individuals from flocks at different times during autumn and winter. The results clearly show that the number of non-territorial birds available to replace removed birds decreases rapidly in late autumn and early winter and suggests that no floaters remain by midwinter. After Hogstad (1989a).

winter flocks with a stable dominance hierarchy, some individuals are 'flock switchers', a strategy whereby an individual regularly moves between flocks to monitor if a vacancy arises, whereupon they join the flock, typically at a high dominance position (Smith 1984, 1987). In that respect they are different from the traditional floaters that tend to be non-territorial subordinate individuals. When Susan Smith removed high-ranked birds of either sex, all were replaced by flock switchers, while removals of subordinate chickadees did not cause flock switchers to fill the vacancies (Smith 1987). As in Hogstad's (1989a) experiment with willow tits, replacement black-capped chickadees were evicted once the original territorial bird was released, confirming that floaters are subordinate individuals. Because of her careful and detailed field work, Susan Smith could document that flock switchers—after having been driven out—did not revisit the flock for an average of 15 days. On the other hand, all experimentally removed birds were replaced within less than 30 hours by switchers who took over both the dominance position and the partner.

In her long-term study Smith observed, in some breeding seasons, unpaired individuals indicating that a small number of surplus birds survived to the breeding season (Smith 1989). Summer floaters were typically birds that had been part of a winter flock (a few were flock switchers), but did not breed because they either could not find a partner or obtain a territory. When a breeding adult died it was immediately replaced by one of the summer floaters, indicating that they were capable of breeding. In that sense these floaters were different from those observed by Hogstad in that they had been established as part of a winter-territorial flock, while Hogstad's birds were non-territorial.

Another North American study in which some replacement of removed chickadees was observed late in the season was by Jeremiah Groom and Tom Grubb in a series of woodlots in an agricultural matrix in Ohio (Groom and Grubb 2006). They removed 102 Carolina chickadees from 25 woodlots in January 2002. In all woodlots some replacement birds settled but in 13 of the woodlots these birds were single and did not breed. Furthermore the median time to replacement was 30–40 days suggesting that replacement birds most likely were not floaters, in the strict sense, but rather subordinates from flocks in other locations or flock switchers. A possible reason why the time to replacement was so long could be related to the fact that the woodlots were relatively isolated from one another.

I must qualify the above conclusion about winter limitation, though, by addressing the question to what extent this conclusion is valid in areas with less extreme climates. All experiments described in Table 3.1 were carried out in areas with cold to very cold winters where over-winter survival was relatively low. Over-winter survival rate of crested tits in Belgium, for example, where winters are much milder than in Scandinavia, was much higher than in Sweden: 90+% in Belgium versus 50% in Scandinavia (Lens and Wauters 1996). Because the average group size at the beginning of winter was four, the high over-winter survival rate in Belgium resulted in surplus birds being present on many territories at the beginning of the breeding season. This led to severe conflicts between same-sex flock members, leading to the exclusion of birds from the breeding territories. The excluded birds became summer floaters. Similarly, bridled titmice in Arizona had a very high over-winter survival in that mild climate, resulting in cooperative breeding (Christman and Gaulin 1998). In both these latter populations, at least, summer density limitation can also be important in limiting the size of the breeding population, although spring removal experiments are needed to confirm this.

Further research should address to what extent winter group-territorial species, in which at the onset of the breeding season there is not enough space for all birds to establish pair territories, may have led to cooperative breeding (Christman and Gaulin 1998; Cimprich and Grubb 1994; Tarboton 1981). One question that can be asked in this context is why, in some of these species with cooperative nesting, delayed dispersal has evolved so that offspring cooperate with their parents in raising their siblings (Bridled titmouse; Christman pers. comm., Tufted titmouse), whereas in other species the breeding pair is assisted by unrelated adults ((Black tit (Tarboton 1981))?

In summary, the multiple removal experiments show that the extent and timing of space limitation in tits (and in other resident bird species) is driven, at least in part, by when the birds defend territories, whether they be pair territories or group territories. If birds only defend territories shortly before and during the breeding season, it is more likely that competition for space will result in summer limited populations. If territories are defended year round (or at last starting in the autumn) many individuals may be unable to establish themselves before winter. In this case they may have a hard time surviving that stressful period locally. Such populations are often winter limited.

Tropical species often occupy pair-territories year-round. In such situations floaters arise when all suitable habitat is occupied. Susan Smith described how rufous-collared sparrow floaters, being space limited, live in a well structured 'underworld'. They monitor a number of territories and replace a same-sex vacancy as soon as it arises (Smith 1978).

It is obvious, though, that space is not the only factor limiting population sizes, as is discussed in the following two chapters.

3.4 Interspecific territoriality

3.4.1 Tits do not defend interspecific territories among themselves

When resources are limiting some species defend their breeding territories not only against conspecifics, but will also exclude breeding pairs of different species from their territory (Orians and Willson 1964). Such interspecific territories can limit the size of the breeding populations of both species involved, or of the subordinate species only (see Chapter 8). Although titmice may be aggressive towards individuals of other tit species, there is no evidence that they defend interspecific territories against other tit species or against other bird species with which they coexist, with rare exceptions (see Subsection 3.4.2).

Nevertheless there is another way in which tits can impact upon each other. In some regions certain tit species do not coexist in the same habitat, while in other parts of their range the same species exist side by side. This implies that in some regions, but not in others, they exclude one another. Under the assumption of competition one expects that different species can coexist in the same habitat when food is sufficiently abundant, whereas they would be segregated by habitat when food is less plentiful. The case of the willow and marsh tit is an interesting one. In western Europe willow and marsh tit can have overlapping territories both in deciduous (Hinsley *et al.* 2007) and in coniferous habitat (Löhrl 1976; Ludescher 1973). In Sweden, however, willow tits primarily occupy coniferous habitats and marsh tit deciduous habitats. In experiments with captive birds Rauno Alatalo and Arne Lundberg showed that while, as expected, marsh tits prefer to forage on oak branches, willow tits have no preference for either oak or pine branches (Alatalo and Lundberg 1983). Surprisingly, willow tits were more efficient at finding food than marsh tits on both substrates. Alatalo *et al.* (1985a) compared habitat use and tree species choice in mainland Sweden (Upland) and on Åland, an adjacent island in the Baltic where marsh tits are absent. They found that the tree species willow tits used for foraging during winter shifted significantly from a strong preference for conifers on the Swedish mainland to no clear preference for either habitat on Åland. None of the other species of the parid guild differed in habitat preference between the island and the mainland, suggesting that the expansion of willow tit habitat use towards the niche space released by the absence of the marsh tit was caused by a release from interspecific competition, rather than by differences in habitat composition between the two locations. They also showed experimentally that, although during the breeding season territories of willow and marsh tit could overlap, marsh tits were almost as aggressive towards willow tits as towards conspecifics, suggesting that the absence of willow tits from deciduous habitat was largely the result of interspecific interference competition.

Another subtle way in which species can interact is through song, as proposed in the 'Acoustic Competition Hypothesis' (Marler 1960; Miller 1982). Territories of great and blue tit overlap. Nevertheless there exists some potential for agonistic interspecific interactions between these two species as illustrated by the work of Claire Doutrelant and Lieve Gorissen (Doutrelant and Lambrechts 2001; Doutrelant *et al.* 2000a; Doutrelant *et al.* 2000b; Gorissen *et al.* 2006). Blue tits sing two kinds of song types: longer song types that end with a repeated trill and are not sung in strophes, and shorter song types without trills that are repeated in strophes with brief pauses between the phrases (Bijnens and Dhondt 1984). A trill is a series of identical notes in the last subphrase of a strophe repeated rapidly. Great tits normally only sing song types without trills, although they can be induced to sing such song types when confronted with playback of trilled blue tit song (Gorissen *et al.* 2006). Claire Doutrelant and collaborators found that that the proportion of blue tit songs with a trill increased significantly with relative great tit density, both at a local and at a macrogeographic scale. Blue tits rarely sing song types with a trill in populations where great tits are absent or rare (Doutrelant and Lambrechts 2001). They concluded that this observation is consistent with the prediction of the Character Shift Hypothesis. Playback experiments showed that great tits responded equally strongly to playback of great tit songs and untrilled blue tit song, while responding significantly less to blue tit songs with a trill (Doutrelant *et al.* 2000a). This suggested that the smaller blue tit used song with trills to avoid interspecific interference with the larger great tit. The recent arrival of the large ground finch on Daphne Island, Galápagos, illustrates how such a shift in the song produced can occur. After the arrival of the socially aggressive large ground finch, the song of medium and cactus ground finch changed, so as to become more different from that of the new immigrant. Sons sang faster songs than their fathers, so that their songs differed more than that of their fathers from the song of the large ground finch (Grant and Grant 2010).

That species song can be influenced by their acoustic environment was shown in the study of Christopher Naugler and Laurene Ratcliffe. They found a significant negative relationship between the complexity of the acoustic environment and song type richness among male American tree sparrows (Naugler and Ratcliffe 1994). In communities with 7–8 sympatric passerine species all male tree sparrows sang the same song type, while in species-poor communities, although each male only sang a single song type, up to 5 different song types were produced by 10 males.

Subtle effects generated by acoustic interference can be the result of interspecific competition in different bird species, although this is not often studied.

3.4.2 Interspecific territoriality between tits and other species

I know of only one example in which interspecific territoriality between a tit species and another bird species has been proved. Tim Reed studied great tits on small Scottish islands (Reed 1982). While on the adjacent Scottish mainland territories of great tit and chaffinch overlapped, the species defended exclusive territories against one another on the island. Reed first tested responses of the two species to playback of conspecific and of heterospecific song. Both species responded equally to conspecific song on island and mainland. Neither chaffinch nor great tit responded to heterospecific song on the mainland site, and neither species responded to the song of willow warbler, used as a control, in either location. On the island, however, both species responded strongly to heterospecific song. Reed then removed chaffinches from part of the island causing great tits to expand their territory into former chaffinch territory; in an island control site great tit territories did not change.

In Chapter 8 I will discuss interactions between tits and other species in the broader context of interspecific competition.

3.5 Conclusions

There are many examples indicating that space is a limiting resource and a few suggesting it is not. Even when all individuals can obtain a breeding territory the quality of these territories will vary, thereby reducing reproductive success and hence population growth. Depending on the species and its social organization, this limitation can be brought about at different times of year. In winter-flocking tit species (great tit, blue tit) intraspecific spring territorial behaviour often limits the size of the breeding population, as is the case in many migratory species. This is not the case in winter-territorial species in cold climates. In species that defend group territories the ability to join a territorial flock increases survival and can be essential for survival to the next breeding season. Individuals that cannot join a flock are food stressed and possibly also more exposed to predators. In warmer climates, though, survival of winter group-territorial tits is often so high that some adults are unable to obtain a breeding territory. These individuals may stay around as non-breeding floaters, but for others this may have led to cooperative breeding. If this scenario is correct we should discover more cooperative breeding in the still poorly known tit species that live in the tropics. In birds in general, ecological constraints are one of the factors leading to cooperative breeding (Koenig and Dickinson 2004).

Effects of space limitation can vary, but all reduce fitness:

- birds may not survive to breed;
- birds survive to breed, but are forced to breed in a low quality site;
- birds survive to breed, but cannot obtain a breeding territory; in the latter case birds can either breed as an intruder or guest pair, help an established breeding pair, or dump eggs in the nests of conspecifics as happens frequently in wood ducks (Nielsen *et al.* 2006; Semel *et al.* 1988);
- birds cannot obtain a breeding territory until they are at least 2 years old. The delay in the onset of reproduction reduces Lifetime Reproductive Success (Dhondt 1989a).

Tits do not defend interspecific territories against other tit species with whom they currently coexist. Space, therefore, does not play a role in interspecific competition among tits coexisting in the same habitat, in contrast to the importance of space in intraspecific competition. This coexistence can be uneasy, as illustrated by heterospecific song-matching in great and blue tit, and as illustrated by shifts in blue tit song types relative to great tit abundance.

Nevertheless, comparing regions where food is presumed to be less or more abundant suggests that interspecific competition for adequate space can lead to a more constrained use of different habitat types, or even to interspecific territoriality in species whose territories normally overlap (Reed 1982). Some observations suggest that the mechanisms involved include agonistic interactions between similar species leading to the exclusion of subordinate species from optimal habitats. It is likely that other mechanisms are involved in other examples that include geographic variation in the extent to which habitat use of closely related species overlaps. Thus Eurasian blackbird and song thrush coexist in the same habitats in western Europe, while in Scandinavia blackbirds are limited to deciduous habitat and song thrushes are limited to conifers. This is probably more the result of competition for food than of agonistic interactions. David Lack's book on ecological isolation in birds is full of such examples (Lack 1971) implying that this phenomenon is of general importance.

4
Food as a limiting resource

4.1 Introduction

In this chapter I explore the extent to which food is limiting. If food is limiting, competition for food is likely.

During the non-breeding season birds need to maximize their chances of survival and optimize their physical and physiological condition in preparation for the following breeding season. Winter food limitation, therefore, can impact survival rates and subsequent breeding population size, as well as the ability to start reproduction at the right time and in optimal conditions. During the breeding season food limitation can affect both nestlings and breeding adults, and adults may need to trade-off their own survival against that of their nestlings. Each species' life-history strategy will drive this trade-off. When food conditions are really inadequate birds may simply skip a breeding season, indicating that food is insufficient to even attempt reproduction. This has been observed in cases of extreme drought (Bolger *et al.* 2005, 2002; Southern 1970), for owls in low rodent years (Southern 1970), for chaffinches in a year with a very cold spring (Mairy 1969), or for black-capped chickadees in very young regenerating forest (Fort and Otter 2004a, 2004b). In all these examples, low food abundance was the direct cause of non-breeding. In some cases males that defend individual territories with insufficient food will remain unmated, as experimentally shown for European robins by Tobias (1997). Males in such territories are so aggressive to prospective partners that they drive them away and food thus indirectly impacts their ability to reproduce.

I will discuss separately the evidence that food is limiting during the breeding and during the non-breeding season. Note that food can be limiting both by amount or by quality, and during the breeding season food can be limiting by the timing and the duration of the 'food peak' (Blondel *et al.* 2006; Visser *et al.* 2004; Visser *et al.* 1998). Note also that many birds hoard food in the autumn that they will consume during winter (Pravosudov 2006). Hoarding is a derived trait suggesting it evolved in the face of food shortage (Dhondt 2007) (see also Smulders (1998)). Hoarding can sometimes intensify intersexual competition for food (Steer and Burns 2008).

Although Ian Newton has reviewed in detail the extensive, mostly circumstantial evidence that food influences numbers (Newton 1998), I will in this chapter explore the link between food limitation and intraspecific competition. I will first discuss an example of correlative evidence that suggests the existence of an effect of winter food abundance on survival and breeding density in the following year, underlining once again that correlation does not equal causation. I will then move to the numerous field experiments that test the extent to which winter food manipulations actually do impact over-winter survival and population size in tits. These experiments (listed in Table 4.2) show that the impact of supplemental food often depends on other factors such as the amount of natural food, climate conditions, or predator pressure. This will lead to exploring how birds respond at the individual level to variations in conflicting demands of food and predation, especially as regards how much fat they carry. Finally I will briefly list some studies that document that bird predation during winter does reduce prey abundance. I will finish this section on food as a limiting factor by summarizing studies that have tested for an effect of food supplementation during the pre-laying period and the relatively small number of studies that have manipulated food during the breeding season.

4.2 The classical case of beech mast: correlation is not causation

During winter seeds are, next to arthropods, a major source of food for tits and other seed eating passerines. Beech masting, or the large-scale synchronous production of fruits from the beech tree, has played an important and recurrent role in the study of variation in numbers, primarily of great tits in western and northern Europe. Staffan Ulfstrand described how tit foraging behaviour differed strongly between years in which the beech trees masted abundantly and years in which there were none or very few beech nuts (Ulfstrand 1962). When beech mast was heavy, Swedish great tits would forage almost exclusively on the nuts during winter. He also showed a close correlation between the migratory behaviour of great tits and masting: in years with no beech nuts and a high great tit autumn population, tits would migrate and breeding populations would decline after that. Because masting occurs simultaneously across western Europe this causes great tit populations to fluctuate in parallel across the whole region (Lack 1966) and significant positive correlations have been shown between the extent of masting and the change in the size of the breeding population (Perdeck *et al.* 2000; Perrins 1966; Van Balen 1980). When beeches mast abundantly tit survival is good and population sizes increase. When beech mast is low tit survival is lower and population sizes decrease.

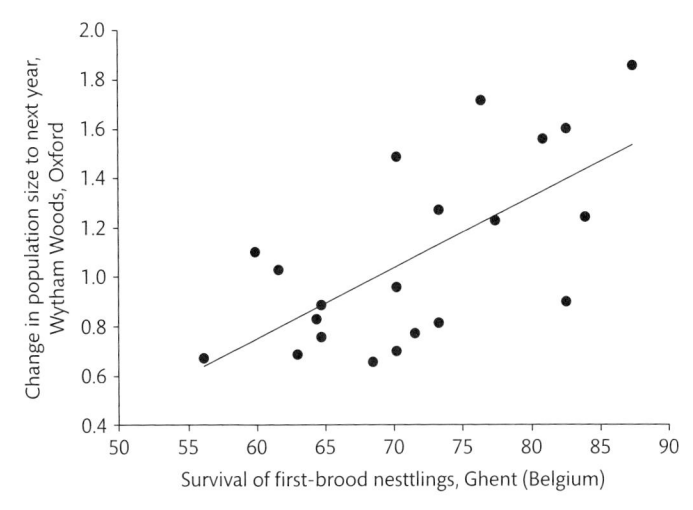

Fig. 4.1 Relationship between great tit first-brood nestling survival in Ghent (Belgium) and the change in population size to the next year in Wytham Woods, Oxford (UK): when first brood great tit nestlings survived well in Ghent, great tit populations increased in the following year in Wytham Woods, Oxford 360 km away (r^2= 0.44; P= 0.001). After Dhondt (1987b).

Although beech mast most likely plays an important role in great tit population dynamics, this may not be the whole story. Perhaps the simplest way to document the fact that we cannot readily infer that great tit breeding populations increase after a good beech mast year because there is a lot of food available in winter, is given in Fig. 4.1. In this graph I show a significant positive correlation between the nesting success (fledglings per egg) of first-brood young at Ghent, Belgium, and the change in the breeding population size for the next year in Oxford, 360 km away. Clearly, the amount of beech mast present in winter cannot explain the survival of great tits in Ghent in the previous breeding season. Chris Perrins, of course, was fully aware of this and, in his 1966 paper, after noting that the critical juvenile mortality occurs before beech mast is available for the birds to feed on, he wrote: '*It seems very unlikely, therefore, that seed crops themselves are responsible for the late summer mortality. It is still possible, however, that some common factor affects both the beech crop and the survival of the young great tit.*' (Perrins 1966). While the causation of beech masting is still obscure, a recent paper links masting in beech to climate variables in the two previous years both in Europe and North America (Piovesan and Adams 2001). So Chris Perrins could be right.

This is why, in the next section, I discuss field experiments in some detail, as they make it possible to understand why the effect of food supplementation is variable.

4.3 Experimental evidence that food does actually influence winter survival or the size of the following breeding population

4.3.1 Interactions between natural and supplemental food

A nice example of contrasting results appears when we compare the experiments carried out by John Krebs in Oxford and by Hans Källander in Sweden (Källander 1981; Krebs 1971). Both tested the possible effect of winter food supplementation on the change in great and blue tit breeding numbers between years. After providing supplemental food in the winter 1968–69, Krebs observed that breeding numbers of blue but not of great tit increased in the area with experimental food while numbers decreased in the control site. Källander found an opposite result. He did not observe a change in blue tit population size between treatments in multiple plots, but great tit numbers increased from 1969 to 1970 in both fed plots, while numbers decreased in the two unfed areas. Over the following winter (1970–71) numbers increased strongly both in the one fed and in the three control areas with no supplemental food. Källander explained the between-year difference as being the result of the large beech mast crop in the second year: there was so much natural food that supplemental food did not have a measurable additional effect. Given that variation in beech mast is synchronous across most of Europe we can use Perdeck's beech mast data for the Netherlands (Perdeck *et al.* 2000) to test whether Krebs's lack of effect of supplemental feeding on Oxford great tits can be explained by an abundant beech mast crop in that year. Perdeck *et al.* confirm Hans Källander's observation that in 1970–71 the beech mast crop was high, and low in 1969–70, but also report that in 1968–69, the winter during which John Krebs performed his feeding experiment in Oxford, the beech mast crop was intermediate. These experiments suggest that providing additional food in winters when natural food is abundant does not have an extra effect on the increase of the breeding population to the next year. As discussed in detail by Hans Källander, the one puzzling result is John Krebs's observation that blue tit numbers increased in a plot after providing winter food.

The interaction between natural and artificial food is best illustrated by Hans Van Balen's long-term—28 control and 7 experimental years—great tit study in the Netherlands (Van Balen 1980), with additional analyses of the data in Perdeck *et al.* (2000). They compared years (not sites) during which additional food had been provided in autumn and winter to years without experimental winter food, concluding that the data supported the Winter Food Limitation Hypothesis. Adult survival and local recruitment of juveniles increased with beech mast crop. In years in which beech trees did not mast but additional food was provided, both adult survival and juvenile local recruitment were higher than in years in which no additional food was provided. That was not the case when the beech mast crop was heavy. High food levels in winter, be they natural (beech mast) or artificial (additional food in non-beech mast years) led to increased breeding populations, indicating that great tit numbers are strongly influenced by winter food (Perdeck *et al.* 2000; Perrins 1966; Ulfstrand 1962; Van Balen 1980), but that additional food on top of a large amount of natural food does not further increase numbers.

In a similar vein two large-scale studies documented that when natural winter food was less abundant birds used artificial food more frequently, indicating winter food limitation. Chamberlain *et al.* (2007) related the annual beech mast crop to weekly winter occurrence rates of 40 bird species at garden feeders in the UK between 1970–71 and 1999–2000. Seven species that commonly feed on beech mast showed significantly lower occurrence in gardens in years of highest beech mast abundance: great spotted woodpecker, woodpigeon, great tit, coal tit, nuthatch, jay, and chaffinch. McKenzie *et al.* (2007) similarly took advantage of the synchronous variation of the Sitka spruce cone crop across the UK. They found that both siskins and coal tits were mistnetted more often in gardens in western Scotland in years with poor cone crops and, using the BTO Garden BirdWatch data, were able to confirm that the local result could be extrapolated across the UK. Both siskins and coal tits switched to supplementary food in gardens more often in years with few cones than in mast years, underlining that in years without masting food was limiting.

4.3.2 Interactions between food and winter temperatures

One of the largest-scale experiments linking changes in breeding population size to the provision of artificial food during winter comes from the long-term nest-box studies supervised by Rudolf Berndt in Germany. In Table 4.1 I summarize the effects of an extremely cold winter on the combined number of pairs breeding in nest-boxes in 12 plots where no food was added and in 4 plots where additional food was provided. The data, which essentially confirm a similar data set describing the effects of the extreme winters 1928/29 and 1939/40 (Berndt 1941), show that access to supplemental food during the extremely cold winter of 1962–63 reduced the decrease in the breeding population size in the next year in great, but especially in blue, tits. After the cold winter great tit populations with access to additional food had decreased by only 6% whereas populations without access to food had decreased 32% since the previous breeding season. Blue tits were much more affected by the cold winter than great tits. Even with extra food blue tit population size decreased by one third, but without it the decrease was 60% (Table 4.1). Coal tit breeding populations collapsed after cold winters, whereas marsh tits numbers were reduced to a lesser extent. Note that

Table 4.1 Effect of supplemental food during an extremely cold winter on tit population size. The percentages represent the change in population size from 1962 to 1963. The numbers in brackets are the number of breeding pairs in 1962, before the cold winter.

Supplemental food	No	Yes
Species		
Great tit	– 32% (619)	– 6% (165)
Blue tit	– 60% (334)	– 34% (112)
Marsh tit	– 23% (22)	– 33% (6)
Coal tit	– 69% (48)	

marsh tits are a hoarding species. The effect of winter feeding on coal and marsh tit cannot be evaluated because of small sample sizes.

Cawthorne and Marchant (1980) analysed changes in bird abundance in British woodlands resulting from the cold winter 1978–79. The data from the Common Garden Census indicated that the cold winter caused a severe reduction in abundance of species weighing less than 10.5 g, while heavier birds showed only small changes; none of the 100+ g species in the woodland index decreased in 1979.

Using the appropriate capture-mark-recapture techniques Doherty and Grubb (2002; 2003) showed that the effect of supplemental food on survival and body condition varied with forest fragment size but only in Carolina chickadees and not in tufted titmice, white-breasted nuthatches, and downy woodpeckers, emphasizing that the extent to which food is limiting during winter varies according to the conditions that the birds are exposed to (Cimprich and Grubb 1994; Pravosudova *et al.* 1999).

4.3.3 Interactions between food and predation

The effects of winter feeding on winter-territorial tits are diverse. The careful experiment of Jansson *et al.* (1981) showed that in southern Sweden the survival rate of both willow and crested tit with artificial winter food improved substantially compared to the control populations without extra food. Winter flocks with access to extra food had a survival rate between November and March of 0.82 (willow tit) and 0.76 (crested tit), whereas flocks with no access to supplemental food had survival rates of 0.45 and 0.39 respectively. This resulted in a two-fold increase in the size of their breeding populations compared to control sites without supplemental food. What was remarkable in this study was that, although the main source of winter mortality for these winter-territorial birds was predation by pygmy owls, providing artificial food to the tits increased their survival because fewer birds were taken by the owls. This underlines the interaction between starvation and predation. After concluding that their data supported the Winter Food Limitation Hypothesis they wrote: '*Obviously, the predation risk of willow tits and crested tits is related to food availability. Food density and foraging efficiency can influence predation by determining the amount of time that can be diverted from foraging to predator scanning.*'

To make matters even more complex, Andrew Dolby and Tom Grubb (2000) showed how risk-taking behaviour during foraging can be influenced by the presence of other species, proving that foraging in mixed species flocks is adaptive by allowing birds to use more food sources. In Ohio woodlots, winter-resident bark-foraging birds usually forage in mixed species flocks. Tufted titmouse and Carolina chickadee are the flock-leading nuclear species, while downy woodpecker and white-breasted nuthatch are follower species. Dolby and Grubb hypothesized that if nuthatches join the nuclear species to reduce predation risk, they should be more reluctant to visit an exposed feeder in the absence of tits than in their presence. They therefore first removed all chickadees and some tits from ten woodlots so that in each plot two titmice, two to four downy woodpeckers, and two nuthatches remained. They then removed either the two titmice or all the woodpeckers to observe nuthatch behaviour in the presence or absence of titmice while keeping flock size constant. Nuthatches were more willing to visit a feeder at 16 m outside the forest in the presence than in the absence of tufted titmice. Given that lone pairs of nuthatches in woodlots, from which parids had been removed, exhibited increased vigilance and reduced nutritional condition compared to pairs with parids, Dolby and Grubb's (1998) results further underline the benefits of mixed species foraging. The same amount of food can thus be more or less limiting depending on predation risk.

4.3.4 Increased survival does not always lead to increased breeding populations

Whereas Jansson *et al.* (1981) found that food supplementation increased both over-winter survival (through reduced predation) and the density in the following breeding season for two tit species, Samson and Lewis (1979) found that food supplementation did not change tufted titmice flock size nor breeding density. Black-capped chickadees autumn–winter flock size did increase substantially, which led to an increase in breeding density from 7 to 10 pairs, compared to no change in the control site.

Experiments on black-capped chickadees by Margaret Brittingham and by André Desrochers documented an improved over-winter survival rate in plots with additional food, which did not lead to an increase of the size of the breeding population (Brittingham and Temple 1988; Desrochers *et al.* 1988). Because Brittingham and Temple's experiment ran over three winters they were able to document an interaction between winter temperature and artificial food on survival. Birds on fed plots always survived better than birds on unfed plots, but during extreme cold weather the advantage of feeding increased tremendously (Fig. 4.2).

When considering the combined effect over the winter (Oct–Apr), survival in the control sites was 0.37 as compared to 0.69 in the experimental sites, a difference comparable to that of the Swedish experiment. It is, therefore, surprising that this had no effect on breeding density. Breeding density was measured as relative abundance obtained by point counts, so was perhaps not very accurate. The only explanation the authors offer for this lack of an effect on breeding abundance is '*A seasonal movement of chickadees apparently occurs on experimental sites in spring, as birds move away from areas of higher densities near feeders and become more evenly dispersed over the available habitat*'. Although they could not document that effect, we can speculate that overall breeding density in an area larger than the study site would have increased, so that feeding would have increased total breeding numbers.

André Desrochers observed, however, as Susan Smith had earlier (Smith 1967), a sharp decline of population size at the onset of territoriality in spring. This was mainly caused by the disappearance of subordinates of each sex. Removal experiments in April and May resulted in partial replacement in one year with a mild winter but not in another year with a very cold winter (Desrochers *et al.* 1988). This suggests that in these sites black-capped chickadee numbers are limited by

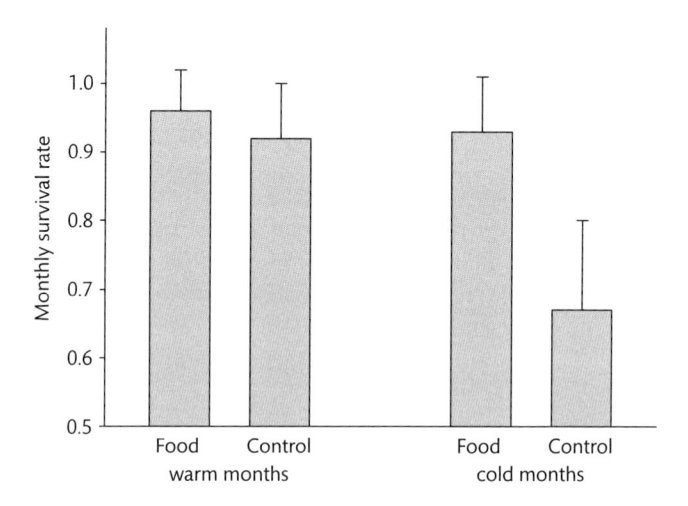

Fig. 4.2 Mean monthly winter survival rates (± SD) of black-capped chickadees on food supplemented plots (food) and on control plots without additional food (control). Added food significantly increased survival rates only in cold months. After Brittingham and Temple (1988).

spring-territorial behaviour (intraspecific competition for space) and/or by winter food (intraspecific competition for food), depending on winter conditions. In either case, intraspecific competition limits the size of the breeding population.

The experiments listed in Table 2 show that in most experiments in which food was supplemented during winter the birds benefitted (increased survival and/or increased density). Effects, however, often interacted with other factors such as amount of natural food available, the coldness of the winter, predation risk, or age and gender.

4.3.5 Large-scale effects and use of food supplementation as a conservation tool

So far the experiments I have listed have tested the effect of food supplementation on one or a few species in a limited area only. Some feeding experiments have attempted to determine the effect of food provisioning on entire bird communities in entire landscapes. Studying 24 landscapes during 2 winters, Yves Turcotte and André Desrochers asked if there would be an interaction between landscape structure (degree of fragmentation = landscape integrity) and food supplementation in Quebec during winter (Turcotte and Desrochers 2005). Their result was not intuitive. In control landscapes, in which there was no food supplementation, species richness declined more strongly over the winter in less fragmented landscapes, while the opposite happened in the food-supplemented landscapes. As a result, the difference in species richness between supplemented and control landscapes increasingly diverged over winter as forest integrity increased. They argued that in more forested landscapes, winter emigration of juveniles and transient birds increased, and that conversely in the highly deforested and fragmented control landscapes, birds became 'gap-locked' when rigorous winter climatic conditions exacerbated already existing movement constraints.

In two even larger-scale experiments that covered the whole of Great Britain, Siriwardena *et al.* (2007) tested the hypothesis that by providing supplemental food to farmland birds their decline could be stopped or even reversed. They analysed results from a three- and a two-year food supplementation experiment at several hundred sites across Great Britain and concluded that supplemental feeding had a positive effect on seven of thirteen species. Species that did not respond to the feeding experiment were species whose abundance was increasing nationally. They therefore concluded that the results provide evidence for winter resource limitation in the populations that responded to the supplemental food, and suggest that supplementing winter food can produce landscape-scale, positive effects and can thus be used as a conservation measure. Effective supply to enough individuals, however, is critical in such an experiment, and many

Table 4.2 Results of experimental food addition during winter on survival or density in following breeding season in passerine birds (with/without stands for treatments with or without supplemental food).

Species	Location	Experiment	Replicated	Interaction with	Effect on survival	Effect on density	Author
great tit, blue tit	N. Germany	site with/without	yes	winter cold	not measured	yes	(Berndt and Frantzen 1964)
great tit	England	site with/without	no	none	not measured	no	(Krebs 1971)
great tit	S. Sweden	site with/without; reversed	yes	none	not measured	yes	(Källander 1981)
great tit	Netherlands	time; (site)	yes	age; beech mast crop–adults: weak effect of added food when beech mast good	yes; especially on juvenile recruitment	yes	(Perdeck et al. 2000; Van Balen 1980)
great tit		site with/without	yes (by location) yes (11 years)	none	no	not measured	(Schmidt and Wolff 1985)
blue tit	England	site with/without	no	none	not measured	yes	(Krebs 1971)
blue tit	S. Sweden	site with/without;	yes (sites reversed)	none	not measured	no	(Källander 1981)
marsh tit	N. Germany	sites with/without	yes	winter cold	not measured	no	(Berndt and Frantzen 1964)
willow, crested tit	S. Sweden	site/time	fall; winter	predation	yes	yes	(Jansson et al. 1981)
Song sparrow	W. Canada	site with/without	no	age	no	not according to authors	(Smith et al. 1980)
willow tit	N. Finland	site with/without (territory)	yes (by territory)	Age, gender	yes	not measured	(Lahti et al. 1998)
black-capped chickadee	PA, USA	site with/without (territory)	no	none	yes	yes	(Samson and Lewis 1979)
black-capped chickadee	PA, USA	site with/without	no	none	yes	not measured	(Egan and Brittingham 1994)
black-capped chickadee	ME, USA	site/period with/without	yes	none	not measured	not measured	(Wilson 2001)
tufted tit	PA, USA	site with/without (territory)	no	none	no	no	(Samson and Lewis 1979)

confounding effects can be present (such as competition for the food with rats or game birds) which decreases the amount or availability of the supplemental food provided. Nevertheless these results suggest that '*appropriate enhancements of food availability, such as through agri-environment measures providing seed resources when they are most needed during the winter, have realistic potential to halt, or at least to slow, population declines at the landscape scale*' (Siriwardena et al. 2007).

4.4 Behavioural responses to winter cold and predation risk: costs and benefits of flocking

Many bird species forage in flocks with conspecifics or in mixed species flocks, both during the non-breeding season and (rarely) in the breeding season (Tubelis 2007). The two benefits of flock foraging are an increased rate of food intake (because of reduced individual vigilance) and a reduced predation risk (because of the 'many eyes' scanning for predators and often convergence in alarm calls). Both should lead to increased survival. In a literature review on survival rates of tropical birds, Jullien and Clobert (2000) did find that obligate flocking species had a higher survival rate than species that did not flock or only occasionally flocked. Blake and Loiselle (2008), however, could not confirm this in a capture-mark-recapture analysis of 31 species in Ecuador. A large number of studies found an increased food intake rate linked to reduced vigilance (reviewed by Beauchamp (1998)) or a reduced predation risk when flocking (reviewed by Lima (1986, 1993) and Pulliam and Caraco (1984)). On the other hand, as flock sizes increase, intra- and interspecific competition increases as well, as does scrounging (McCormack *et al.* 2007). Intraspecific competition increases when conspecifics join a flock. The number of conspecifics can reach an optimal number, after which the presence of more conspecifics reduces the advantage of flocking because the costs of intraspecific competition outweigh the benefits of flocking ((e.g. Fernandez-Juricic *et al.* (2007)). Since the cost of interspecific competition should be less than that of intraspecific competition, birds should prefer to associate with individuals of other species when large flocks are advantageous. Some good, detailed studies document how the birds carefully trade-off their food intake needs against predation risk by short-term variation of the group sizes in which they associate. Luc Lens, studying group-territorial flocks of four crested tit in Belgium, found that as the temperature increased during the day the number of tits foraging in the same tree gradually decreased from four—when cold in the morning—to two, or even one, when warmer in the afternoon (Lens and Dhondt 1992). In Norway, Olav Hogstad documented how temperature influenced flocking by willow tits. At warmer temperatures willow tits foraged in single species flocks. At colder temperatures the number of willow tits in a flock decreased, but they associated with coal tits so that the total flock size was larger. As the metabolic needs of the birds increased, they could allocate more time to foraging due to improved predator detection by many eyes, and suffered less from intraspecific competition (Hogstad 1988). Similarly Kubota and Nakamura (2000) found that varied tits foraged in monospecific flocks when additional food was provided, but joined mixed species flocks when food was removed or temperatures decreased. Szekely *et al.* (1989) also observed the relation between food availability and flocking behaviour. When provided with extra food, mixed species flocks were significantly smaller and birds were more often observed alone or in flocks with only conspecifics than occurred in the control situation.

As energy demands increase because of, for example, cold weather, birds change foraging strategies so as to increase their food intake rate while attempting not to increase their predation risk. As in flocks there exists a clear dominance structure, these strategies should differ between individuals of different rank, as illustrated by Kimmo Lahti's *et al.* experiment. They provided extra food to selected groups of willow tits in northern Finland. Individuals in groups with extra food survived considerably better (72.4% from autumn to spring) than those in control groups (39.1%). The effect was especially strong amongst the yearlings (9.5 to 59.4%) that are subordinate in the winter groups Lahti *et al.* (1998).

4.5 Individual responses to managing body fat reserves in the context of food availability and predator presence

The main risks to which birds are exposed on a daily basis during the non-breeding season are the risk of starvation and the risk of predation. The risk of starvation can be mitigated by carrying more fat reserves, especially in the evening (to survive the night and have some reserves in the morning in case conditions have suddenly deteriorated), while the risk of predation can be mitigated by optimizing the ability to escape predators. Unfortunately for the birds, there is a conflict between these two strategies: the ability to escape a predator is related to the take-off speed and take-off angle of ascent, which is dependent on body mass (Krams 2002; Kullberg *et al.* 1996). Lower mass leads to increased take-off speed (and hence an increased probability to escape a predator), but on the other hand lower mass increases the probability of starvation when conditions deteriorate. Hence birds need to optimize body mass, the optimal solution being the mass that minimizes the joint risk (Lima 1986).

The amount of fat birds carry in winter depends on the predictability of access to food, which is influenced by many factors:

- *Foraging site*: Rogers and Smith (1993) found that ground foragers carry more fat than tree-feeding species, but only in harsh winter conditions when ground foraging becomes difficult and unpredictable.

- *Temperature:* body mass varies seasonally. Birds are heavier in winter, although it is not easy to determine which temperature they respond to because one can expect them to respond both to short-term and long-term changes in temperature (Rogers 1995). Using field data collected over a 17-year period Andy Gosler observed both strategic fattening (by optimizing fat reserves in response to temperature during the past few hours) and constraints on mass gain caused by the more immediate temperature, which could impede a bird's ability to achieve optimum mass (Gosler 2002).
- *Predator presence*: in a longitudinal study of body mass of great tits Gosler *et al.* (1995) observed that in areas and periods where sparrowhawks were rare or absent great tit winter body mass was higher than when hawks were present. The response to predator pressure can vary with type of predator. Thus Krams (2000) showed that, in an area with both sparrowhawk and pygmy owl, birds carried more fat in the evening, compared to an area with only sparrowhawks. This suggests they perceived this double threat as generating more unpredictability in feeding conditions. Macleod *et al.* (2005) demonstrated experimentally that birds, in response to heightened perceived predation risk, manipulate their diurnal body mass gain strategy. When 'attacked' by a plastic sparrowhawk model, which swooped over the foraging site, great tits delayed mass gain until later in the day, as predicted by mass-dependent predation risk theory. Furthermore, the results are consistent with changing flight performance rather than changing exposure time to predators being the driving force for mass-dependent predation risk.
- *Dominance position in a group*: many studies have shown that, in general, subordinate individuals carry more fat than dominant ones both in winter-flocking species such as the great tit (Gosler 1996; Krams 2000, 2002) and in winter-territorial species such as willow and crested tit (Haftorn 2000; Krams 1998). In a study of coal tits in a subalpine coniferous forest, Broggi and Brotons (2001) found that transient coal tits (who are subordinate) carry more fat than resident ones, suggesting that food conditions for resident birds are more predictable. In a very elegant experiment, Indriķis Krams showed a causal relationship between dominance position, predation risk, and fattening strategy. He first documented that crested tits are dominant over willow tits, and forage in sites that are safer from predation (Krams 1996). In both crested and willow tits the body mass index (a measure for mass corrected by size) was lower in dominant birds, illustrating how the role of intraspecific competition for foraging sites influences fattening strategy. Furthermore willow tits, being subordinate to crested tits, carried more fat than the crested tits (Krams 1998), illustrating the impact of interspecific competition.

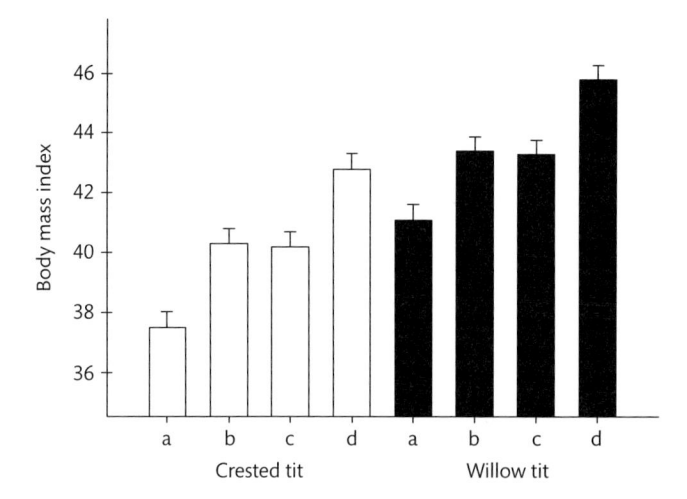

Fig. 4.3 Body mass index (BMI) in Latvian tits in mixed species flocks. Crested tits are dominant over willow tits, adults over juveniles, and males over females. More dominant individuals carry less fat and hence have a lower BMI: a: dominant males; b: subordinate male; c: dominant female; d: subordinate female. After Krams (1998).

Krams then removed all crested tits from five mixed species flocks, and all willow tits from five other flocks. Removal of crested tits resulted in a change in the foraging position of willow tits to sites that were safer from predation and a decrease of the body mass and fat reserves of willow tit α-males, while there was no change in foraging site nor in body mass of crested tits foraging by themselves (Krams 1998). Carrying less fat is safer when being attacked by predators because different fattening strategies lead to differences in take-off speeds (Krams 2002) and hence the ability to escape from predator attacks. The change in willow tit fattening strategy after crested tit removal was thus adaptive and underlines the impact of interspecific competition for foraging sites. Their non-optimal behaviour when foraging in mixed species flocks was thus the result of interspecific displacement by the more dominant crested tits.

- *Foraging success.* Will Cresswell found a direct correlation between foraging rate when alone (a measure for foraging ability) and mass gain for Eurasian blackbirds in winter (Cresswell 2003). Blackbirds with the lowest foraging rate put on about 19 g over the winter, compared to about 9 g for the birds that had the highest foraging rate. This result, although non-intuitive, is as predicted by the Mass-dependent Predation Hypothesis (Lima 1986; Witter and Cuthill 1993) which predicts that for poor foragers conditions are always less predictable, and they therefore need to carry more reserves. The best foragers can thus maintain a relatively low mass.

Although not strictly experimental the interaction between winter food limitation and predator pressure is illustrated very nicely by Andy Gosler's work in Wytham Woods, Oxford (Gosler 1996; Gosler *et al.* 1995). He took advantage of the effect of DDT on predator abundance and the effect of beech masting on the amount of winter food (Fig. 4.4). The graph makes two points: great tits were heavier in years with no sparrowhawks than in years when hawks were increasing; and they were heavier when sparrowhawks were increasing than in the period when hawks were abundant. An increased predation risk shifted the trade-off point between the risk of starvation and the risk of predation by increasing the risk of starvation. A second effect on mass comes from the presence or absence of beech mast. When sparrowhawks were absent, great tit winter body mass was not affected by the presence of beech mast; they simply carried a fair amount of fat, probably to reduce the risk of starvation. When hawks were present, the birds carried less fat overall but were heavier in non-mast years than in mast years. The optimal body mass in years with sparrowhawks is lower when beech mast is present, because food resources are more predictable and birds can afford to carry lower food reserves thereby reducing their risk of being taken by a hawk.

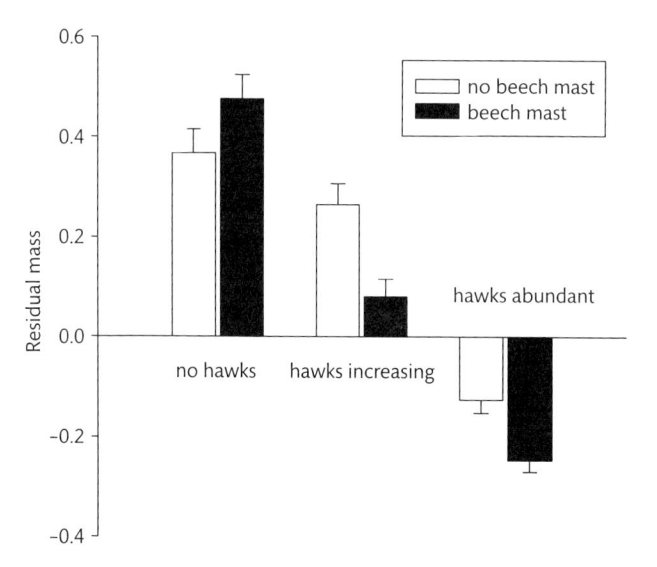

Fig. 4.4 Mean residual mass ± SE of great tits in Wytham Woods, Oxford (UK) when sparrowhawks were absent (no hawks), present in small numbers (hawks increasing), or fully reestablished (hawks abundant); and in years with a good (dark bars) or poor (open bars) beech mast crop. Birds carried more fat, and residual mass was therefore higher, when there were no hawks. When hawks were present birds carried more fat in years when beeches did not mast. Data from Table 1 in Gosler (1995).

While all factors listed so far relate to winter survival strategies, the presence of sparrowhawks in the breeding season has a similar effect. It has long been accepted that heavier fledglings survive better (Dhondt 1971; Perrins 1965). Much to their surprise Adriaensen *et al.* (1998) found that in the presence of sparrowhawks this was no longer the case. This is probably because after fledging, when predation can be intense (Geer 1981, 1982), the very heavy young are an easy prey for the hawks. Optimal fledging mass can thus vary with the presence of a predator (Fig. 4.5).

The results discussed in this section make two points clearly: Lack's 'Winter Food Limitation Hypothesis' (Lack 1966) gathers a lot of support. Food is often, but not always, limiting during the non-breeding season. The extent to which food is limiting, though, varies with other conditions (winter cold, predator pressure, amount of natural food available), but varies also between species and individuals. Dominant individuals suffer less than subordinates; adults suffer less than juveniles; males suffer less than females; resident or established birds suffer less than floaters or transients. Note also that although a fair number of

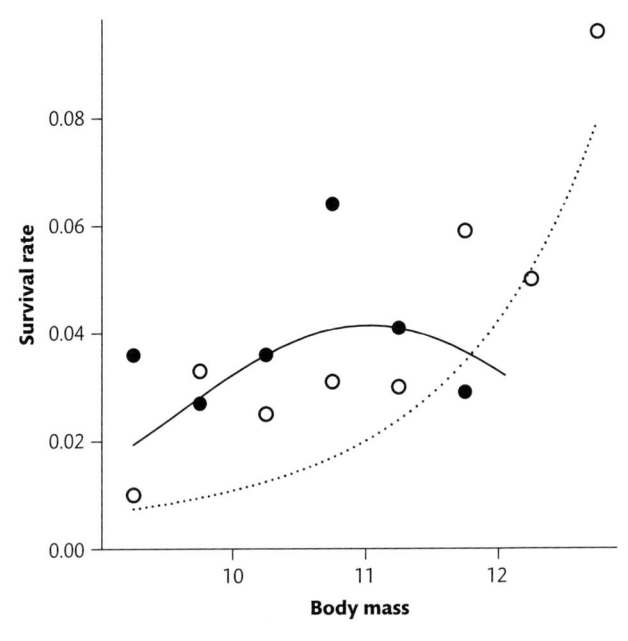

Fig. 4.5 Local recovery rate of blue tit fledglings in the next breeding season as a function of fledgling body mass. Symbols: proportion of surviving birds per 0.5 g mass class (at least three recoveries per class); open symbols: years without sparrowhawks breeding in plot; filled symbols: years with sparrowhawks breeding in the plot. Lines calculated using logistic regression analysis. After Adriaensen *et al.* (1998).

experiments have been performed, the majority have tested tits, and many have not been replicated during the same study. Because, however, many experiments have used the same species (by different scientists at different locations), the experiments have actually been replicated in time and space. This gives more trustworthy information than had a single scientist carried out multiple replications at the same location, because the experiments are independent from one another and because they can allow a broader generalization across space and time, although they also underline geographic variation in effects. Thus experiments using great tits showed food limitation in Germany during extreme winters (Berndt and Frantzen 1964), but not during normal years (Schmidt and Wolff 1985); when beech mast was absent, but rarely when it was present (Källander 1981; Perdeck *et al.* 2000; Van Balen 1980).

A second generalization we can make is that Lima's (1986) conclusion that birds optimize body mass very precisely to trade-off the risk of starvation and the risk of predation is largely supported.

4.6 Pre-breeding food supplementation effects on reproduction

Multiple food supplementation experiments have been carried out during the pre-laying period in order to determine to what extent additional food at that time can advance laying and increase the number or the quality of the eggs laid. Given that most experiments (Schoech and Hahn (2008) mention more than 100) find that food supplemented females lay earlier than control birds and some lay a larger clutch or larger eggs, we must conclude that at this critical time of year food is limiting. Nager *et al.* (1997) reviewed food supplementation experiments and carried out within-species pairwise comparisons for all species with at least two years of food supplementation. They found that in the less favourable year the difference between food supplemented birds and controls was larger than in the favourable year, and this both as concerns the extent to which laying was advanced and the size of the clutch produced (Fig. 4.6).

Perrins (1970) argued that birds lay as early as they can obtain sufficient food to form the eggs, and the fact that most food supplementation experiments cause advanced laying would support Perrins' *Constraints Hypothesis*. On the other hand the response to supplemental food before laying depends on broodedness (Dhondt 2010a; Dhondt *et al.* 2002; Svensson 1995). Species that normally lay a single clutch in a season respond less strongly or not at all to supplemental food (mean of 12 species 3.8 days ±1.01), while multi-brooded species respond more strongly: nine double-brooded species advanced laying on average 7.6 ±1.16 days, while two triple-brooded species advanced on average 18.4 ±2.46 (Dhondt 2010a). This modulated response to supplementary feeding is most

56 | Interspecific Competition in Birds

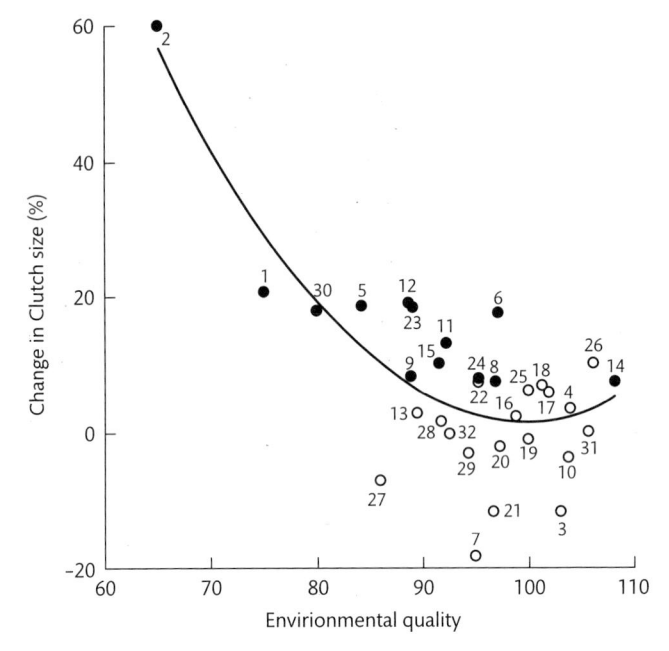

Fig. 4.6 Effect of supplemental feeding on clutch size in relation to natural environmental quality. Each point represents a different study. Results presented in graph involved 22 different species. Supplementary feeding resulted in significant increases in clutch size (filled circles) at below-average environmental conditions but not (open circles) at average or better conditions. The line shows the overall relationship between increase in clutch size of food supplemented birds and environmental quality ($r^2 = 0.64$, $P < 0.001$). From Nager et al. (1997), with permission from John Wiley and Sons.

likely adaptive and illustrates that species that will only be able to raise, at best, a single brood in a season should be particularly careful in responding to environmental cues. They should make their decision to start laying by using multiple cues, not food abundance alone. This group of species follows Lack's *Anticipation Hypothesis* that states that birds time their breeding season so that the period of peak food supply overlaps with the time of peak demand by nestlings. Birds, on the other hand, that will be able to raise multiple broods in a season should start laying as soon as conditions permit (Perrins' *Constraints Hypothesis*).

In the single equatorial study of food supplementation (a stonechat population in Tanzania) the birds advanced laying by 44 days (Scheuerlein and Gwinner 2002). Scheuerlein and Gwinner argued that in a system in which natural food conditions vary strongly between years, birds should respond strongly to supplemental food, especially as close to the equator day length cannot be used to make the decision on when to start laying. They proposed that in their study this strong response to supplemental feeding is adaptive under environmental conditions that are relatively constant within a given year but may vary considerably between years.

In a recent large-scale winter feeding experiment which stopped in early March, six weeks before breeding started, blue tits not only laid 2.5 days earlier but also increased productivity in the subsequent breeding season because of increased fledging success, an as yet unsuspected long-term effect of winter feeding (Robb *et al.* 2008). A similar, but much smaller scale, experiment in Korea (Park *et al.* 2004) found that in a plot in which hog fat was supplied from November to February marsh, great and varied tits laid a larger clutch consisting of bigger eggs.

The fact that most studies show an advance in laying date when food is supplemented means that competition for food at that time of year must be intense, although the extent to which food is limiting varies between years, locations, and species. In larger species that lay eggs at intervals of two or more days, supplemental food often results in shorter intervals between successive eggs (Aparicio 1994), suggesting again food limitation during the egg-laying phase.

4.7 Food manipulations during the breeding season

During the breeding season birds not only need food for themselves but also to raise their young. Although birds breed at a time of year when food is very abundant, adults may still have to make allocation decisions between self-maintenance and investment in the offspring. David Lack (1966) showed that many species time reproduction so that peak food demand for the nestlings coincides with the natural food peak. Birds can be food limited because of: insufficient quantity (not enough food to raise all young successfully); insufficient time to collect the food because it is too hard to find; mistiming (birds miss the food peak); food quality being inadequate; or some essential element being lacking. Species that do not obtain sufficient calcium in their normal diet to either form eggs or to provide to the young will forage specifically for this resource (Dhondt and Hochachka 2001). There are extensive data showing that the number of young raised is correlated with the amount of food available at the particular time and place of nesting (e.g. Blondel *et al.* (2006)). This does not necessarily mean that food is limiting. It could be that birds have evolved adaptations to adjust clutch size to the amount of food that will be available, although neither great nor blue tits (Dhondt *et al.* 1990b) nor starlings (Granbom and Smith 2006) adjusted clutch size to food availability in the breeding season. In the case of the tits, clutches were larger than optimal in poor quality habitats and smaller than optimal in high quality habitats, while in the starling example clutch size did not differ between habitats of different quality. In the following chapter, I will provide ample evidence for density-dependent effects on reproduction. If food is limiting, then less food will be available for each pair and reproductive success will decrease.

In a recent review, Robb *et al.* (2008) summarized the proportion of experimental studies that documented a significant effect of supplementary food on various components of reproduction (Fig. 4.7). It is clear that in general supplementary food had a positive impact on one or several aspects of reproduction, but that different components were not affected equally, which implied that the exact period when food was limiting, and hence intraspecific competition occurred, varied. They also found some studies in which supplementary feeding had an adverse effect on reproduction. A recent careful and well-replicated study on great and blue tits confirmed that this can be the case (Harrison *et al.* 2010). The exact mechanism through which this comes about is unclear.

In the following sections I will limit myself to experimental effects which have complex, surprising, or novel effects. Effects of insufficient summer food can vary from delay in clutch initiation, hatching failure, nestling mortality or quality, to reduced adult condition and survival.

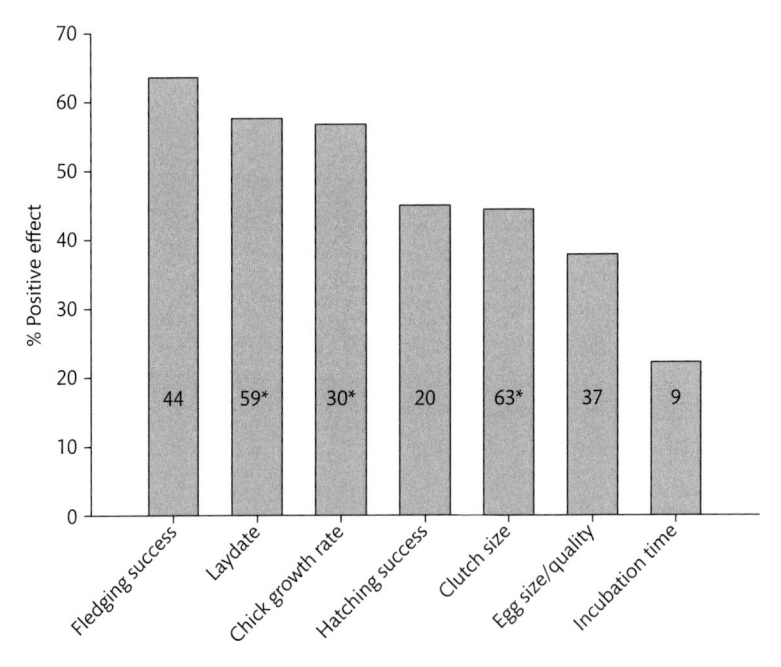

Fig. 4.7 Percentage experiments in which a positive effect of supplementary feeding during the breeding season was observed on various components of breeding success. Numbers in the bar represent sample size. A * next to the number indicates that in one of the experiments the effect of supplementary food was negative. Effects of supplementary food differed on different components of the reproductive cycle. Based on Table 2 in Robb *et al.* (2008a).

4.7.1 Food reduction experiments yield variable results

A small number of food reduction experiments have been carried out during the breeding season, either by spraying the foliage with *Bacillus thuringensis* (which kills caterpillars when they ingest it) or by spraying with insecticides, usually aimed at particular taxa of arthropods. In a two-year, large-scale caterpillar reduction experiment, Marshall *et al.* (2002) found a delay in laydate of red-eyed vireos, underlining the importance of food abundance on the initiation of breeding (Food Constraints Hypothesis). The absence of differences between three 30-ha experimental and six control plots in nesting success, clutch size, hatching success, nestling mortality, or adult mortality (that could be attributed to the experimental reduction of caterpillars) did not support the idea that during the nestling stage food was limiting. Rodenhouse and Holmes (1992) performed a similar experiment with black-throated blue warblers in three successive years. In none of the years could between-plot differences in laydate be attributed to food reduction caused by spraying. In only one of three years was there any effect. In 1983, when caterpillars were very abundant, the birds on the experimental plot, in which caterpillar abundance had been reduced, made significantly fewer nesting attempts and the diets of nestlings included fewer caterpillars. Clutch size, hatching success, and number of young fledging per nest did not differ among the food reduction site and controls. Also, the reduced number of nesting attempts per pair in the food reduction site in 1983 did not significantly lower production of young per pair. In both other experimental years natural caterpillar abundance was much lower than in 1983 and spraying had no effect. Based on natural variation in caterpillar abundance Rodenhouse and Holmes (1992) concluded that neotropical migrant bird species are probably limited periodically by food when breeding in north-temperate habitats, a conclusion later confirmed in their study site by Sillett *et al.* (2000).

In a food reduction experiment with the chestnut-collared longspur, a grassland species that primarily (> 85%) feeds its young with grasshoppers, the experimental reduction of grasshoppers by more than 90% through spraying resulted in the birds switching to alternate food sources, but without an effect on mass nor size of the young at fledging (Martin *et al.* 1998). Another insecticide treatment, causing food reduction in shrub-steppe habitat, caused effects on nestling size in one year out of two in each of two species studied, but no effect on number of fledglings nor nestling mortality (Howe *et al.* 1996).

All in all the few examples of experimental food reduction, although logistically very demanding, had only limited effects. There are several possible, not mutually exclusive, reasons for these negative results which include, but are not limited to, the following: food reduction is largely ineffective, meaning that food abundance is not reduced below a critical threshold for a sufficiently long period, as might have been the case in the red-eyed vireo study; natural food abundance varies between plots and years in unpredictable ways, as seems to have been the case in the black-throated blue study and in the red-eyed vireo study in a year with an unusually cold spring; as a result effects might be discerned in some years but not in others; birds are flexible and able to shift their diets opportunistically as clearly was the case in the chestnut-collared longspur study, and could have been the case in several other studies in which the treatment targeted caterpillars only.

4.7.2 Food addition: effects on nestlings, parents, both, or neither

Garcia *et al.* (1993) studied possible trade-offs between future and current reproductive success in mountain bluebirds that were supplemented with mealworms during the nestling period. Supplemented young were heavier at fledging compared to young from control nests. Food supplementation had no effect on adult male mass, while it did have an effect on adult female mass: supplemented females maintained their body mass while unsupplemented adult females lost about 10%. The authors conclude that adult females allocated the additional energy to self-maintenance rather than to increased investment in current offspring while males did not. The authors assume that this behaviour should impact survivorship, and that therefore food during the breeding season would limit lifetime reproductive success in this species.

Davis *et al.* (2005) provided artificial food to a declining population of parasitic jaegers in the Shetland Islands once the young had hatched. This food represented about 20% of the nestling's assimilated protein but birds with additional food did not grow larger. Adult female, but not male, attendance was significantly higher in food supplemented pairs, and food supplemented pairs raised 71% more nestlings than control pairs. Finally, supplemented adults showed a significantly higher return rate the following year than the control birds. In this jaeger population, which suffers from a declining food stock, food limitation affected both current and future reproduction.

In a similar experiment Ritz (2007) provided supplemental food to South Polar skuas. As in the previous study, the author found no effect on chick growth but found an effect on adult attendance. In this study, however, males but not females responded to supplemental food by increasing nest attendance, and male but not female condition increased during the nest rearing stage. Ritz concluded that food was not very limiting in his population.

As in the previous section the results underline the variable effects of food manipulation experiments, implying that the intensity of intraspecific competition for food might vary from absent to intense. Furthermore, the effect of competition for limited food resources will depend on a species' life history.

4.7.3 Interactions between food supplementation and predation

As during winter the effect of food supplementation during the breeding season can interact with other factors. In very elegant experiments, Liana Zanette and colleagues showed a synergistic effect on reproductive output in song sparrows between additional food and predator presence: with additional food, reproductive output increased by a factor of 1.1; in the absence of predators reproductive output increased 1.3 times; but when both food was supplemented and predators were absent, song sparrow annual fecundity increased almost two-fold over what was expected if the factors acted independently and their effects were simply additive (Zanette *et al.* 2003; Zanette *et al.* 2006). The authors suggested that *'this more than additive result most likely stems from indirect predator effects: lower predator abundance means parents can spend more time foraging, taking full advantage of the added food and thereby substantially reducing partial clutch/brood loss.'* In a similar experiment, Preston and Rotenberry (2006) found that the combined treatment of food supplementation and a reduction in predation pressure was slightly less than additive. They suggest that whether effects of food and predation on population parameters are synergistic or additive depends on the life history of the species and the environment in which they live (Preston and Rotenberry 2006). In either case these experiments on passerines suggest that food can be limiting during the nestling stage. It is important to point out that food limitation at that time, and hence the possible existence of intraspecific competition, can depend on other factors, such as the presence of predators.

4.7.4 Interactions between food supplementation and habitat quality

The effect of supplemental feeding in goshawks depended on the natural food abundance in a territory: extra food reduced nestling mortality in low-quality territories, but not in high quality territories (Byholm and Kekkonen 2008). In contrast to this result, Granbom and Smith (2006) also showed a positive effect of food supplementation on the growth and survival of starling nestlings in habitats of different quality, but did not find a significant interaction between habitat quality and supplemental feeding. This suggests that food was equally limiting in poor as in good habitat, although clutch size did not differ between them.

4.7.5 Surprising effects of food supplementation

Food limitation during the chick rearing period will result in an increase in sibling competition, which can generate some surprising results. Fabrizio Grieco

(2003) observed that providing supplemental food to nesting blue tits resulted in reduced asymmetry of nestling tarsus length. Nilsson and Cardmark (2001) manipulated marsh tit brood size, thereby increasing sibling competition in augmented nests. In these nests the mass of the heaviest young was the same as in nests of reduced size, but the mass of the lightest nestling was considerably less. The latter redirected the energy provided by its parents to growth of the wing at the expense of general size, so that they were able to fledge with their siblings. In house sparrows rates of extra-pair paternity were five times lower among pairs nesting at sites continuously provided with supplemental food compared to nests at sites without extra food (Vaclav *et al.* 2003).

In the marsh and blue tit studies supplemental food reduced the intensity of intraspecific competition between sibs leading to increased overall fledgling quality. In the house sparrow example supplemental food shifted the outcome of intersexual competition in the favour of the male, who spent more time with his partner and could thereby more effectively control access to the female by other males.

4.7.6 Food supplementation and immune competence

In recent years field tests of immune competence have become more commonplace. Researchers regularly measure responses to injections of Phytohaemagglutinin (PHA; to measure T-cell immune competence), sheep red blood cells (to measure humoral immune competence), or other proteins. This permits the detection of subtle effects of food supplementation in birds, when the more straightforward measurements (survival, mass) do not show any effects.

Thus Saino *et al.*(1997) found in barn swallows and Gasparini *et al.* (2006) in kittiwakes increased immuno-competence in food supplemented nestlings, although in swallows T-cell immuno-competence improved (humoral not tested), while in kittiwakes humoral but not T-cell immuno-competence improved in the experimental group.

A particularly interesting experiment using magpies showed a significant interaction between the effect of food supplementation and parental quality on T-cell mediated immune response of the nestlings (De Neve *et al.* 2004). They provided a high-calorie paste, enriched with essential micronutrients, to half of the nestlings in each nest every other day, and measured T-cell response after a PHA challenge at 16 days. The T-cell response was significantly stronger in food supplemented nestlings, but the difference between food supplemented young and control nestlings increased as nest size decreased. Since in magpies nest size reflects parental quality, this result shows again that the effect of supplemental food is modulated by other factors in such a way as to have a stronger effect when conditions are more constraining.

4.7.7 Effects of specific food supplements: carotenes and calcium

Whereas we have so far mostly looked at effects of the amount of supplemented food, some experiments found that specific elements had an impact. Thus Fenoglio *et al.* (2002) found that providing nestling moorhens with beta-carotene-rich diets caused these birds to have increased immuno-competence and improved behavioural performance. Cucco *et al.* (2006) observed improved growth and cell-mediated immunity, but no change in behavioural performance, in grey partridges supplemented with beta-carotenes. Interestingly, providing carotene supplements to the females before laying had no effect on chick physiology, suggesting that it would only be during the nestling stage that birds could compete for carotene-rich food.

In an important field experiment Jaap Graveland proved that the reason why great tits breeding on calcium-poor soils in the Netherlands produce eggs with thin and porous shells (that often do not hatch), or even do not lay at all, reflects calcium deficiency in their diet (Graveland and Drent 1997). This conclusion has now been confirmed in numerous studies, summarized by Patten (2007).

4.8 Predation by birds and other taxa can reduce food availability and thus have indirect effects

Food supplementation or food reduction experiments vary the amount of food available. As described in the previous sections many of these experiments show that birds do better when provided with additional food, and may do less well when natural food is reduced. What is further needed to determine the extent to which food can be limiting are experiments testing whether bird predation does actually cause a measurable reduction in natural food availability, thereby possibly exacerbating the intensity of intraspecific competition for food. Some experiments tested if predator exclusion increased food abundance in exclosures compared to unmanipulated areas, while a series of studies explored if exclusion of birds had indirect effects and influenced the extent to which damage to leaves or fruits increased.

Arthropods are an important food source for tits during winter. Very few studies have measured natural food availability during winter and even fewer have related this to bird numbers. Using exclosures, a Swedish study documented that birds take a high proportion of the arthropods present during winter. At the end of winter enclosed branches, from which the birds had been excluded, had 50% more large spiders than the branches on which the birds had been foraging (Askenmo *et al.* 1977), implying that during winter tits reduce arthropod prey considerably, and hence gradually increase intraspecific competition for food. Although they did not relate food abundance to bird numbers, they nevertheless argued that food reduction during winter through bird predation might cause a density-dependent mortality due to starvation among the birds. Taking advantage of the fact that coal tits in pinewoods rely heavily on the larvae of *Ernarmonia conicolana* (Heyl.) that live in pine cones, John Gibb documented that up to 80% of these larvae were taken by the tits and found that, over a five-year period, at the *end* of winter there existed a close correlation between the abundance of the food and the numbers of coal tits remaining (Gibb 1960).

In one of the first breeding season exclosure experiments, Dick Holmes and colleagues placed crop protection netting over 10 areas of about 36 m² understory shrub to measure the impact of bird predation (Holmes *et al.* 1979). For most arthropod taxa (Arachnida, Coleoptera, Homoptera, and Hemiptera) they did not find a difference in abundance between the exclosures and the control sites. They did count, however, significantly more Lepidoptera larvae inside than outside the enclosures. They estimated that birds removed an average of 37% of the caterpillars from understory foliage. Other experimental studies also found significant effects of birds as predators on the abundance of their prey. These studies include predation of reed buntings on scale insects (Kaneko 2005), of shorebirds on chironomid larvae (Sanchez *et al.* 2006), and of birds on spiders (Gunnarsson 2007). In recent years more sophisticated experiments have tested for the possible effects of multiple predators. In one such very interesting experiment, the effects of predation by spiders and birds on caterpillars was shown to be non-additive (Hooks *et al.* 2003). They measured arthropod abundance, leaf-chewing damage, and final plant productivity associated with broccoli in four treatments: predation by birds only, predation by spiders only, predation by both spiders and birds, and no predation by either group. They found a very strong effect of bird predation on caterpillar numbers which translated into a very low level of defoliation. Spiders had some effect but the effect of the two predators was not additive.

Rosa *et al.* (2008) similarly tested for the effect of predation by birds and by nekton (crustaceans and fish) on the abundance of a polychaete worm *Hediste diversicolor* in an estuarine setting by creating exclosure plots that excluded birds and nekton, nekton only, and neither, for a nine month period. At the end of the experiment the average density of *H. diversicolor* in the plots protected from all predators was eight times greater than in those without any protection.

More recently a series of studies has explored not only direct effects of bird predation but also its indirect effects. By reducing herbivore arthropods, birds can indirectly reduce damage to fruits (great tit in orchards, Mols and Visser (2002)), or foliage (Koh 2008; Van Bael and Brawn 2005; Van Bael *et al.* 2003; Van Bael *et al.* 2007). The importance of such trophic cascades varies between habitats and years (Mazia *et al.* 2009; Van Bael and Brawn 2005).

Since most of these experiments document a considerable impact of bird predation on prey abundance one can infer that prey abundance is often limiting for the bird predators and that, therefore, competition for food can be important.

4.9 Food supplementation experiments as a conservation tool

Reintroductions are a common tool in conservation. When introducing captive-bred individuals they are usually provided with supplemental food so that they can gradually learn to provide for themselves. This does not provide any information on the extent to which food can be limiting. Thus the routine long-term provision of food for species such as old world vultures at 'vulture restaurants' is simply assumed to be beneficial, although usually not tested. Daniel Oro and colleagues have analysed the effect of long-term feeding on the Spanish population of the bearded vulture using capture-mark-recapture techniques (Oro *et al.* 2008). They concluded that survival of pre-adults but not of adults increased because of the use of artificial feeding sites, suggesting that food is limiting for the younger but not for the older age classes. This indicates that competition between age classes impacts population size.

Another study on a reintroduced vulture population found that griffon vulture group size increased with food biomass, leading to increased aggressiveness, suggesting that interference competition at food sites was important (Bose and Sarrazin 2007). They recommended not only increasing the number of carcasses but also the number of feeding sites. This would induce dispersal and reduce intraspecific competition.

In New Zealand reintroduction of endemic bird species on islands is a relatively common practice, often after removing the non-native mammalian predators. As of March 2002 there had been 224 introductions of native species to New Zealand islands (Armstrong *et al.* 2002). In one such example the population dynamics of the reintroduced populations were carefully monitored for at least five years during which the effects of experimental food addition were tested (Armstrong *et al.* 2002). The team reintroduced New Zealand robin and stitchbird (or hihi) on Tiritiri Matangi, a 220-ha offshore island near Auckland, and saddleback and stitchbird to Mokoia, a 135-ha island in Lake Rotorua. Both islands were free of mammalian predators, but had vegetation that was regenerating after clearing. Food, therefore, could have been limiting. The populations of New Zealand robin and of saddleback grew without management, but the stitchbird populations required supplementary feeding in order to maintain themselves. Removal of additional food sources in one year caused the population to decline. On Mokoia, however, even with supplementary feeding stitchbirds could not survive because they suffered from aspergillosus caused by the pathogenic fungus *Aspergillus fumigates*. After having determined that adult survival rates were too low for the population to survive, Armstrong and his team removed the 12 remaining stitchbirds from Mokoia and combined them with a reintroduced population on another island which was growing in response to supplementary feeding (Armstrong *et al.* 2007).

As a final example, I want to summarize the work of Shane Heath and colleagues who tried to save the San Clemente loggerhead shrikes, a critically endangered subspecies that has been the recipient of an extensive and expensive recovery effort. In their approach they combined experimental food supplementation with rodent control, targeting individual pairs. Pairs receiving rat control alone produced an average of 1.1 more fledglings than pairs receiving no management during the same period, but adding food on top of rat control increased productivity by an additional 1.4 fledglings (Heath *et al.* 2008). As effects of food supplementation were most pronounced during dry years (when food was most likely less abundant), the intensity of intraspecific competition for food varied with environmental conditions.

By combining reintroduction experiments (important for conservation) with quality experimental science the researchers not only gained insight into their systems, but also contributed to the basic understanding of processes influencing population dynamics in general. The examples presented above show that in many, but not all, studies food was a limiting factor. As reiterated in this book, interactions between various factors also need to be considered.

4.10 Conclusions

Experiments whereby the amount or the quality of food is manipulated at different times of the year have yielded mixed results, suggesting that food is limiting throughout the year and that, therefore, competition for food may be common.

Effects of food supplementation in the pre-breeding season usually caused the laydate to advance and sometimes clutch or egg size to increase, suggesting that many bird populations are food limited at that time of year. Competition for food during the pre-laying has both direct and deferred effects. Direct effects are delayed laying, causing indirect effects such as reduced clutch size (if clutch size decreases with first-egg date), nest success, fledgling condition, or reproductive rate. The probability of initiating a second clutch is also higher in pairs that fledged their first brood earlier. Deferred effects are caused by effects on fledgling date (young that fledge later usually have reduced survival) and of laydate on fledgling mass. Competition for food during the pre-laying season, therefore, increases between-individual differences in reproductive success or in fledging survival after independence.

Effects of supplementary feeding during the breeding season are summarized in Fig. 4.7. The large number of experiments performed, and the detailed analyses of the results, illustrate the mostly positive, but variable, effects of supplementary food. This suggests that the intensity of competition for food can vary during the breeding season. A surprising result was that sometimes supplemental feeding has an adverse effect on reproductive success (Harrison *et al.* 2010).

Some authors performed food reduction experiments. These are logistically much more demanding than food supplementation experiments, because large areas have to be treated with insecticides or with *Bacillus thuringensis*, and because of the need to measure treatment effects on arthropods at regular intervals. Given the large proportion of food supplementation experiments that had a positive effect on reproduction, one would expect that food reduction would have the opposite effect. That was not the case. Most food reduction experiments had no or minor effects; and in replicated studies, not in each year. Two of the studies found that food reduction resulted in the birds switching to alternative foods not affected by the treatment, perhaps explaining why the effects of the experiments were small.

Food supplementation during winter had positive effects, especially in years and locations in which natural food was low, temperatures extreme, or predation intense. Two good examples of how, at the population level, the response to food addition is modulated by predation are changes in great tit body mass in the presence and absence of sparrowhawks in beech mast and non-beech mast years (Fig. 4.4), and the increase in reproductive success of song sparrows to food supplementation in areas with abundant or few predators. Individual responses to food supplementation can vary strongly with dominance position in species that live in dominance-structured single species or mixed flocks. Subordinate individuals benefit more, although they still have to take more risks (De Laet *et al.* 1985), implying that in dominance-structured systems effects of intraspecific competition are unequally distributed between individuals. Recent work has documented that winter feeding not only impacts over-winter survival, but can also have delayed effects on reproduction (Robb *et al.* 2008).

Experiments with reintroduced or threatened populations underline the extent to which food availability can limit success, and the need to take into account interactions between food availability and other environmental factors.

Given that food is often limiting, a necessary condition for the existence of intraspecific competition is met, and we can now test to what extent interspecific competition for food also exists. Note that intraspecific competition does not necessarily occur when food is limiting, and that many other biotic (predation, pathogens) and abiotic factors (cold winters, rainy breeding seasons, drought) will influence population size. However, as illustrated in this chapter, food limitation does in many cases lead to competition for this limiting resource.

5
Nest sites as a limiting resource

5.1 Are nest sites limiting in cup-nesting species?

It is difficult to determine directly if nest sites of non-cavity-nesting species are limiting, although some authors suggest that limitation of high quality nest sites can limit populations of endangered species. A number of observations and experiments provide circumstantial evidence that this might be the case.

Predation on bird nests is usually very high and birds attempt to build their nests at the safest sites. Tom Martin formulated and tested two not mutually exclusive, mechanistic hypotheses that related to factors that influence the risk of nest predation. The Total Foliage Hypothesis states that *'predation risk decreases with increases in total vegetation in the nest patch because greater foliage density inhibits transmission of visual, chemical and auditory cues by prey'*. The Potential Nest Site Hypothesis states that *'increases in the density of plants of the type used by prey reduces the probability of predation because it increases the number of potential prey sites that must be searched and can cause the predator to give up before finding the occupied site'* (Martin 1993). Of 36 studies relating nest concealment to predation 29 found predation rates to be lower in nests with greater concealment (Martin 1992), supporting the Total Foliage Hypothesis. Foliage structure, per se, thus often seems to play a role in nest safety. Recently Chalfoun and Martin (2009) experimentally tested the Potential Nest Site Hypothesis and found that birds preferentially nested in habitat patches containing greater densities of potential nest sites than the average available throughout the landscape. This reduced predation probability independently from foliage concealment. These findings underline the likelihood that predator search behaviour can limit the number of quality sites where birds can build their nests and successfully raise young. Furthermore, as the activity of different predators can vary across the season the nest locations that are most at risk will also change. This is nicely illustrated by a study of snake predation by Texas ratsnakes on nests of black-capped vireos and golden-cheeked warblers in central Texas (Sperry *et al.* 2008). Both species, but especially the vireos, were more at risk when snakes were more active: seasonal variation in avian nest success was driven by seasonal variation in predator activity. The authors proposed that the difference in the strength of the relationship between snake activity and nest predation between the two bird species was best explained by the difference in nesting habitat between vireo and warbler. Snakes would use vireo habitat mostly when foraging, while snakes would use warbler habitat also when resting or digesting so that more nests were taken opportunistically.

There are many examples in which birds vary the placement of their nests during the season as a function of changes in vegetation or changes in the activity of various predators. Open nesters often increase the height at which they build their nests as the season progresses. Field sparrows, for example, increased the mean height at which they built their cup nest from 26 cm in May, to 38 cm in June, and 47 cm in July and August (Best 1978). Given that 90% of the nests failed, this suggests that good sites were scarce. A similar observation was made for Eurasian blackbirds: Ribaut (1964) observed that throughout the season the height at which nests were built gradually increased. As leaves developed on deciduous trees, higher nests were better concealed and nest sites protected from predation became more numerous. This underlines again that good nest sites for this cup-nesting species were in limited supply, especially early in the season. In this study also nest failure rate was 86%, mostly caused by predation.

Tom Martin carried out an interesting set of experiments to test the idea that a possible reason why nest locations vary between different open nesting species (thereby further reducing the number of possible locations for successful reproduction and increasing the intensity of competition) is related to the very strong predation pressure they suffer. He placed eggs in one, three, or seven experimental nests sited in a clump. The number of intact nests after three days was 10% if all seven artificial nests contained eggs, but 80% if only one of the seven nests contained eggs. This indicates that predators changed their search behaviour after a success, as described in detail by Jamie Smith for European thrushes foraging on earthworms (Smith 1974). In a second experiment Tom Martin compared predation rates on artificial nests either all placed in a similar location in a tree (simulating a single species assemblage) or distributed over four different locations (simulating a four species assemblage) (Martin 1988). He found in all replicates that more nests remained intact in the multiple species treatment, and concluded that predation behaviour may represent a process that favours coexistence of species that partition nest sites (see Fig. 5.1). This again emphasizes that heavy predation on nests severely limits high quality nest sites, and illustrates an interesting example of a complex form of interspecific competition. This may lead to dominant species keeping subordinate species from using preferred nest sites (see Fig. 5.2). Thus, when orange-crowned warblers were removed, Virginia warblers shifted their nest sites to locations very similar to those used by the orange-crowned warblers, implying that optimal nest sites, at least for the Virginia warblers, were limited through interspecific competition (Martin and Martin 2001).

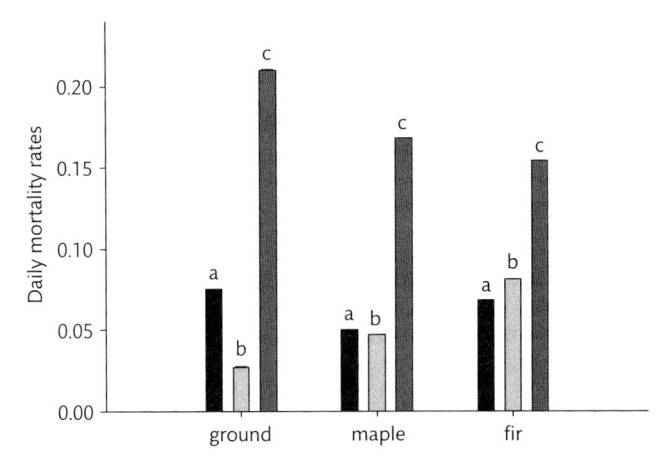

Fig. 5.1 Daily mortality rates at the egg stage in three different habitats. In all habitats predation rates (expressed as daily nest loss) were lower when (a) artificial nests were partitioned among multiple sites (mimicking nest placement of different species), than (c) when nests were located at similar sites (mimicking nest placement by a single species). Mortalities in the multiple nest site treatment were similar to those of natural nests that belonged to different species. Based on data in Table 1 in Martin (1988).

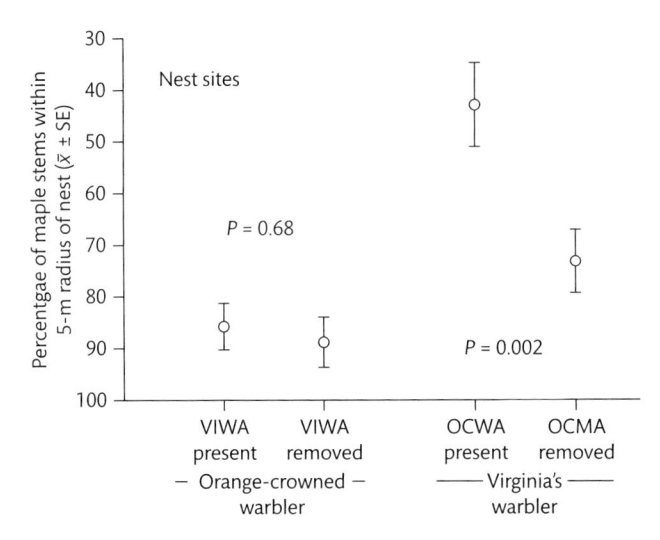

Fig. 5.2 Experimental evidence for the exclusion of subordinate Virginia's warblers (VIWA) from preferred nest sites by the dominant orange-crowned warblers (OCWA). When OCWA are experimentally removed VIWA select nest sites that are more similar to those used by OCWA. Behavioural observation documents exclusion of VIWA from preferred nest sites through direct aggressive interference of OCWA, supporting the hypothesis that high-quality nest sites are in short supply, even in open nesting species. From Martin and Martin (2001) with permission of the Ecological Society of America.

5.2 Are cavities limiting for cavity nesters?

Among the 103 bird families for which they provide information, Bennett and Owens (2002) list 33 families of which some or all species breed in holes or burrows. Cavity nesters can be found not just in forests but in a great diversity of habitats. The very unusual Tibetan Hume's ground tit, for example, only recently placed in the Paridae (James *et al.* 2003), breeds in a hole dug in the ground. Many seabirds either use existing burrows (including nest-boxes; see Chapter 8), or excavate a burrow themselves; some use cracks in or under rocks.

Some species excavate their own holes (primary cavity nesters), others can only breed if they find an adequate existing cavity (secondary cavity nesters). Some excavators will use an existing hole if available, probably because excavating a hole requires a lot of energy and takes a lot of time. Only one group of birds, the woodpeckers, have special adaptations which make it possible to excavate in living trees. Some woodpeckers even excavate separate cavities for nesting and for roosting. Although one would expect that excavator species would not be limited by nest sites in appropriate habitat, this is not always true. Thus the density of Gila woodpeckers increased with increasing numbers of large saguaro cacti, in which they excavate their holes; the species defended its holes against other species both in direct agonistic interactions and through interspecific territoriality (Brenowitz 1978).

There is an active discussion as to whether or not cavities limit nest sites and hence numbers. On the one hand Tomasz Wesołowski (2007), in a review of the important work he and his colleagues carried out over many years in Białowieza Forest in Poland, is quite explicit: there is no shortage of holes in this forest during the breeding season, and hence cavity nesters do not compete for holes. He points out that the conclusion that cavities are limiting is based on the fact that most studies of cavity nesters have been performed in temperate secondary forest, where the trees are relatively young and therefore contain few or no natural cavities. He also concludes that woodpeckers do not always play an important role in creating cavities, because in Australia there are many cavity-nesting species but no woodpeckers. He claims that this conclusion is generally valid in natural woodland habitat around the world and that in primeval forest conditions, hole-nesting non-excavating birds are not, as a rule, nest site limited. He admits that he does not know why not all species have adopted cavities for nesting, since nest success in cavities is usually higher than in open nests (Wesołowski 2007), nor is it clear why some tit or nuthatch species have developed the ability to excavate nest cavities (a derived trait) while the ancestral taxa are secondary cavity nesters (Dhondt 2007). The traditional view is that excavating a cavity offers protection against predators: such cavities are tailored to the body size and needs of the species, and using a different cavity each year reduces the chance of being found by a predator. Observations on crested tits in Belgium (Lens and Wauters 1996) and on black-capped chickadees in the USA (Christman and Dhondt 1997) do not support this idea—at least for titmice—since 70% of crested tit nests and 62% of chickadee nests (both excavators) were depredated. Nest predation in eight studies of five species of excavating parids varied between 14 and 70% averaging 37% (Christman and Dhondt 1997; Mahon and Martin 2006). In a four-year study Mahon and Martin (2006) observed extreme variation (10–64%) in predation rates on chestnut-backed chickadee nests linked to the presence of other foods for the red squirrel. In mast years cone abundance was high and nest predation on chickadees low. I suggested that the smaller taxa among tits (and nuthatches) excavate their own nest because interspecific competition forces them to do this (Dhondt 2007). Nest usurpation by larger species of cavities used by smaller species is common, but the smaller species can protect themselves against this by making cavities with entrance holes that exclude the competitors. It is probably for this reason that some nuthatch species reduce the size of the entrance hole of the nesting cavity with mud (Matthysen 1998).

5.3 Are cavities in natural forests superabundant?

5.3.1 Experimental manipulation of resource availability

In contrast to Wesołowski's conclusion, there is a considerable body of experimental work that suggests or documents that cavities do limit the population size of cavity-nesting species, because when nest-boxes are added or removed there is usually a concomitant increase or decrease of their breeding population: Ian Newton lists 35 experimental studies involving 26 bird species and in all but a few a change in the number of cavities (either increase or decrease) resulted in a concomitant change in abundance of secondary cavity nesters (Newton 1998). These experiments, though, cannot really be used to evaluate Wesołowski's conclusion because most have some shortcomings.

First, only two had replicate control and experimental sites, and are thus experimentally valid (Brawn and Balda 1988; Gustafsson 1988b). On Gotland, Lars Gustafsson used nest-boxes in all study plots, but changed the number of nest-boxes available either to all species, or to the collared flycatcher (Gustafsson 1988b). Although he found clear evidence of strong interspecific competition for nest cavities between tits and flycatchers, the experiment does not test for a possible effect of the addition of nest-boxes in natural forest situation and hence does not prove that cavities in a natural forest are limiting. In a ponderosa pine forest at about 2,000 m in Arizona, Jeffrey Brawn and Russ Balda added nest-boxes to three plots, and kept two control sites unchanged. They collected data on bird density for all sites over three breeding seasons, and also during four pre-treatment years (Brawn and Balda 1988). The problem with this experiment is that the three treatment sites differed in silvicultural history. Two of the three plots had been thinned about 10 years before the experiment; the third plot had not been thinned for at least 60 years. The results were clear: in the plots that had been thinned recently the density of cavity nesters increased strongly with nest-boxes, both compared to the control sites and compared to the pre-treatment years. In the third plot (that had not been thinned recently) bird density was the same in pre-treatment years (1.03 pairs ha^{-1}) and in years with nest-boxes (1.08 pairs ha^{-1}). The conclusion, therefore, was that in thinned forest cavities were limiting, but that there was no evidence that this was also the case in more natural forest.

Second, all experiments listed by Newton, except one in Venezuela, were carried out in temperate regions: 20 in northern or western Europe, 12 in the US or Canada, 1 in Japan, and 1 in New Zealand, reflecting the bias in distribution of ornithologists. These experiments are therefore inadequate to generalize to conditions in natural areas in tropical regions.

5.3.2 How many suitable cavities are there in natural forests?

A different approach to determine if natural cavities are limiting is to count all cavities present in a plot and either compare that number to bird density or to occupation rate. Most of these studies concluded that in natural forest, and in contrast to forest in which trees had been removed, cavities were present in surplus amounts. In a deciduous forest in Sweden, Carlson *et al.* counted 60 cavities ha^{-1}, about half of which were limb holes (Carlson *et al.* 1998). In two different years bird nested in 9.1 and 5.3% of the cavities respectively. Cavities occupied by bird nests had narrower entrances, were located higher up, had smaller volumes, thicker walls, and a smaller circumference of the stem at the hole compared with unoccupied cavities. Furthermore the cavity characteristics differed between bird species. Nevertheless the authors concluded that bird density in this forest was not limited by cavities. As Ken Otter pointed out (*in litt.*) one cannot simply say that any crevice in a tree is a potentially usable nest. Many cavities that chickadees start are abandoned as soon as they hit hard wood, and the resulting hole is likely too small for a practical nest for any species. Yet, these often get classified as 'available cavities'. Realistically, one has to classify the cavities as being of the correct dimensions and potentially also having the correct insulative properties to make them acceptable to birds for nesting. Such cavities may well be limited, but wouldn't appear so if one counted every concavity in a tree as a potential 'cavity'.

A small number of studies have counted cavity density in primary forests in the tropics, and several of them were based on counts from the ground without systematic verification of whether the cavities were suitable (sufficient depth, overall size for nesting). Typically these counts found large numbers of cavities.

Boyle and colleagues found an average of 112 cavities ha^{-1} (29% in snags) in the primary forest at La Selva, Costa Rica (Boyle *et al.* 2008). Cavity density in different plots ranged from 9–131 ha^{-1}. Pattanavibool and Edge (1996) studied the effect of single-tree selection silviculture on cavity abundance in Huai Kha Khaeng Wildlife Sanctuary in Thailand where more than 30% of vertebrate species use cavities. They counted 407 cavities ha^{-1} in unlogged forest but only 189 in selectively logged areas. A suitable cavity was defined as one with a horizontal depth of more tham 7.5 cm and an entrance of more than 3 cm. They calibrated their ground-based counts by climbing 120 trees of the 3 main tree species. It is unclear though what really constituted a suitable cavity for the birds in their system, given that they did not measure cavity depth.

Donald Brightsmith counted cavities in primary forest in Manu National Park, Peru. He asked the question if competition or predation had been the main selective pressure causing cavity nesters to breed in termite mounds, a trait that developed independently at least eight times (Brightsmith 2005). He counted and inspected natural cavities present up to a height of 15 m, in a forest in which the forest canopy height is 35–40 m. He was careful to determine if cavities were suitable for secondary cavity nesters of less than 200 g using the following criteria: to be suitable the cavity had to be at least 4 cm deep and at least 7 cm wide, or if it was a horizontal tube at least 14 cm long. Although in his study site about 100 species are suspected to breed in cavities, none of the 29 cavities were used in 1996, and 11 of 64 in 1997. The occupants were one secondary cavity nester, four primary cavity nesters, three mammals, and three bees or wasp nests, leaving 84% of the available cavities empty. Of the 60 nest-boxes he provided in a 10 km^2 area only one contained an active nest in one year, and another might have been also used in the other year. Brightsmith concludes that apparently there is very little competition for nest cavities among small (less than 200 g) subcanopy, cavity-nesting birds at his site in Manu National Park. He does underline though that the unused cavities and the nest-boxes might not really have been suitable. Given the high number of cavity-nesting species in his study site, and the very low number of nests he found, his data suggest that counting cavities up to a height of 15 m only might have been insufficient. Furthermore, he does mention that canopy macaws fight over cavities and are therefore possibly nest site limited.

James Gibbs and colleagues, on the other hand, suggested that in the tropics cavities might be limiting precisely because the proportion of secondary cavity nesters among the bird species breeding in holes is high, and because the number of cavities in snags is lower in tropical and subtropical forests than in temperate and boreal regions (Gibbs *et al.* 1993).

In their review of cavity-nesting birds, Cintia Cornelius and colleagues suggested that in degraded or exploited neotropical forest, cavities are limiting but believe it might not be the case in primary tropical forest (Cornelius *et al.* 2008).

They underline that unused cavities are measurably different from active ones, suggesting that many of the existing cavities are not really suitable for secondary cavity nesters. They conclude that '*although cavities may be abundant, cavities of the right size and characteristics may be in short supply for many bird species, showing that it is important to consider cavity quality when assessing cavity availability and nest site limitation*'. More recently, Natalia Politi and colleagues found that in subtropical montane forests of the Andes, cavities used for nesting were a non-random subset of all available suitable cavities. Cavity suitability was verified with a pole-mounted camera. Birds selected cavities that were relatively high above the ground, had smaller entrances, and were excavated by woodpeckers (Politi *et al.* 2009). Overall, while the large numbers of cavities counted and the low proportion occupied suggest that secondary cavity nesters would not be limited in unmodified tropical forests, several authors emphasized that although cavities might seem superabundant, cavity quality might limit bird abundance and diversity (Camprodon *et al.* 2008; Lohmus and Remm 2005; Remm *et al.* 2008). Given the very limited information on cavity requirements for most tropical birds it is difficult to evaluate cavity suitability for most species.

Recent work by Cockle and Bodrati (2009) on nesting of the Planalto woodcreeper, a 60 g bird, illustrates this. Cockle and Bodrati found that natural nests of the species have an opening of 5–7 cm and are about 45 cm deep. When given a choice of nest-boxes in subtropical forest in Argentina the birds preferred deeper boxes and did not use any of the boxes less than 40 cm deep. The size of the entrance hole did not influence box choice. In a review of raptor species that breed in the dry tropical deciduous forests of Tikal National Park, Guatemala, Gerhardt reported that among the four species of Falconidae and the three species of Strigiformes that nested primarily in tree cavities, only the ferruginous pygmy owl used cavities excavated by woodpeckers. The other species used non-excavated cavities (Gerhardt 2004). Gerhardt concluded that cavities suitable for the owl species appeared abundant and not limiting. Cavities suitable for nesting by falcons were more specialized and rare, thereby possibly limiting numbers and enhancing intra- and interspecific competition for the scarce resource.

Ken Otter commented that the extent to which a particular species uses nest-boxes offered by researchers can differ between habitats. For years he has been trying to get black-capped chickadees to nest in boxes. Within his study sites in Northern British Columbia success rates are really low: in one year, 1 of 50 boxes was used, despite having birds breeding in every territory in which the boxes occurred, and there being an average of about three boxes per territory. His forests are largely an aspen/conifer mix, and aspen provides ample soft, dead wood for chickadees to burrow into. However, these same nest-boxes are used in the harder-wood deciduous forests of the Carolinian belt of the eastern US with much greater success (Otter, *in litt.*).

These results underline the importance of providing the correct boxes with the correct internal dimensions and entrance hole size. If 'wrong' boxes are offered the fact that they are not used cannot be invoked to conclude that cavities are superabundant. Small tropical bird species seem to like deep cavities.

Finally, effects of cavities can come about in indirect ways because birds often use boxes not just for nesting but also as roost sites. I found, for example, that blue tit breeding density increased markedly when small-holed nest-boxes (with an entrance hole excluding the larger great tits) were provided, and that this increase was independent of the number of large-holed nest-boxes present in which they often bred (Dhondt and Adriaensen 1999) (see Fig. 9.4). The mechanism underlying this increase in blue tit density was that when small-holed cavities were available in winter, large numbers of blue tits used them for roosting, and these birds then stayed in the area to breed. When no adequate roosting sites were present (small-holed boxes blocked in winter) blue tits did not stay in the area during winter, and small-holed boxes were mostly used for nesting by immigrant yearling birds in the breeding season.

5.4 Studies of nest web communities

Manipulating the density of available cavities does not always have direct effects, such as providing more or fewer nest or roosting sites. Indirect effects can be quite important, as illustrated in Blanc and Walters' study of red-cockaded woodpeckers (Blanc and Walters 2008). Red-cockaded woodpeckers excavate cavities in live trees. Northern flickers use and enlarge these cavities, which is detrimental to the woodpecker, but makes them suitable for large secondary cavity-nesting species such as the eastern screech owl and the American kestrel. By using metal restrictor plates, Blanc and Walters excluded flickers from red-cockaded woodpecker cavities. This did not cause a significant reduction in nest abundance of flickers who simply switched to using snags, but resulted in a (non-significant) reduction in nesting abundance of the large secondary cavity nesters (American kestrel and Eastern screech owl).

What the above example makes clear is that in order to understand effects of cavities on animal communities one needs to look at all animals that use cavities and not just at one or several bird species. Most experiments discussed so far have only looked at part of the problem: how many cavities are present? How do occupied and not occupied cavities differ? Does the density and species composition of breeding birds change when cavities are added or removed? Kathy Martin and John Eadie propose that it is more meaningful to study cavity-nesting bird communities as structured in nest webs (analogous to food webs) where both the cavity producers (primary cavity nesters) and the cavity consumers (secondary cavity nesters) are studied in their entirety (Martin and Eadie 1999; Martin *et al.* 2004).

Kathy Martin and her group at the University of British Columbia have not only looked at the entire nest web (as illustrated in Fig. 5.3), but have also carried out a series of long-term before-and-after-control-impact experiments in which they increased or decreased the number of cavities available, and this both in temperate and tropical unexploited forests.

In a study in mature aspen groves in the Cariboo-Chilcotin region of British Columbia, Canada, Kathryn Aitken and Kathy Martin tracked cavity use for two years, then blocked the high-quality nest sites (those used in both of the previous years) (Aitken and Martin 2008). They reduced the number of cavities from 27 to 15 per ha. Over the two years when cavities were blocked the abundance of cavity-nesting species decreased by over 50% and the number of bird species decreased from 9 to 7. The dominant secondary cavity nester, the European starling, decreased by 89%, but the mountain bluebird increased significantly, while tree swallow abundance did not change. The result showed a complicated response

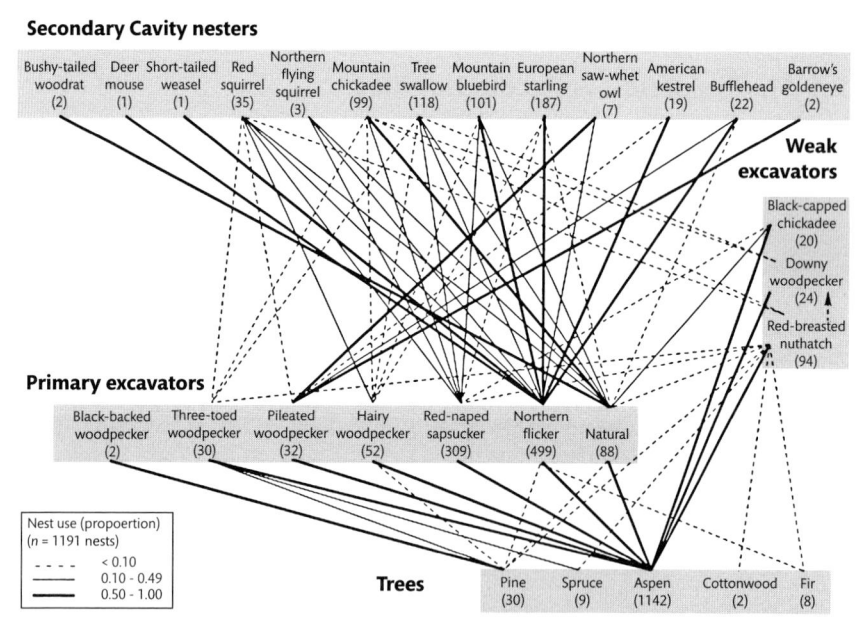

Fig. 5.3 A diagram of a nest web representing the resource flow (cavity or tree) through the cavity-nesting vertebrate community in interior British Columbia, Canada. Resource use in the nest web shows links between species using nests (secondary cavity nesters and excavators) and the excavator or tree species that provided the resource. Numbers under each species indicate the number of occupied nests for which there was information on the excavator or tree species used. Links for species with fewer than 15 occupied nests are considered preliminary findings. From Martin et al. (2004), with permission from Cooper Ornithological Society.

to the manipulation, whereby a subordinate cavity nester (bluebird) benefitted from the reduction in the more specialized and choosy species (starling), illustrating that the outcome of interspecific competition is not always predictable from known dyadic interactions.

In a more recent paper, they reported on a study in mature mixed conifer forest in central British Columbia in which they added nest-boxes. The experiment lasted a total of eleven years: six pre-treatment years, two treatment years during which nest-boxes of two sizes were added, and three years following nest-box removal. The pre-treatment observations suggested that cavities should not be limiting because, although the density of natural cavities was very low (< 2 ha^{-1}), cavity occupation rate was less than 10% in each year (Aitken and Martin 2011). In the literature such an observation is seen as evidence that cavities are not limiting. Aitken and Martin tripled cavities by adding nest-boxes on three experimental sites, while continuing to study cavity nesters on four control sites. The effects on the numbers of different cavity nesting species was quite variable (Fig 8.2). Mountain chickadee nests increased nearly nine-fold on treatment sites during the two years in which boxes had been provided, but declined by half the year following box removal to return to pre-treatment levels the following year. Red-breasted nuthatch density doubled during the treatment years and returned to pre-treatment density following box removal. Surprisingly only one nuthatch used a box for breeding. Red squirrel and northern flying squirrels increased strongly when nest-boxes were present: the red squirrels preferentially used large boxes, while the flying squirrels preferred small boxes. Thus both bird and both mammal species responded positively to the nest-box addition experiment but to a different extent and in a different way. As in previous examples, some of the effects were indirect, as shown by the increase in the density of red-breasted nuthatches (who often excavate nests), although only one used a nest-box for breeding. This could be an example of heterospecific attraction (see Chapter 8).

A recent experiment carried out by Kristina Cockle and Kathy Martin in subtropical Atlantic forest in Argentina is particularly interesting (Cockle *et al.* 2010). They studied cavities and cavity-nesting birds both in unexploited and in logged forest, and added nest-boxes in replicate plots. They found only 0–7 suitable cavities ha^{-1}, which was 19% of all potential cavities counted from the ground. When they climbed up to these potential cavities some of them were not cavities at all but a slight indent in the tree with no real cavity (Cockle *in litt.*). Based on their research the definition of what represented a suitable cavity was much more restrictive (> 12 cm deep) than in the study by Brightsmith (2005) (4 cm deep). The nest-boxes used in this study were 60 cm deep (as compared to 20 cm in Brightsmith's experiment) and were added at a high density (15 boxes ha^{-1}). Although the proportion of boxes occupied in unexploited forest was not very high (8–9%), this was sufficient to more than double the density of cavity nesters on the treatment plots, implying that it is very difficult to determine a priori what represents a suitable cavity, and that a low occupation rate of natural or artificial cavities cannot be used to infer that cavities are superabundant.

5.5 Conclusions

In this chapter I have attempted to evaluate the data that address the question of to what extent nest-sites are limiting. As regards open nesting species, indirect arguments indicate that, because nest predation is very high, quality nest sites are in limited supply, causing both intra- and interspecific competition for high-quality nest sites. The evidence concerning nest site limitation in secondary cavity-nesting species is confusing. It is very hard to determine how many suitable cavities are available for several reasons. First, most cavity-nesting species have different requirements, and cavities that are quite acceptable for one species may not be for others. Especially in tropical and subtropical regions, these requirements are not or poorly known, making it difficult to define suitability. Counting cavities from the ground, without later verification, is insufficient. Second, observing that only a low proportion of apparently suitable cavities are occupied is not sufficient evidence that cavities are available in surplus and thus not limiting, as illustrated by the experiments of Aitken and Martin (2011). Third, cavities are used not only for nesting but also for roosting. If birds roost by themselves, more cavities are required for roosting than for nesting. Fourth, birds are not the only users of cavities: mammals, reptiles, and insects also use cavities. Each probably has different suitability requirements, and in many case preemptive occupation of a cavity by one species removes the cavity from the real estate market.

What is clear, though, is that exploitation of forests for timber, even when removing trees selectively, has a major impact on the availability of suitable cavities, thereby posing a threat to the persistence of cavity-nesting species. Especially if large trees are selectively logged this impacts cavity-nesting species and will increase both the intensity of intra- and interspecific competition.

In order to determine the extent to which cavities are limiting one needs to (1) know the optimal dimensions of the cavities for each species; (2) verify the dimensions of all natural cavities; (3) study the spatial distribution of cavities of different dimensions in the context of the distribution of the territories of cavity nesting species (Newton, *in litt.*). If, for example, one tree per 10 ha contains 10 cavities suitable for a given species, and exclusive territories are 5 ha, then only one pair will be able to breed, although mathematically the number of suitable cavities per ha would not be limiting.

The objective of the last three chapters was to evaluate the evidence that space, food, or nest sites are actually limiting, and if so to what extent. To discuss this I preferentially emphasized field experiments. The evidence suggests that all three of the resources discussed can be and often are limiting, although none seem to be always limiting. When one resource (winter food for example) does not limit population size, another one (space, as influenced by territorial behaviour) or the availability of high-quality habitat can limit breeding population size and therefore population growth.

The general conclusion of these three chapters is that the necessary conditions for the possible existence of interspecific and intraspecific competition are fulfilled and that interspecific competition can exist over space, food, or nest sites as each of these resources is often limiting. The direct evidence for interspecific competition will be discussed in detail in Chapter 8.

6
The effect of intraspecific competition on population processes

6.1 Intraspecific competition seems to be generally important in birds

As explained in Chapter 2, a necessary condition for the existence of competition *between* species is that is that *conspecific* individuals compete among each other for limiting resources, and that this intraspecific competition has an effect on per-capita growth rate or on one or more of its components. In this chapter I want to explore which demographic parameters or population processes are affected by resource limitation, how frequently and in what conditions this occurs, and what mechanisms translate resource limitation into intraspecific competition.

I will first discuss this in general terms and illustrate intraspecific competition in birds using a handful of examples from long-term year-round studies. I will then discuss a small number of experimental studies that shed light on this and one specific mechanism that can result in density-dependent reproduction. Finally, I will review in detail density-dependence in tits and explore what resources cause this.

The most common way to document the existence of intraspecific competition is to document density-dependent effects, that is to show that as population size increases one or more demographic parameters change or that per-capita growth rate approaches zero. Thus a decrease in reproductive, immigration, or survival rate or an increase in emigration or mortality rate with increasing density would be evidence for the existence of intraspecific competition. Note however that density-dependent mortality, for example, can also be brought about by an increase in predation pressure through a numerical or functional response of the predator as prey density increases, or by the impact of parasites or pathogens in social species (Hochachka and Dhondt 2000) rather than by limitation in the resources discussed in Chapters 3–5. Note also that a reduction in reproduction with increasing population density could be the result of adaptive restraint rather than food limitation, as proposed by Simmons (1993), although this idea, which goes back to Wynne-Edwards (1962), is controversial and usually not supported. It is therefore useful to identify the resources over which intraspecific competition occurs and the mechanisms that cause the density-dependent relationship.

Careful analyses of time series data can provide strong powers of inference. An analysis of 1,198 times series data of both invertebrates and vertebrates showed that in about 75% of them per-capita growth rate was density dependent, implying that negative feedback mechanisms exist that regulate numbers (Brook and Bradshaw 2006). They concluded that their analysis '*on an expansive empirical dataset of* 1198 *species abundance time series provides a convincing and broad-scale reinforcement of the theory that density dependence is a pervasive ecological process*'.

6.2 Case studies show variation in what processes are affected by density-dependence

In the next sections I summarize a few examples of studies that have attempted to identify if and when density-dependence occurs and what population parameter is affected. They are among the few studies that comprehensively look at all demographic variables that can influence population growth rate so that it is possible to understand which demographic parameters explain variation in growth rate. This is necessary if we want to identify which resources are limiting and which mechanisms operate that cause density-dependence. Describing such studies briefly will underline (1) that intraspecific competition seems to occur in many populations, be it often for different resources at various times of the year and impacting very different population processes; (2) that populations need to be studied year-round in order to understand what is really happening; and (3) that including movements in and out of study populations is essential as free-living populations are normally open, and these movements could be density dependent. Erik Matthysen's (2005) review of density-dependence of dispersal in mammals and birds shows that few studies set out to test specifically the existence of density-dependent dispersal, that about one third of bird studies find it, and that when an effect is present in birds dispersal increases with density.

To explore the occurrence of intraspecific competition in bird populations I have chosen three detailed, long-term studies on passerines that combined observations and experiments: Jamie Smith's work on Mandarte Island song sparrows, a resident isolated population with very little immigration; Jan Ekman's year-round study of mostly resident willow tits in Sweden in which dispersal was important; and Dick Holmes' long-term studies of the migratory black-throated blue warbler in which intraspecific competition both on breeding and wintering grounds was studied; I also discuss two less detailed studies of long-lived birds that will illustrate how variable the effects of intraspecific competition can be. These brief reviews are not intended to be exhaustive, but to identify primarily when intraspecific competition occurs, what resources are limiting, and which demographic parameters are affected. The many density-independent factors influencing numbers, although clearly important to understand population fluctuations, will often not be mentioned.

6.2.1 Song sparrows on Mandarte Island (Arcese *et al.* 1992; Smith *et al.* 2006)

Jamie Smith started his research on the song sparrow population on Mandarte Island in 1974. Mandarte Island is a 6 ha island near Vancouver Island, B.C., Canada, whose breeding population in the period 1975–2006 varied between 4 and 74 females. Very few immigrants (< 3% of all breeders and between 1 and 4 per year (1.2 on average)), join the breeding population. The number of fledglings produced per female per year is strongly density-dependent, varying between 1 at high densities and 5 at low density. Two variables cause this variation in fledgling production. The first, nest failure by cowbird predation, is density-independent. The second, variation in clutch size, is closely correlated with population density and the result of intraspecific competition. As density increases, territory sizes become smaller and females have less food to produce eggs. The existence of food shortage influencing reproduction was supported by feeding experiments (Arcese and Smith 1988). While adult survival is not density-dependent, juvenile recruitment is. The percentage of independent young that recruited into the breeding population varied between 13% at high densities and 70% at low density. Since the Mandarte song sparrows defend territories year-round, intraspecific competition for space plays an important role in juvenile recruitment, and the density of same-sex individuals may play a more important role than density overall. Intrasexual competition affects male fitness by limiting access to territories and mates, and affects female fitness by limiting access to male parental care (Smith *et al.* 2006). As regards the effect of cowbirds on reproduction, it was shown experimentally that cowbirds as predators have both a direct (destroying eggs) and an indirect effect on song sparrow reproduction (Smith *et al.* 2002; Zanette *et al.* 2006) (see also Subsection 4.7.3). Density-dependent production and recruitment of offspring are sufficient to regulate the population, but extreme conditions during winter sometimes resulted in a collapse of the population. Furthermore the importance of competition changed over the course of the study as described in Chapter 4 of Smith's book:

'Strong regulation characterized the Mandarte song sparrow population through 1989 despite two crashes in numbers. After 1989, however, the population behaved somewhat differently…. About 59% of the annual variation in the per capita production of young per female was accounted for statistically by density to 1989, but only 35% of this variation was explained by density over the entire study' (Smith *et al.* 2006).

This is a particularly interesting observation underlining how the relative importance of competition depends on the role of other factors, in this case possibly predation and parasitism by cowbirds.

When studying song sparrow dispersal in a suite of small islands near Mandarte Island, Wilson and Arcese (2008) found important variation in immigration rates between islands depending on their size and degree of isolation but also varying with sex. Immigration rates of males declined as population density increased, and immigration rates of females increased as the sex ratio became biased towards an excess of adult males. The effect of density on immigration was supported by the observation that annual immigration rates varied inversely with population density on nine focal islands over seven years, in contrast to the lack of density dependence found on Mandarte Island. On small islands immigration sometimes rescued the population from extinction and added genetic diversity. Whereas dispersal is caused by intraspecific competition for space or partners, island isolation can still result in the costs of exploration of dispersers being so high that local populations temporarily go extinct or are composed of birds of one sex only.

6.2.2 Willow tits in Sweden (Ekman 1984a, 1984b)

Jan Ekman studied very closely a population of individually colour-ringed willow tits in an area of about 6 km^2 (an area about 100 times larger than Mandarte Island) in south-west Sweden over a six-year period. He found a suite of density-dependent effects: between breeding density and clutch size; between number of fledglings and both juvenile and adult summer survival (Fig. 6.1); between the number of yearlings in February and their late winter survival. Most if not all of these density-dependent effects were related to intraspecific competition for food (during winter in relation to predation risk) and/or for space, as shown by numerous field experiments in which either food was added or birds removed (Cederholm and Ekman 1976; Ekman *et al.* 1981; Hogstad 1989a, 1989b; Lahti *et al.* 1998). Like song sparrows, willow tits are year-round territorial, but unlike the song sparrows juveniles after dispersal join non-kin adults during their first summer or autumn in dominance-structured flocks. This social organization buffers fluctuations in the overall size of the breeding population against winter mortality in a severe environment.

6.2.3 Warblers in New Hampshire and the Caribbean (Holmes 2007)

The study of the ecology of forest birds (especially black-throated blue warbler and American redstart) at the Hubbard Brook Experimental Forest in New

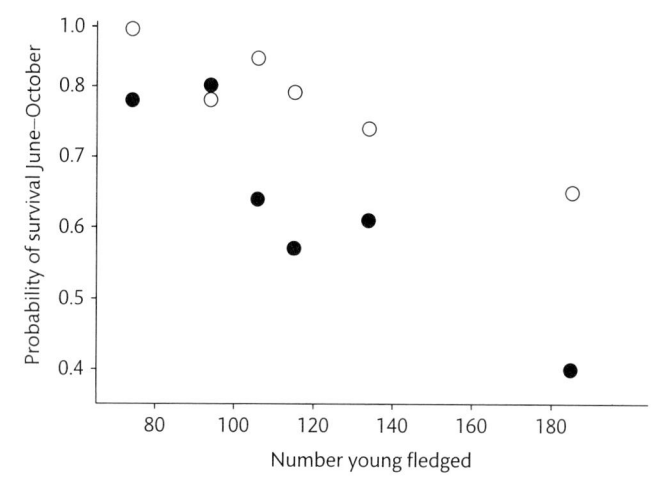

Fig. 6.1 Summer survival of juvenile (filled symbols) and adult (open symbols) willow tits in Sweden. Note that adults survive better than juveniles, but that summer survival of both age groups declines with increasing numbers of young fledged. After Ekman (1984b).

Hampshire, USA, which Dick Holmes started in 1969, is still ongoing. Although these species are long-distance migrants, Holmes and colleagues have successfully identified effects on the population both at the breeding grounds and on wintering sites in the Caribbean, generating unusually interesting insights. On the breeding grounds the data for the black-throated green warbler are most detailed, while on the wintering grounds those for the American redstart are more complete. The 37-year long time series of the black-throated blue warbler showed strong density-dependence. The factors generating dampened fluctuations of the population size stem both from what happens during the breeding season and what happens in the non-breeding season. Both the number of young fledged per territory and the proportion of male fledglings that recruited into the breeding population the following year were negatively related to breeding density (Fig. 6.2).

Furthermore there was also a strong and statistically significant positive relationship between recruitment and mean annual fecundity of black-throated blue warblers in the previous breeding season, illustrating the importance of fecundity in maintaining local populations even for a species that spends more than 8 months of the year away from the breeding grounds. As in the studies described above, field experiments made it possible to translate correlation into causation as concerns the link between food and density-dependence. In this warbler study both reduction in food through aerial spraying (Rodenhouse and Holmes 1992), reduction in local competition through removal of neighbours (Rodenhouse

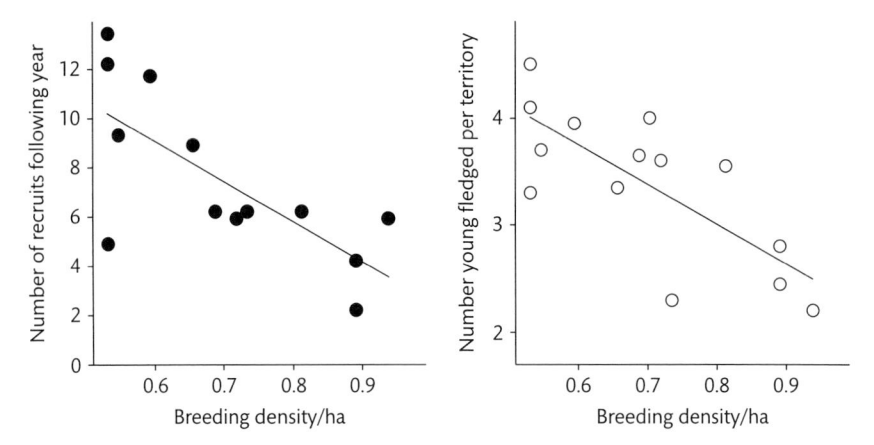

Fig. 6.2 Density-dependent relationships of black-throated blue warblers at Hubbard Brook, New Hampshire, USA: on left, relation between the number of yearling males recruiting in the study plot and breeding density ha^{-1} (P=0.005); on right, relation between number of fledglings per territory per season and breeding density ha^{-1} (P=0.001). After Holmes (2007).

et al. 2003; Sillett *et al.* 2004), and food addition experiments (Nagy and Holmes 2005) confirmed the importance of intraspecific competition for food. Holmes (2007) concluded that '*food clearly limits reproductive output in this species, and this limitation probably occurs to at least some extent in most breeding seasons*'. Other factors that impact reproductive success severely, such as nest predation and overall fluctuations in food abundance through variation in the SOI (Southern Oscillation Index), act in a non-density-dependent manner. During the non-breeding season intraspecific competition for food is important in American redstarts and causes limitation in population size. On the wintering grounds the birds defend individual territories in habitats ranging from second-growth scrub to mangroves to wet forest as well as in agricultural habitats such as coffee and citrus plantations. Adult males, the more dominant individuals, occupy the best sites which contain most food and subordinate juvenile females occupy the poorest territories. The unique aspect of Holmes's study was the link between conditions on the breeding and wintering grounds. Conditions on the wintering grounds carry over to breeding success both at the individual level (females on poorer winter territories depart later and in poorer condition and hence will be less successful at breeding) and at the population level. SOI fluctuations affected food conditions both in the breeding and in the wintering grounds, increasing food shortage (and hence intraspecific competition for food) adversely in El Niño years when rainfall is reduced in eastern North America: both adult survival and fecundity were lower in El Niño years and higher in La Niña years (Sillett *et al.*

2000). In these warblers intraspecific competition for food and for quality space is important and impacts population size.

6.2.4 A stable and a recovering sparrowhawk population (Newton 1988; Wyllie and Newton 1991)

In much of eastern England sparrowhawks had disappeared by 1960 because of the use of organochlorine pesticides. Immigrants from outside the study area in Rockingham Forest, Northamptonshire, began to settle as breeding birds in the late 1970s and three new nests were found in 1979 when the study started. The dynamics of this expanding population in an area of about 48 km^2 were compared to a stable population in south Scotland. In Rockingham Forest population size increased rapidly and by 1989 84 nests were found. The rate of annual increase was density-dependent. As density increased the dispersal distance between birth site and breeding site increased implying intraspecific competition for space (Fig. 6.3). In the increasing English population, age at first breeding was lower than in the stable Scottish population, and recruitment rate as well as adult survival rate were higher, but reproductive rates did not differ between the populations. In a separate analysis of the stable Scottish population, Ian Newton also found that the proportion of young birds that recruited

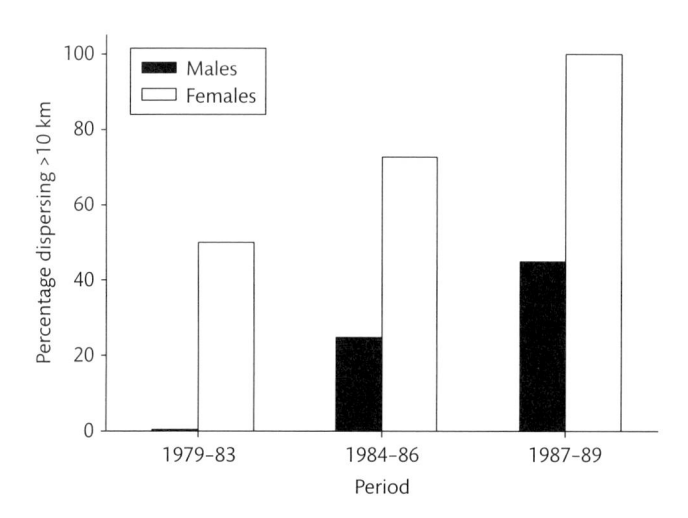

Fig. 6.3 Percentage of nestling sparrowhawks recovered at > 10 km ringed in different periods of population recovery. As population size in Rockingham, Northamptonshire, UK, increased, nestlings settled further and further away from their birth place, indicating density-dependent post-fledging dispersal. Based on data in Table 8 in Wyllie and Newton (1991).

into the population was density-dependent, while reproduction was not (Newton 1988). Similar results as regards younger age at first breeding in low density populations were also observed in other birds and mammals leading Wyllie and Newton (1991) to conclude that change in age at recruitment to first breeding may be a widespread response to density change among long-lived animals. This is most likely linked to intraspecific interference competition for space.

6.2.5 Increase in population size in the colonial Audouin's gull in Spain

Giacomo Tavecchia and colleagues studied demographic parameters of an Audouin's gull colony in the Ebro delta, north-east Spain, using appropriate capture-mark-recapture techniques on individually marked birds (Tavecchia *et al.* 2007). Their study population grew exponentially from 1981 to 1997 and fluctuated after that at around 10,000 pairs. They showed that the per-capita growth rate ($\log_e$ of N_{t+1}/N_t) decreased not linearly but logarithmically with density; that reproductive rate did not decrease with density but was strongly influenced by food abundance; and that whereas age at first breeding did not change as colony size grew, adult survival, recruitment rate, and adult emigration rate were all density-dependent, though to different degrees, suggesting the importance of interference competition.

6.2.6 The resource for which competition occurs varies between populations

The five examples presented show that the resources for which intraspecific competition takes place and the time of year when it occurs vary strongly between studies. In the three passerines reproduction was density-dependent, implying intraspecific competition for food during the breeding season, although the warbler study also showed an effect of habitat heterogeneity (see below). The song sparrow work showed the role of an interaction between predation and food, illustrating once more that one should not look at factors by themselves. In the longer-lived species (sparrowhawk, Audouin's gull), reproduction was not density-dependent. This suggests that the impact of intraspecific competition might differ between species with different life-history strategies. In all studies juvenile survival and/or recruitment into the breeding population was inversely related to density implying that both exploitation competition for food and interference competition for space play a major role in population regulation. The reason why not many studies report this result is that it is very difficult to measure juvenile survival and dispersal accurately.

6.3 Density-dependence in introduced populations

One way to test experimentally for effects of intraspecific competition is to carefully follow the fate of introduced populations. In Section 4.9 I described in some detail one of the many reintroduction experiments performed in New Zealand (Armstrong *et al.* 2002). I here discuss the data concerning the successful reintroductions of New Zealand robin on Tiritiri Matangi, a 220-ha offshore island near Auckland, and that of the saddleback to Mokoia, a 135-ha island in Lake Rotorua. Careful monitoring and modelling of the data led to the conclusion that, as concerns the New Zealand robin, neither adult survival nor reproduction were density-dependent, although reproductive success varied between forest patches and

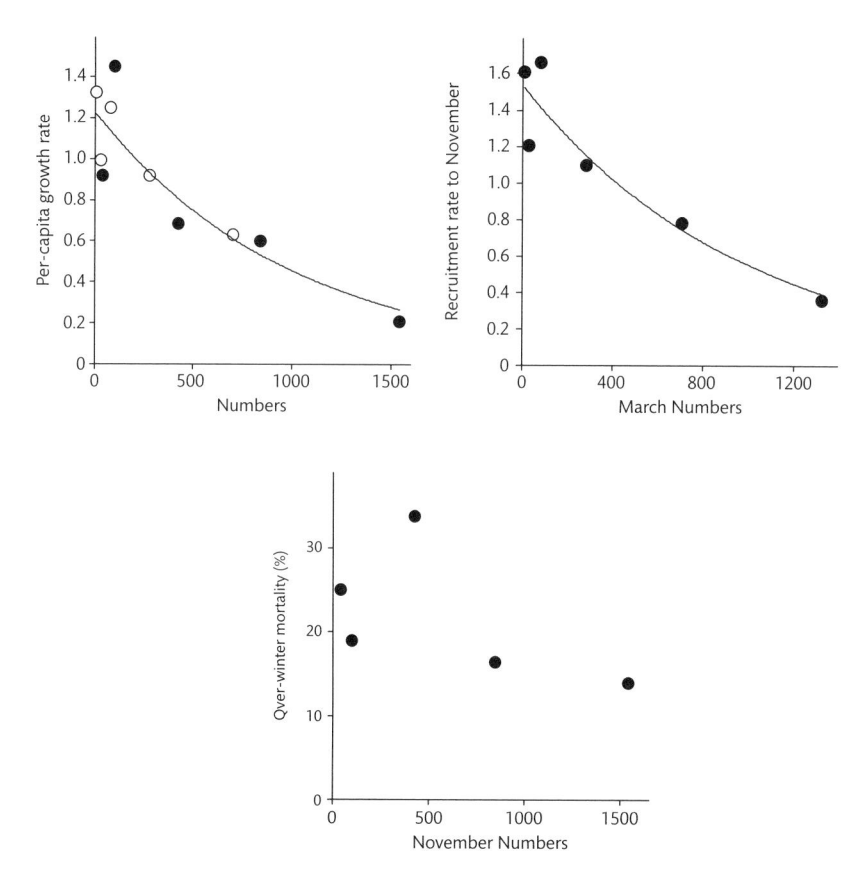

Fig. 6.4 Demography of the pheasant population introduced into Protection Island, Washington, USA. In the top panel open symbols represent per-capita growth rates calculated from autumn to autumn, while closed symbols are values calculated between successive springs. After Einarsen (1945b).

was lower in first-year birds. Juvenile mortality, most of which occurred during the initial two months after fledging, was density-dependent. In the reintroduced population of saddlebacks both juvenile survival and reproduction were density-dependent. What is interesting about these reintroduction studies of highly endangered endemics is that the models formulated based on the data led the investigators to conclude that they could 'harvest' birds, and use individuals from the successful reintroductions for further reintroductions.

The objective of the introduction of pheasants on the 159 ha Protection Island, Washington, USA in 1937 was to determine how rapidly a small introduced population would grow, if it would be viable, and how heavily it could be exploited through hunting (Einarsen 1945a, 1945b). Two cocks and six hens were introduced in spring 1937, and numbers were counted every November and March for five years. The population was not hunted until the island came under military control in April 1943. This classical data set documents very clearly that per-capita growth rate was density-dependent, and that this came about through density-dependent factors operating in spring and summer (Fig. 6.4), while over-winter mortality did not vary with density (Dhondt 1985b). Arthur Einarsen commented that the decrease in population growth in spring and summer was primarily the result of egg and chick loss. Males harassed females, and an increasing number of nests were abandoned as the population increased, implying that the mechanism causing density-dependence in recruitment was primarily interference competition (Einarsen 1945b).

6.4 Mechanisms resulting in density-dependence: the importance of habitat heterogeneity

Traditionally, density-dependent reproduction has been explained by intraspecific competition for space (see Buffer Hypothesis in Chapter 3) and, within a habitat, for food: as population size increases each female has less food to invest in reproduction and hence produces a smaller clutch, so that the mean reproductive rate in the population decreases. Implicit in this concept is that all individuals are average and live in a homogenous environment. Following Andrewartha and Birch (1954), who believed the environment to be non-uniform, Dhondt *et al.* (1992) proposed another way by which mean reproductive rate within a study plot can decline as density increases. They assumed that there exist quality differences between territories and that birds settle preferentially on the best available territory in which reproductive rate will be highest. As the population grows the proportion of individuals breeding in high-quality sites decreases (because their number is limited) and the proportion of the population occupying poorer quality sites (in which reproduction is less successful) increases. Even if reproduction in high-quality sites does not change, the mean reproductive rate in the population will decrease as the proportion of individuals with reduced reproduction will increase because they occupy poorer territories. Dhondt *et al.* (1992) showed that during five years in which they experimentally increased blue tit population density, mean clutch size was lower than in five years with low density. In low density years most of the low quality territories remained vacant, whereas the entire study plot was occupied in high density years. In the high-quality sites mean clutch size did not differ significantly between high and low density years, but females breeding in low quality sites, on average, laid significantly smaller clutches than those occupying high-quality sites. Females that shifted from low- to high-quality sites increased clutch size more than females moving from high- to low-quality sites. Using six study sites in The Netherlands, Christiaan Both (1998) found that in only one of them did clutch size of individual females not change with changes in breeding density, and that therefore the *Individual Adjustment Hypothesis* was supported rather than the *Habitat Heterogeneity Hypothesis*. The point he missed is that the two hypotheses are not mutually exclusive and that actually both can be true simultaneously, a point illustrated in Fig. 6.5 in which I present my earlier results in a different way. In order for habitat heterogeneity to cause density-dependence, habitat heterogeneity must be at the level of a territory (Dhondt *et al.* 1992).

The Habitat Heterogeneity Hypothesis was renamed 'site-dependent regulation' by Rodenhouse *et al.* (1997). Whatever its name, results consistent with predictions of the Habitat Heterogeneity Hypothesis were reported in studies of species as diverse as various eagle species in Spain (Bonelli's eagle (Martinez *et al.* 2008), booted eagle (Ferrer *et al.* 2006), Spanish imperial eagle (Ferrer and Donazar 1996)), goshawk in Germany (Krüger and Lindström 2001), sparrowhawk in Scotland (Newton 1991), great bustard in Spain (Martinez *et al.* 2008), black-throated blue warbler in north-east USA (Rodenhouse *et al.* 2003), and nuthatch in Sweden (Nilsson 1987). Interestingly, even in colonial species such as the griffon vulture in Spain (Fernandez *et al.* 1998), common guillemot in Scotland (Kokko and Lopez-Sepulcre 2007), and lesser sheathbill on the subantarctic island of Kerguelen (Bried and Jouventin 1998) observations supported predictions of the Habitat Heterogeneity Hypothesis. This mechanism—intraspecific competition for high-quality sites—resulting in density-dependence therefore occurs generally. The fundamental difference between the two mechanisms causing density-dependence is that in the Individual Adjustment Hypothesis intraspecific competition is primarily exploitation competition for food, while in the Habitat Heterogeneity Hypothesis competition is preemptive, transient interference competition for high quality space. This space can be a specific nest site in colonial birds or a breeding territory in a specific patch in the habitat.

Note that habitat heterogeneity at the territory level can also operate outside the breeding season as illustrated by the elegant experiments with American

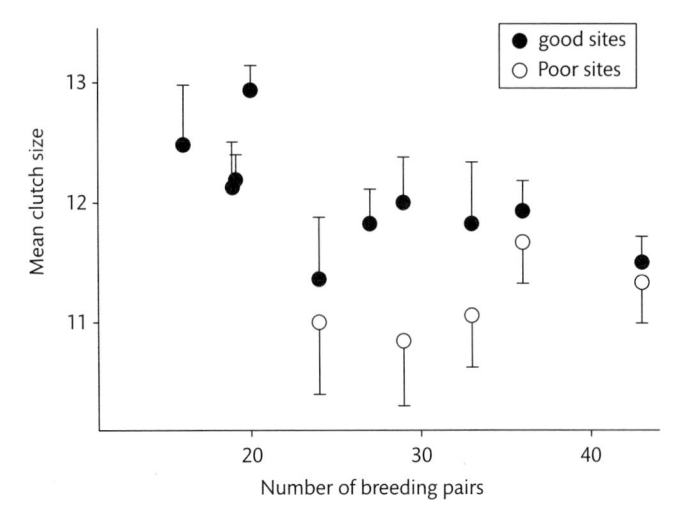

Fig. 6.5 Relationship between blue tit abundance and first-brood clutch size in plot T (Antwerp, Belgium) in which density was low during five years and experimentally increased during the five following years. 'Good sites' are nest-boxes in which, during the initial five years at least, one clutch of 12 eggs or larger was found. Most 'poor sites' were not occupied during the initial five-year low-density period. Using good sites only the decrease of clutch size with density was marginally significant (r = −0.62; P = 0.06). Using all data, clutch size decreased significantly with density (r = −0.68, P < 0.05). Data from Dhondt *et al.* (1992) and unpublished.

kestrels (Ardia and Bildstein 1997) and American redstarts (Studds and Marra 2005). In winter kestrel females—the larger sex—occupied individual territories in preferred habitat and maintained a better body condition than males that occupied less preferred habitats (Ardia 2002). Removal experiments showed that areas previously held by females were reoccupied more frequently than those previously occupied by males and that birds of both sexes preferentially reoccupied vacated female areas. Given that birds in poorer territories were in poorer body condition, this behaviour should result in density-dependent mortality, especially of males, through intraspecific competition for high quality territories (Ardia and Bildstein 1997).

6.5 Density-dependence in titmice

There is ample evidence for effects of intraspecific competition on demographic parameters and population processes. These are usually caused either by competition for space or food. Having established in the previous chapters that resources are often limiting I here limit myself to review density-dependence in tits, as they illustrate the general imbalance between the number of studies on different variables. Sinclair (1989) reviewed studies reporting density-dependence at different life stages. A large number of studies explored to what extent clutch size decreases with density, while very few studies attempted to determine if dispersal or survival also decrease, although they are bound to have a much larger impact on population size.

Let us take a look at how density-dependence relates to intraspecific competition and what resources birds compete for in each of the cases. Most studies report correlations between density and a demographic variable; some have manipulated density experimentally and compared demographic variables at high and low densities. A few have used k-factor analysis to identify density-dependence. In the following section I discuss these results per demographic variable. In each case I try to identify if density-dependence is the result of intraspecific competition and if so over what resource. I discuss dispersal in a bit more detail given its general importance in regulating numbers.

6.5.1 Laying date

I was surprised to find so few studies that asked the question of whether the laying date could be influenced by breeding density. This is the more surprising as many studies have documented that supplemental food advances laying, showing therefore that food at that time of the breeding cycle must often be in short supply (Section 4.6). At higher densities, therefore, intraspecific competition for food should be more intense. I assume the reason why so few people have looked at the effect of density on laying date is that the laying date varies so much between years because of variation in spring temperature, and that the warmth sum explains a high proportion of between year variation in laying date (Kluijver 1951). The only paper showing an effect of density on laying date is one in which blue tit density was manipulated in one study plot but not in the other. During the five-year period when blue tit density was low in both plots laydate in the more rural site (Plot SO) was 6.4 days later than in the suburban site (Plot MA). In the five-year period when blue tit density had been experimentally changed to 'high' in the suburban site but remained unchanged in the rural site the difference between the two plots decreased significantly to 3.1 days (Dhondt and Adriaensen 1999) (see Appendix 3, Table 3). Wilkin *et al.* (2006) used a different approach to determine the possible effect of density on demographic variables. They estimated the territory size of great tits in Wytham Woods, Oxford, UK, between 1965 and 1996 by calculating Thiessen polygons around the nest sites. They found no effect of polygon size on laying date unless they capped the polygon size at 0.9 ha. When they did that, the laying date was significantly earlier in larger polygons, although there was no association between laying date and the number of birds breeding in Wytham after controlling for spring temperature.

The effect of density on laying date is most likely the result of competition for food at a critical time when food is in short supply. Another factor that could contribute to this effect is that at higher density the proportion first-time breeders—who usually lay later—is higher.

6.5.2 Clutch size

In the four species of tits in which this has been studied, clutch size decreases with increasing density, although this is not found in all studies or study plots (great tit: (Both 2000; Both and Visser 2000; Dhondt 1977, 2010b; Kluijver 1951; Krebs 1970; Lack 1966; Orell and Ojanen 1983; Perrins 1965; Sasvari and Orell 1992; Wilkin *et al*. 2006); blue tit (Both 2000; Dhondt 1977, 2010b; Dhondt and Adriaensen 1999; Dhondt *et al*. 1990a; Lack 1966); coal tit (Both 2000; Lack 1966; Löhrl 1974); willow tit (Ekman 1984b; Jansson *et al*. 1981)). Crested tit clutch size was not smaller when breeding density had been experimentally increased (Jansson *et al*. 1981). Food supplementation experiments not only resulted in birds laying earlier but sometimes also caused birds to lay a larger clutch (Nilsson 1991). The strength of that effect depended on the quality of the year/habitat; in poor quality situations the effect was strongest (Dhondt 2010b; Nager *et al*. 1997) (see Chapter 4, Fig. 4.6). It is therefore likely that smaller clutches at higher density are the result of intraspecific competition for food. Alternative hypotheses are that it is adaptive to lay a smaller clutch when breeding density is higher, that competition occurs at the landscape level not at the individual level (Nicolaus *et al*. 2009), or that the male's time budget changes and he spends more time in territorial strife. While the first hypothesis has not been tested experimentally, both others have. Nicolaus *et al*. (2009) compared great tit clutch size over 11 years in a young mixed deciduous forest planted in a reclaimed polder in The Netherlands. They created five plots with very low nest-box densities (0.11–0.83 boxes ha^{-1}) and three plots with higher box densities (1.25- 1.64 boxes ha^{-1}), resulting in plots with very low (around 0.2 pairs ha^{-1}) and relatively higher breeding densities (0.6–1.0 pair ha^{-1}). The negative relationship between clutch size and breeding density across all plots was very strong, although there was no difference in mean clutch size between higher and low density plots. They suggested that food was limited during winter, impacting the female condition and thus clutch size, but not during the breeding season, as nestling mass did not differ between treatments, a conclusion congruent with the results of the winter feeding experiment of Robb *et al*. (2008) (see Section 4.6).

In a study plot with more than 10 nest-boxes ha^{-1} and great tit densities of more than 2 pairs ha^{-1}, Kempenaers and Dhondt (1992) tested the hypothesis that increased interactions at higher densities would result in smaller clutches. They presented a caged male close to the nestbox in which a female had started laying during seven consecutive days for four to seven hours each day, finding that this had no effect on clutch sizes of either great or blue tit. The increased time the males spent in aggressive encounters did not result in smaller clutches. It is thus likely that increased intraspecific competition for food at higher densities or in smaller territories has a direct effect on clutch size.

Although some hypotheses have received support (see Section 4.6, 6.4) the exact mechanisms that cause females to adjust their clutch size to breeding density between years still largely remains a puzzle.

6.5.3 Egg mass

Christopher Perrins and Robin McCleery found that great tit egg mass was inversely correlated to great and to blue tit breeding density (Perrins and McCleery 1994). This result was contradicted by a newer analysis of the same data by Wilkin *et al.* (2006). They wrote that Perrins and McCleery (1994) did not control for confounding factors such as female age which in the newer analysis had a strong negative effect on the mass of the eggs. At present it seems that great tits reduce clutch size but not egg mass as density increases and intraspecific competition for food becomes more intense.

6.5.4 Percentage second broods

In great tits the percentage of pairs that produce second broods declines with both conspecific and heterospecific breeding density (Dhondt 1977). The mechanism underlying this trend is related to weight loss while raising the first brood. Females start forming the eggs for the second clutch while still feeding their first brood (Kluyver *et al.* 1977) and weight loss while raising their first-brood nestlings is higher in females that did not start a second brood compared to females that did (4.9% of body mass compared to 2.1% over an 8-day interval). Blue tits, which rarely undertake second broods, decreased body mass by 8.6% over the same 8-day period, double the 4.0% decrease of great tits (De Laet and Dhondt 1989). This information, combined with Van Balen's (1973) finding that second broods are more frequent in habitats that have more food later in the breeding season, clearly implies that intra- and interspecific competition for food influences the ability of females to start a second clutch.

6.5.5 Nest success/nestling mass

There are three components to reduction in nest success. One is hatching failure of individual eggs in a clutch; a second is complete failure caused, for example, by predation; the final one is nestling mortality. Typically death of individual nestlings is caused by starvation, while death of entire broods is caused by predation or nest abandonment. Neither hatching failure—sometimes caused by inbreeding (Kempenaers *et al.* 1996)—nor complete nest losses through predation can be explained by competition among tits, while mortality through starvation can. Even if nest failure through predation was density-dependent in Wytham great tits, this was clearly not directly related to competition between tits, but simply reflected a change in predator behaviour in response to prey abundance (Dunn 1977). On the other hand, given that nestling mortality was correlated to fledgling weight (Dhondt *et al.* 1990b) and varied seasonally and between habitats with food abundance (Van Balen 1973) and with density, nestling mortality is certainly influenced by intra- (and interspecific) competition for food during the breeding season. The winter feeding experiment carried out by Jansson *et al.* (1981) increased crested tit density 2.33 times and willow tit density 1.61 times. This had no significant effect on crested tit reproduction, but halved the number of fledglings per clutch in willow tits, primarily through a reduction in nest success of almost half.

All these observations suggest that intraspecific (and interspecific) competition for food can occur during the nestling stage at least in great, blue, and willow tit.

6.5.6 Reproductive rate

Given that various components of reproductive success are density-dependent and impacted by competition for food, it is not surprising that overall reproductive rates decrease with density. Note, however, that the intensity of intraspecific competition varies with habitat quality. The effect of additional breeding pairs on clutch size, nest success, or reproductive rate is higher in poor quality (suburban parks, beech wood) than in high quality (mature oak forest) habitat, suggesting that in high-quality habitats space rather than food becomes limiting (Dhondt 2010b) (see Fig. 9.6).

We can thus conclude that in various tit species (but not in all, and not always) intraspecific competition for food occurs during the breeding season, be it to a different extent in different species and in habitats of different quality.

6.5.7 Dispersal and survival: the important role of intraspecific competition for space

After fledging, young stay with their parents for up to a few weeks and then disperse, in part because of increased parental aggression (Holleback 1974). This initial dispersal has been described as being 'explosive' in great tits (Goodbody 1952). Dispersal stops earlier in males than in females and density seems to influence its duration (Dhondt 1971, 1979; Ekman 1984a; Goodbody 1952; Nilsson and Smith 1988). Rapid establishment is essential because juveniles that settle first tend to win future encounters (Drent 1984; Nilsson and Smith 1988). Piet Drent's careful observations of interactions between juvenile great tits in summer showed that the loser of a conflict would move away, while the winner would initiate a new conflict at the same site, thereby emphasizing its local dominance. Intraspecific competition at that time is thus for space mediated by conflicts over dominance position. This is probably generally true in resident bird species (Drent 1984).

In winter-territorial species juveniles rapidly leave their parents and attempt to settle in a high-quality site. An extreme example is that of the female Eurasian nuthatch who, within a week of fledging, had already settled on her own territory with a partner while she regularly returned to her parent's territory to be fed (E. Matthysen, pers. comm.). In marsh tits this pressure to rapidly occupy a good territory causes young males to depart from the natal territory earlier than females, and causes them to settle closer to home (Nilsson and Smith 1988). In winter-territorial tits in saturated environments juveniles may not disperse and become helpers ((tufted titmouse, Grubb and Pravasudov (1994)), or breed jointly with their parents on the same territory ((bridled titmouse, Christman and Gaulin (1998)). Ekman (1979) showed that the tendency to disperse increases with the total number of juveniles fledged and is thus density-dependent (Fig 6.1). If groups are too large in late winter there may be a new episode of territorially-induced dispersal that may lead to some individuals being excluded from breeding altogether (Desrochers *et al.* 1988; Lens and Wauters 1996; Smith 1967).

In the great tit, a winter-flocking species, there are a series of dispersal episodes. During the establishment of dominance hierarchies in flocks of juveniles the birds display a special 'molesting' behaviour: an aggressive individual directs a one-sided wing flapping towards a conspecific, which usually results in the displaying individual pursuing the other one in flight (Kothbauer-Hellmann 1990). Subordinate birds are forced to disperse through intraspecific competition for dominance position and space, as shown by autumnl removal experiments (Drent 1984). The general importance of numbers of birds present during this dispersal period on the tendency to disperse is shown by the fact that a higher proportion of second-brood birds move away from their birth area (Dhondt and Hublé 1968; Dhondt and Olaerts 1981), and that they do not do this when first-brood young are experimentally removed (Kluyver 1971). A second dispersal period occurs when moult is over and autumn-territorial behaviour resumes. This can lead to massive movements that may generate irruptions (Cramp 1963; Cramp *et al.* 1960; Newton 2006; Perrins 1966). Two irruptions of coal tits in Europe in 1989 were estimated to involve more than two million birds (Van Gasteren *et al.* 1992). The authors concluded that one irruption, which originated in the taiga zone, was the result of food shortage, while the second, which originated in Northern Germany and Poland, was the result of a very high population density. In either case, birds could have been subjected to food shortage and intense competition. Although in some populations great tits are regular migrants, a larger number of birds migrate in better condition in high density years, suggesting that large population size rather than food shortage triggered the mass movements (Korner-Nievergelt *et al.* 2008). There may be no effective difference between a large population and a scarce food supply, because in both cases food may be hard to get. It is the imbalance between the birds and their food supply that causes mass emigration, and this is dependent on both food supply and population level (I. Newton, *in litt.*).

Usually, short-distance autumn dispersal simply generates a large number of immigrants in sites in which winter resources (roosting sites, food) are plentiful. At the end of winter when territorial behaviour gradually resumes, subordinate individuals of both sexes are forced out of high-quality plots and either move to lower quality breeding sites (Dhondt 1971) or are prohibited from breeding altogether as shown by spring removal experiments (see Chapter 3). Intraspecific competition through aggressive behaviour thus influences dispersal at different times during the year: in late summer, in late autumn, and again in late winter. The extent to which this leads to movements depends on bird abundance at that time (Dhondt 1971). It is therefore surprising that so few studies have attempted to document density-dependent dispersal (Matthysen 2005). In the Antwerp research

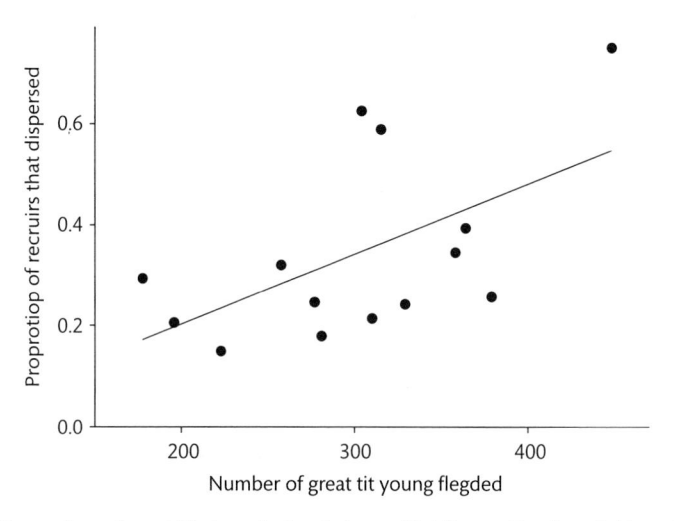

Fig. 6.6 Proportion of great tits born in the Antwerp Plot B recruiting in neighbouring plots. The proportion increases with number of fledglings (1979–93, r^2 = 0.31; P = 0.036). Dhondt (unpublished).

we studied several plots close together and hence recaptured many of the birds that dispersed and bred in another plot. This made it possible to calculate the proportion of birds among those that were found breeding that had dispersed. As can be seen in Fig. 6.6, when breeding density was high and hence many young fledged, up to 75% of great tits were found breeding outside their plot of birth.

6.6 Conclusion

The numerous studies documenting density-dependent effects on various demographic parameters show that in titmice intraspecific competition for food and for space occurs regularly. I will illustrate this by following a young tit from the day of its birth using documented effects.

In the nest the hatchling competes with its siblings for food. This intraspecific competition for food is shown both by the fact that fledgling weight and quality decrease with brood size and with breeding density, and by the observation that nest success decreases with breeding density. The incidence of intraspecific competition at that time increases with decreasing habitat quality. After fledging, survival and dispersal are strongly affected by density in all species. Intraspecific competition is for food, for dominance position in the flock, and for space. The effects for males and females must be different because, on average, female dispersal continues for longer and they thus move further. What happens next differs between juveniles from winter-flocking and winter-territorial species. Juveniles from winter-territorial species try to join an adult pair to become a member of the group. They thus compete for access to flocks in the best possible territories. This is illustrated by willow tit males dispersing less far and settling sooner than females (Nilsson and Smith 1988), and by crested tits from habitat fragments that disperse later, because they fledge in poorer conditions, being forced to join flocks on poorer territories (Lens and Wauters 1996). If they cannot join such a territorial group they will become a floater or a flock switcher and have a hard time surviving until the next breeding season (Chapter 3). Intraspecific competition for space is thus intense at this time. If they do join a flock they will become a subordinate member. The effect of the social situation will influence the winter-fattening strategy of a bird and will therefore influence its risk of being taken by a predator (Krams 2001). Competition is a combination of interference competition (for social status which gives priority access to food) and exploitation competition (more directly for food). In Scandinavia over-winter survival of crested tits and perhaps of other winter-territorial tit species is low enough that by the beginning of the breeding season most (but not all) survivors can breed. Over-winter survival is influenced by density at that time. In other winter-territorial tits surplus birds are forced to disperse and will probably often not be able to breed (Desrochers *et al.* 1988; Smith 1967). During the breeding season competition for food is less intense although often sufficient to influence clutch size and nest success.

If we now return to the winter-flocking species, survival and especially dispersal are strongly influenced by intraspecific competition. This can lead to short or long distance movements. Here again interference competition is important. During winter there is intense competition for roosting sites, again interference competition. Birds that cannot sleep in a cavity survive less well, although in some populations birds do not use cavities for roosting (Dhondt *et al.* 2010). If only few suitable cavities are available for blue tits, a larger number of them will disperse. As for chickadees, subordinate juveniles will be forced out by territorial behaviour and some birds will not breed and remain floaters. During the breeding season competition for food influences clutch size, nest success, and reproductive success overall. In both groups of species intraspecific competition occurs throughout the year, and both exploitation competition for food and interference competition influence individual fitness.

Intraspecific competition can often, but not always, be documented. If we look at other studies often no effects of density on reproduction are found (Sinclair 1989; Wiktander *et al.* 2001).

This does not mean that intraspecific competition is not important because, as was explained in the first section of this chapter and has been narrated in detail for tits, it is not always easy to demonstrate. The effects of intraspecific competition for space on dispersal or for dominance position often play a more important role in the dynamics of tit populations than density-dependent effects on reproduction. Furthermore, and as pointed out by Newton, most researchers have used, when measuring density, the currency of 'birds per unit area'. But many species, including some tits, are exposed to an enormously varying food supply from year to year, such as beech mast (Newton 1998, 2006). It is then not space that birds are competing over, but food. And if bird densities were expressed as 'birds per unit of resource' rather than 'unit of space', more density-dependent relationships may emerge. This is probably why irruptions have been associated with either scarcity of food or high population density: both produce an imbalance between birds and food, a situation that could be resolved if densities were expressed as 'birds per unit of food'. The same presumably applies for any other resource that varies greatly from year to year. Without taking account of this, we are measuring density against a continually changing baseline, so no wonder we cannot always detect density-dependence and food competition when it is surely there (I. Newton, *in litt.*).

Depending on the type of competition (exploitation or contest) the mechanism that causes the density-dependent effects varies. Short and long-lived birds seem to differ to some extent in what demographic variables are most affected, although density-dependent variation on especially juvenile survival and dispersal seem to be particularly important in all species (Saether *et al.* 2002).

7
Studies of foraging niches and food

Having established that resources often are limiting and that intraspecific competition can often be observed, one final condition needs to be satisfied before we can consider that interspecific competition exists, that is condition 3, that resource use between potential competitors overlaps. The literature on this subject is vast. Again I primarily discuss tits, particularly to provide the historic framework, but generalize to other species when appropriate.

A very large number of studies have described where birds forage and a smaller number what birds eat. Inspired by David Lack's pioneering work on cormorants (Lack 1945) and Darwin's finches (Lack 1947) the earlier studies showed how, on the one hand, foraging niches and diets differed between species coexisting in the same habitat, but on the other hand showed at least some overlap between the foraging niches of coexisting species. In the American literature the study of niches was expanded to altitudinal replacement of closely related species, which is at least partly because habitats change with altitude, and these data were used to support Robert MacArthur's idea that interspecific competition shaped communities.

In studies in which niche attributes (foraging niche, food taken, altitudinal range, micro-habitat use) of more than one species were described, the authors not only described overlap and differences between niches of coexisting species, but also often documented how niches of a given species differed between situations when a putative competing species was present or absent. It is therefore logical to include in this chapter the observational and experimental evidence that niches are influenced by the presence of putative competitors. These data provide evidence that interspecific competition exists by showing how resource use of one species affects the resource use or availability for another (evidence 1). In this chapter, I will describe and discuss those results, provide evidence that in many studies condition 3 for interspecific competition to be considered is fulfilled, give examples of data supporting evidence 1, and critically evaluate non-experimental data.

7.1 The early studies of foraging behaviour emphasized differences between species

The first detailed and longer term field study in a long list of studies documenting how coexisting tit species differ in their foraging niches was by Hartley (1953) at the Edward Grey Institute in Oxford. It is interesting to re-read the motivation for that study. Hartley wanted to determine the effect of predators on prey and of prey abundance on predator numbers. His work was funded by the Agricultural Research Council and influenced by David Lack. Hartley chose to study the foraging niches of tit species that coexisted in the same habitats. He argued that:

'If no distinction could be made between the foods, feeding places and feeding methods... it might reasonably be assumed that food supply was not the factor limiting their numbers. Conversely, a measure of ecological separation between the species would justify the assumption of some degree of interspecific pressure.'

He concluded from his two-and-a-half year study that:

'the five species of titmouse... are separated in three different ways – in foraging at different heights, in searching in different trees and in seeking food in different parts of those trees. Not all species are separable all the time; but no two are identical in their feeding behaviour.'

In the discussion, Hartley stated that the titmice in English woodlands behave very similarly to the Darwin's finches as described by Lack (1947) '... *ecological divergence has mainly taken the form of difference in food habits, in which case the species are not forced to occupy separate habitats, but can exist side by side*'. This paper then concludes, as do many others that have used a similar approach, that closely related bird species can coexist in the same habitat by occupying different foraging niches. In Lack's view, these differences in foraging niche represent the evolutionary outcome of interspecific competition in the past. Hartley concludes (p. 287):

'The influence of food supply upon the populations of different species of titmice has not been proven. But the foraging habits here described are such that their survival value may be traced to the partial separation of each species from the others and the resulting diminution of interspecific competition. Only by the evolution of specialization of feeding habits are several closely allied species able to occupy and to exploit a single habitat. These specializations the English titmice show.'

During the next 20 years or so a suite of publications would essentially confirm Hartley's results, in European tits but also among North American parids (Dixon 1954) and in other bird groups, as illustrated by Macarthur's (1958) classic work on North American warblers.

John Gibb expanded Hartley's work at Oxford of where tits foraged (Gibb 1954) by studying what tits ate and how their food supply changed during winter in pine forest (Gibb 1958, 1960). This approach permitted him to demonstrate directly that food limited the number of coal tits that remained alive at the end of winter. He also observed that coal and blue tits competed for these limited resources. Svein Haftorn emphasized the importance of food hoarding in a number of Norwegian tit species (Haftorn 1954) and Staffan Ulfstrand underlined the importance of beech mast as a winter food (Ulfstrand 1962). He documented how feeding behaviour and foraging niches changed in years when beech mast was abundant. These studies were the onset of the use of observational arguments to support the existence of interspecific competition described in the next section.

7.2 In the 1970s observational arguments were used to document the existence of interspecific competition. These arguments only convinced the believers

Doug Morse studied foraging niches in North American birds, especially of species foraging in mixed species flocks. In his 1967 paper he concluded from many observations of hostile interspecific interactions between pine warblers and brown-headed nuthatches, and from niche shifts in the presence/absence of each other in longleaf pine habitat in Louisiana, that interspecific competition between these species occurred (Morse 1967). He concluded that *'Brown-headed nuthatches and Pine Warblers are important factors regulating each other's distribution in the habitat while in each other's presence'*. In the discussion, though, he underlined the need for experiments to prove conclusively that interspecific competition occurs, but did point out that both Snow (1949) and Hartley (1953) had also observed that subordinate tit species shifted their foraging niche away from that of the socially dominant species when the dominant species was present. He generalized his ideas by summarizing the literature using examples from birds, mammals, fish, insects, and crustaceans and showed that overlap in resource use between dominant and subordinate species *'invariably decreased when together, usually as a consequence of a decrease in niche breath of the subordinate species'* (Morse 1974). In this latter paper, as already in his 1970 paper (Morse 1970), he was much more careful than in his 1967 paper to link shifts in foraging niches to interspecific competition. When I asked him about this he wrote (*in litt.*, 2002):

'At that time there was a violent reaction among many ecologists, including my colleagues at the University of Maryland, against assumptions by MacArthur, then Cody and Diamond, who chose to equate correlation with causality, and then to draw competition out of the hat.... I was by that time backing away rather quickly from assuming competition in the absence of experimental evidence... My 1974 American Naturalist paper also raises the possibility of niche shifts as being a consequence of competition, but treats the matter gingerly'.

It is interesting to note that it would take quite a bit longer for European ornithologists to follow suit and 'treat the matter gingerly', possibly because the discussion as to how interspecific competition could be documented was still more strongly dominated by David Lack, who accepted that shifts in foraging niches were sufficient to prove the importance of interspecific competition. In North America the big debate as to the importance of interspecific competition in structuring communities was just getting underway in the early 1970s and culminated in a series of highly critical papers (Connell 1980; Connor and Simberloff 1979; Wiens 1977). These authors were critical as to the importance of interspecific competition in structuring communities and argued that descriptive observations, which could be interpreted in different ways, could not solve the problem as to its importance.

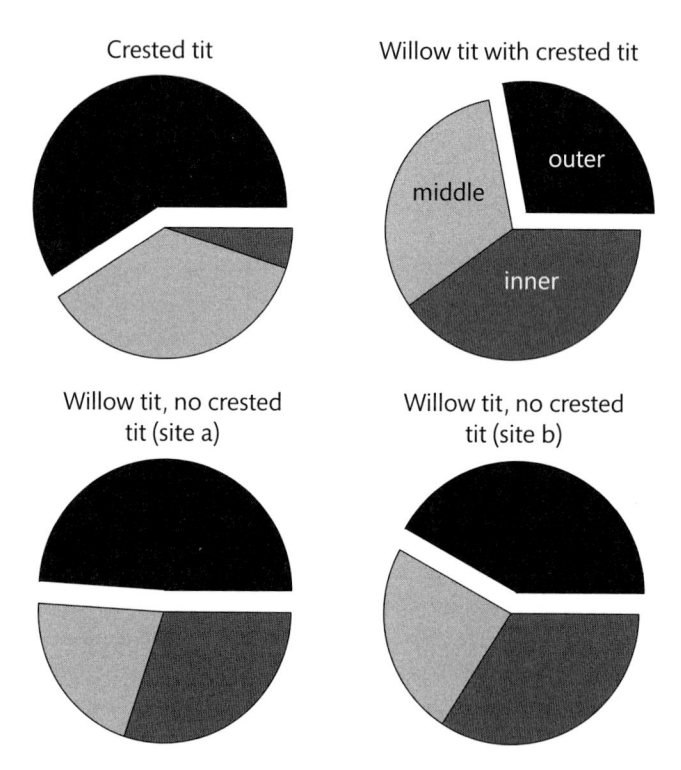

Fig. 7.1 During winter in central Norway crested tits forage mostly on the outer parts (black) of the trees (top left). In the absence of crested tits (lower diagrams), willow tits also forage mostly on the outer parts of the branches (42–49% in two sites). In the presence of crested tits (top right), however, only 22% of willow tit foraging observations are on the outer third of the branches; they are competitively displaced to less preferred parts of the tree. Black: outer third of the branches; light grey: middle third of the branches; dark grey: inner third of the branches. After Hogstad (1978).

On the European scene, a paper documenting niche shifts relative to community composition in tits was not published until 1978 (Hogstad 1978). In the introduction of this paper Olav Hogstad wrote:

'The present study was undertaken to test for the existence of mutual interactions in the winter foraging patterns of the Willow Tit, Crested Tit and Coal Tit. Variations in foraging niche were sought by comparing the foraging zones of the species in different study areas in which only one, two, or all three species, respectively, were present'.

His study showed very nicely how the subordinate willow tit was confined to the inner-lower parts of spruce trees—the less preferred foraging site in spruce in Norway—in the study plots in which crested tits also lived, but expanded their foraging niches to higher-outer parts of spruce trees in the study area in which no crested tits lived (Fig. 7.1). Hogstad attributed the willow tit foraging niche expansion in the absence of crested tits to ecological release resulting from reduced interspecific competition.

A similar, but more subtle, effect of crested tits on the foraging niche of willow tits was reported in pine forest in northern Belgium (Dhondt 1989b; Lens 1996)

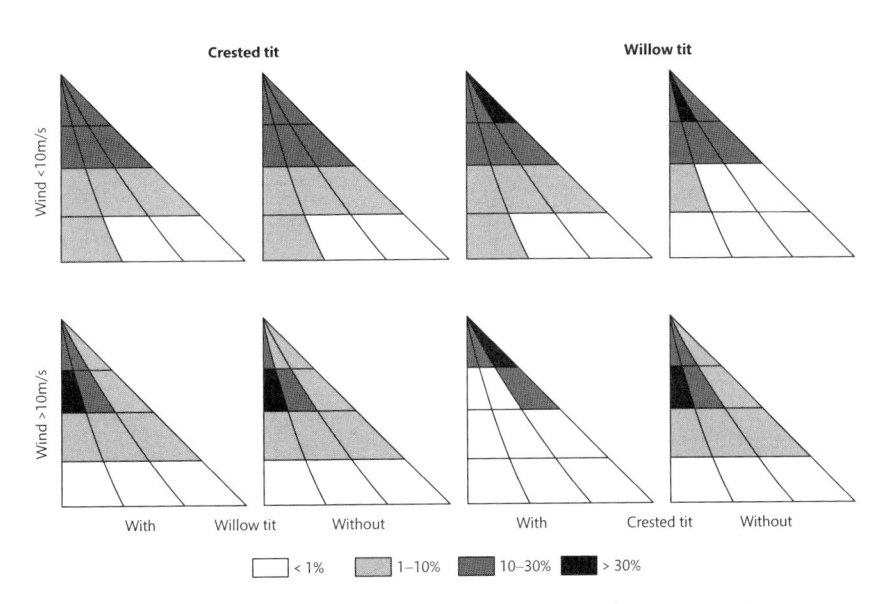

Fig. 7.2 Winter foraging niches of crested (left) and willow (right) tits in a pine forest in N. Belgium. For each species mentioned on top foraging positions are shown in the presence (left) or absence (right) of the other species, and at low (above) and high (below) wind speeds. Note that crested tit foraging is not affected by the presence of willow tits, while willow tit foraging niches are strongly affected by crested tits, especially in high wind conditions. Shading indicates the proportion of observations in different parts of the tree. After Dhondt (1989b) and Lens (1996).

as shown in Fig. 7.2. During winter both crested and willow tits defend group territories of about 10 ha. Some of the willow tit territories overlap with those of crested tits, while in adjacent territories willow or crested tits live without the other species. Which particular areas had territories of both species, and which were occupied by one species only, varied between years suggesting habitat differences did not drive this. Foraging niches of the dominant crested tit did not differ between territories in which crested tits were by themselves and those in which they were together with willow tits. For the subordinate willow tit, though, this was not the case: willow tits were more often observed closer to the trunk—the preferred foraging site in pine trees in Belgium—in the absence than in the presence of crested tits. In conditions of high wind speeds and low temperatures the birds were more energetically stressed. When alone, both species foraged lower and closer to the trunk to seek protection against the wind. Again, crested tit foraging locations were not influenced by the presence of willow tits, but willow tit foraging niches were dramatically affected by crested tits. In their absence they moved to the same locations in the tree that crested tits occupied, but in their presence they shifted to the outer and upper sections of the trees, a location that in high wind conditions would definitely be energetically costly. At high wind velocity, foraging shifts of willow tits decreased the niche overlap with crested tits foraging in the same tree, but there were no such shifts when foraging in neighbouring trees (Lens 1996).

By 1986 Rauno Alatalo and colleagues could evaluate seven studies reporting non- experimental data on foraging niche use of European tits in allopatry compared to niche use in sympatry (Alatalo *et al.* 1986). In the absence of interspecific competition one would predict that by chance an equal number of convergent and divergent shifts in foraging niches should be found. All seven studies, however, reported at least one ecological niche change in the direction predicted by interspecific competition, and none reported convergent changes in foraging niche. Furthermore, niche changes were found only in species that morphologically most closely resembled the species that was absent in allopatry. After excluding alternative explanations for these niche changes, Alatalo *et al.* (1986) concluded that these results provided strong support for effects of interspecific competition on foraging niches in conditions of food limitation (but see Section 7.7).

7.3 Field and cage experiments provided conclusive evidence as to the effect of interspecific interactions on the foraging niches used

However many ornithologists were convinced by the descriptive results alone, the need to carry out experiments to convince non-believers became increasingly strong. Thus, for example, in the paper mentioned in the previous paragraph and in which the case was made that descriptive studies could be used to infer causality, Alatalo *et al.* (1986) nevertheless concluded that '*In general, we believe that well-done experiments, if feasible, are the best way to evaluate alternative hypotheses.*'

To determine to what extent foraging niches were constrained by the presence of other closely related species, Rauno Alatalo and his friends in Sweden therefore carried out a series of very elegant field and aviary experiments. These experiments not only confirmed that in winter the subordinate species (coal tit, goldcrest) expanded their foraging niche to safer parts of the tree when the dominant willow and crested tit had been removed (Alatalo *et al.* 1985b), but also showed that the removal of subordinate species resulted in the dominant species expanding their foraging niche into the more peripheral parts of the trees, which are normally used by the subordinate species only (Alatalo *et al.* 1987). They thus not only confirmed that the subordinate species were, as expected, displaced through interference competition by the larger guild members, but surprisingly also showed that the larger species were affected by exploitation competition of the socially subordinate species.

To make sure that the foraging niche of the smaller and subordinate coal tit was actually the result of interspecific aggression or avoidance rather than a preference for the outer parts of the tree, Alatalo and Moreno (1987) tested foraging efficiency of both coal and willow tits in aviaries. Coal tits tended to be more efficient foragers in any part of the tree. When placed singly in an aviary with a choice of inner and outer branches, both species collected most of their food on the inner branches. When individuals of the two species were placed together the willow tit did not change its foraging pattern, but the coal tit moved (was displaced) to the outer branches. Interestingly, this shift happened with very little overt interspecific aggression.

Although niche shifts as described so far provide strong support for the hypothesis that interspecific competition does occur, one problem remains before one can conclude that interspecific competition really does occur: in neither field experiment did the authors attempt to determine if the experimental manipulations had an impact on a fitness-related trait such as survival or reproduction in the following breeding season. For the foraging shift to be considered a true example of interspecific competition, this would need to have been documented. Although unlikely, it cannot be excluded, for example, that food was not sufficiently limiting and that therefore the niche shift caused by the presence of other species did not result in an increase in over-winter mortality, at least in the subordinate species.

7.4 Measures of fitness-related traits are needed, however, to prove the existence of interspecific competition

Cimprich and Grubb (1994) realized that in order to prove the existence of interspecific competition it is not sufficient to observe that the foraging niche expands when a putative competitor is removed, but that it is also necessary to document that this niche expansion impacts fitness-related traits (evidence 2). After removing tufted titmice at the beginning of winter in small isolated woodlots in Ohio, they studied not only changes in Carolina chickadee behaviour but also possible effects on their survival and nutritional condition. In a very careful study over four winters they confirmed that the subordinate Carolina chickadee expanded its foraging niche towards the foraging niche of the removed species, but did not find a concomitant change in over-winter survival. However, they had another trick up their sleeve. In earlier work, Tom Grubb had documented that the extent to which a pulled feather regrows is influenced by the amount of food a bird can obtain, the 'ptilochronology approach' (Grubb 1989), and that supplementary food during winter for several species, including the Carolina chickadee and tufted titmouse, improved feather regrowth (Grubb and Cimprich 1990). The nice thing about this approach is that, since it takes a titmouse about one month to regrow a feather, the extent to which the feather regrows integrates the energetic conditions to which the bird has been exposed during about one month. Cimprich and Grubb found that the mass of the regrown feathers was significantly lower in the Carolina chickadees living in the control woodlots with tufted

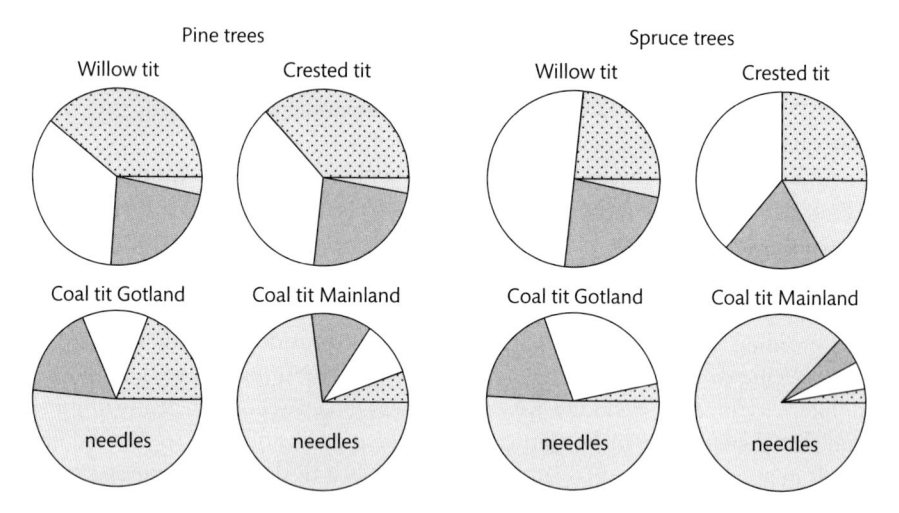

Fig. 7.3 Foraging stations of willow, crested, and coal tits in pine and spruce dominated forest on the mainland of Sweden and on the island of Gotland. On Gotland there were no willow or crested tits, while on the Swedish mainland both species coexisted with coal tits. In the presence of dominant competitors coal tits foraged almost only on needles and rarely on trunks or thick branches. In the absence of the dominants, coal tits expanded their foraging niche towards the centre of the tree. Stippled: trunk and thick branches; white: thin branches; dark grey: twigs; light grey: needles. After Alerstam et al. (1974).

titmice than in the experimental woodlots where these had been removed. Resource availability for chickadees was thus lower in woodlots with tufted titmice than in experimental woodlots without them, proving the existence of interspecific competition for food. By comparing the differences in feather growth between woodlots with results obtained with caged, food deprived birds they estimated that the presence of tufted titmice represented a reduction of about 5–10% in food intake. Cimprich and Grubb did not document that individuals in poorer condition began breeding at a later date and/or were less successful at reproducing, so in the strict sense their experiment did not prove that interspecific competition had an adverse effect on lifetime fitness of Carolina chickadees, but it can be assumed that in stressful conditions the reduced food intake in the presence of another species would have affected fitness.

7.5 The story of the coal tit on Gotland: alternative explanations can be right

On the Swedish mainland great, blue, willow, marsh, and crested tit all coexist with the socially subordinate coal tit. On the Baltic island of Gotland willow, marsh, and crested tit do not breed, and the coal tit is the only species present among the smaller species. Not only has the coal tit expanded its habitat use and foraging niche on the island, but coal tit density on Gotland is also much higher than on the mainland. On Gotland foraging coal tits also use the inner parts of the trees where on the Swedish mainland the absent willow and crested tits forage (Fig. 7.3) (Alerstam *et al.* 1974). This classic example of ecological release has led to an evolutionary change: coal tits on Gotland are considerably larger and have heavier bills than on the mainland (Gustafsson 1988a), adapting them to the novel foraging substrate. The size difference has a genetic basis (Alatalo and Gustafsson 1988). Coal tits on Gotland are a good example of release from interspecific competition and its effects. The underlying assumption is that the socially dominant willow and crested tits have not been able to reach the island of Gotland or were not able to maintain a breeding population there. This is surprising because Scandinavian willow tits regularly migrate. So until the year 2000 the story was as follows: coal tits are socially subordinate to crested and willow tit; coal tits are therefore constrained to forage on the outer parts of the conifers; they are small and nimble and have a fine beak, making them better adapted to forage between the pine needles (Norberg 1979). Where the dominant competitors are absent, as on Gotland, coal tits can exploit the foraging space vacated by the more dominant species, and hence increase in abundance. This increase in abundance enhances the intensity of intraspecific competition and coal tits, especially the males, eventually evolve to become more willow tit-like with a bigger body mass and a heavier beak. The importance of interspecific competition is therefore proved.

A paper by Cecilia Kullberg and Jan Ekman (2000), however, added a twist to this story. They found that on four Scandinavian islands where pygmy owls were breeding, crested, willow, and coal tits coexisted, whereas on five islands where the owl was absent only coal tits bred. Why is this? The two groups of islands do not differ in size, in habitat, nor in distance from the mainland. The explanation proposed by Kullberg and Ekman is that, when no owls are present, the socially dominant species have been competitively excluded by the smaller coal tit! Their reasoning is as follows. In experimental situations coal tits are more efficient foragers than either other species (Alatalo and Moreno 1987). With an average of 14 eggs/female/season, coal tits have a much higher reproductive rate than either of the other species (fertility 6–7 eggs/female/season) (Dhondt 2001). However, because coal tits forage in more exposed sites, they are over-represented among the prey taken by the bird-specialized pygmy owl, whereas both other tit species, because they forage in safer parts of the tree, are under-represented among its prey (Ekman 1986). Owl predation, therefore, keeps coal tit numbers down. When no owls are present, coal tits simply out-compete the other tit species. These become eventually excluded by the more efficient and more numerous coal tit through exploitation competition for food. So the absence of the larger tit species in the absence of pygmy owls, a keystone species, is caused by competitive exclusion by the smaller species.

In both scenarios interspecific competition is involved, but cause and effect are to a large extent reversed between them. In the original scenario interference competition by the competitively superior willow and crested tits narrows the coal tit niche, while under the Kullberg and Ekman scenario exploitation competition by the more efficient coal tit drives the other species to extinction in the absence of pygmy owls. In both cases ecological release of coal tits, as observed on Gotland and some other Scandinavian islands (Sørensen 1997a), is the result of the absence of willow and crested tit. The controversy is about the cause of the absence of the larger species. In the original scenario their absence is the result of weak dispersal capability, while in the Kullberg and Ekman scenario the absence of pygmy owls allows coal tits to competitively exclude the larger species from the islands through exploitation competition.

This interpretation has recently been challenged by Rodriguez *et al.* (2007). They claim that the competition-colonization trade-off (CCTO) explains the observed patterns. The CCTO proposes that early successional species are able to persist, allowing them to temporarily out-perform late successional species because they colonize disturbed habitats before the arrival of late species, an idea already proposed in 1974 by Diamond as the 'supertramp strategy' (Diamond 1974). They document that in forest patches in mainland Sweden the probability of occurrence differs between the three tit species: crested and willow tit (poor dispersers) are more sensitive to fragmentation than the coal tit (a good disperser), because of their difference in dispersal ability. They further state that the islands studied by Kullberg and Ekman (2000) that were inhabited by crested and willow tits or not, differed in their degree of isolation from the mainland by the presence or absence of stepping stones between the island and the mainland. In the absence of stepping stones coal tits would be the only small tit on the island, while when stepping stones connect the island from the mainland crested and willow tit had also been able to colonize the island. The analysis by Rodriguez *et al.* (2007) makes it impossible to determine whether the pygmy owl does play the role of a keystone predator on islands in the Baltic without performing an experiment. Unfortunately it seems unlikely that owls would be introduced on islands where they are currently absent.

Regardless of the exact mechanism involved, the existing differences in morphology, behaviour, and genetics between coal tits on Gotland and on the Swedish mainland are a really nice example of ecological release leading to an evolutionary change (see Subsection 10.2.1).

7.6 Altitudinal replacement of closely related species

Altitudinal replacement of closely related species was used extensively in the 1970s to suggest or conclude that interspecific competition influences community composition and leads to parapatric distributions along elevational gradients (e.g. Terborgh and Weske (1975)). The importance of interspecific competition is inferred when distributions are narrower in the presence of presumed competitors (leading to competitive displacement) than in their absence (leading to ecological release). In the absence of field experiments (such as those performed by Hairston with plethodonthid salamanders (Hairston 1980)) ecological niche modelling makes it possible, to some extent, to test the importance of interspecific competition more rigorously. I here discuss one such example regarding the distribution of two *Buarremon* brush-finches from tropical Central and South America. Remsen and Graves (1995) described the distribution of two species of Andean brush-finch, the stripe-headed brush-finch and the chestnut-capped brush-finch, and observed that they show an interdigitating pattern of distribution in the Andes that they interpret as the result of competition between them. The two species have complementary, parapatric distributions with minimal overlap. Where there is only one species, that species usually occupies the entire elevational range in the region. In two regions the usual relative elevational pattern of the species is reversed so that the stripe-headed brush-finch is found at low rather than at high altitude, and the chestnut-capped brush-finch occurs at high rather at than low altitude. Remsen and Graves believe it therefore unlikely that the altitudinal distribution is governed solely by autecological factors and conclude that therefore interspecific competition caused it. Cadena and Loiselle (2007) revisited this example. Using ecological niche modelling they calculated expected distributions along axes of climatic variation and found that some patterns of elevational range variation did appear to be consistent with predictions of the hypothesis that release from competition underlies expanded elevational ranges in allopatry, while other patterns of expanded elevational ranges in the absence of putative competitors are better explained by a hypothesis related to species' autoecology and geographic variation in the environment. They therefore caution against using elevation uncritically as a dimension of ecological niches.

This kind of argument as to the existence of competitive exclusion is, at best, circumstantial and does not really contribute to the understanding of the importance of interspecific competition today.

7.7 Seasonal variation in niche overlap

In their study of the effects of changes in food availability on niche width and overlap in Darwin's finches, Jamie Smith and colleagues observed that, when food resources diminished in the dry season and birds did not leave the island, food niches became narrower leading to a reduction of niche overlap between species using the same seeds (Smith *et al.* 1978). They concluded that, during periods of limited food availability, the balance between intra- and interspecific competition shifted towards the latter, causing specialization (Fig. 7.4). They reviewed all existing examples of studies comparing niche width and overlap in different seasons and concluded that their observation was typical: in lean times food niches become narrower and niche overlap between competing species decreases. A few years later, and using 30 examples, Tom Schoener confirmed that in most cases this is what happens (Schoener 1982).

A more recent example illustrates how complex such interactions can be. The African black-bellied seed cracker exhibits a non-sex-related polymorphism in beak size that enables the small- and large-billed morphs to utilize different trophic niches. The bill polymorphism between small- and large-billed individuals is under genetic control by a single autosomal locus with the allele for a large bill being dominant (Clabaut *et al.* 2009). Thomas Smith showed that in lean times diet overlap between different morphs decreased, that therefore the intensity of intraspecific competition increased, resulting in an increase in the niche breadth of the species (Smith 1990). When, additionally, large flocks of redheaded quelea arrived during lean times—their diet being most similar to the small-billed morphs, both preferring small, soft seeds—while the amount of soft seed declined, the small-billed seed crackers shifted from soft to hard seeds, illustrating the impact of an increase in interspecific competition.

Jaksic *et al.* (1993), however, found that during lean times niche overlap between ten species of predators (two fox species, four owls, and four falconiforms) was greater than during fat times. In a comment, John Wiens (1993)

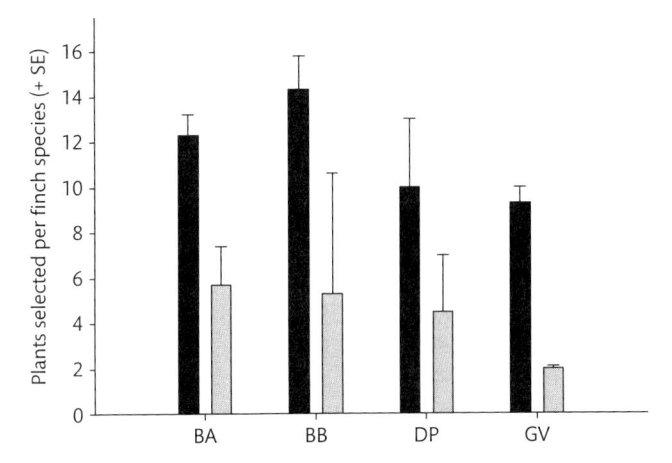

Fig. 7.4 Seasonal variation in the number of food plants selected by Darwin's ground finches in four study sites on three islands in the Galapagos. During the dry season (grey bars), there is less food available, and foraging niches are narrower than during the wet season (black bars), when food is more abundant. Based on Table 5 in Smith *et al.* (1978).

proposed that the niche response to lean times depends on the extent to which resources are limiting. If resources are not limiting, neither in fat nor in lean times, niche overlap should always be large as species should opportunistically use the same abundant resources. If resources are always extremely limited species should be forced to take whatever few resources remain, and resource use should also largely overlap but normally lead to starvation. Only when resources are above the resource-limitation threshold during the fat times and below the threshold during lean times would we expect to find an increase in specialization as resources become more and more limiting driven by competition.

7.8 Effects of migrants on residents

Billions of birds migrate to areas to survive the non-breeding season, joining the many residents that live there. Unless the migrants use resources that are not exploited by residents, or resources really are superabundant, one would expect that their arrival should result in some degree of interspecific competition because of resource use overlap. Remarkably few studies, though, have studied this in detail, and I am not aware of any experimental studies addressing possible interspecific competition. Although there are some studies suggesting the existence of niche overlap and interspecific competition between residents and migrants (Rabol 1987), recent reviews conclude that evidence, especially experimental evidence, is needed before we can conclude that interspecific competition between residents and migrants is generally important (Salewski and Jones 2006; Salewski *et al.* 2003). The detailed work by Dick Holmes and Pete Marra on Caribbean islands clearly shows resource limitation for overwintering parulids, that percolates all the way through to effects on the breeding grounds but does not provide strong evidence for the existence of interspecific competition (Marra 2000; Marra and Holmes 2001; Sillett and Holmes 2002).

7.9 Conclusions

Studies of foraging niches and diet show that food resources are often limiting (condition 1) and that resource use of potential competitors often overlaps (condition 3). The presence of one species may cause a shift in the foraging niche of another species (evidence 1). Note that, in general, food niches are narrower when food is scarce, implying that in such conditions the balance between intra- and interspecific competition shifts in favour of interspecific competition. In some cases ecological release caused by the absence of competitors has led to evolutionary changes, probably caused by a change in the balance between intra- and interspecific competition (see also Chapter 10). Several of the examples presented in this chapter underline, once again, that using non-experimental evidence to infer the importance of competition is usually open to criticism. In the next chapter I will therefore mostly emphasize experimental results.

8
Field experiments to test the existence and effects of interspecific competition

In previous chapters I have set the scene for the evaluation of the evidence that interspecific competition occurs. I have shown that various resources are often limiting (Chapter 3–5), that in many studies intraspecific competition has been documented that impacts a variety of demographic parameters (Chapter 6), and that niches often overlap between species so that the presence of one species affects the niche use of another (Chapter 7). In this chapter I want to discuss direct evidence concerning the existence and effects of interspecific competition, primarily using experimental data.

Experiments that attempt to document the existence of interspecific competition can be separated into three groups, depending on the effect measured. The first group, which looks at possible niche shifts, has already been discussed in Chapter 7. Both other groups of evidence (effects on a demographic parameter; experiments testing for an effect on numbers) are discussed in this chapter as they provide direct and compelling evidence for the existence or absence of interspecific competition.

Experiments to test for the existence of competition can either manipulate a resource (for example by blocking cavities in which birds nest, or by adding or removing food) or manipulate the density of a presumed competitor (by removing or adding individuals).

Most 'older' experiments reviewed by John Wiens had no control, were not replicated, and were continued for a short time only (Wiens 1989). In the last 20 years many experiments have continued over several years and compared multiple study sites in which a manipulation was performed to several control sites. These experiments were thus replicated in time and space and compared experimental and control plots. Although technically all experiments, including field experiments, should be carried out this way, I have included in this chapter less perfect experiments. Although such experiments, by themselves, do not provide compelling evidence, combining the results of several such 'imperfect' experiments carried out by different scientists at different locations that do (or do not) give the same result can provide compelling proof for the existence or absence of interspecific competition. In the following paragraphs I have therefore grouped experiments that have used a similar approach or asked a similar question to determine if the results would support each other or would vary between sites, times, or habitats. If all experiments, regardless of their quality, yielded the same result, it would be possible to conclude that interspecific competition for a given resource was generally important.

Given that both resources and abundance vary in time and space one would expect that the outcome of the same experimental treatment could vary. If different experiments generated contrasting results, one could therefore ask why (quality of experiment; differences between sites, species; lack of general importance of competition) and provide a nuanced answer to the question regarding the importance of interspecific competition. For this reason I have listed all the experiments I could find and have described most in some detail, although this makes this chapter a bit long.

In some experiments the manipulation of a limiting resource resulted in a change in abundance of one putative competitor. The effect of that change was then measured on other competitors. Adding nest-boxes, for example, often causes an increase in the numbers of cavity nesters. This may or may not result in a change in abundance of open-nesting species (see Section 8.4).

When reviewing the experiments it becomes clear that very few experiments with birds have tried to measure simultaneously effects on demographic variables and effects on population size. None of the experiments that manipulated the density of one species and measured its impact on the reproduction of a second species also reported effects on population size of the second species. I remember asking Göran Högstedt about his beautiful experiment documenting the severe impact of jackdaws on magpie reproduction ((see Section 8.1 and Högstedt 1980a) if he believed this would have an impact on magpie population size. He replied that the effect on reproductive success was so strong that it was bound to have an impact, although he did not measure this. Similarly, none of the experiments that tested for a possible effect of a manipulation on population size also measured which demographic variables would cause such a change. In Chapter 9 I will therefore review in some detail my own work on competition in blue and great tits in which I attempted both to manipulate population size and determine how this effect came about.

In Section 8.9 of this chapter I will further explore the somewhat surprising result that experimental increases of secondary cavity nesters sometimes result in increases in breeding density of other bird species caused by heterospecific attraction, as briefly mentioned in Section 8.4.

The experiments described in this chapter are grouped in nine sections. In each section there is a table listing the experiments discussed. In the first column I briefly state what manipulation was performed (for example: block natural cavities; remove/add fish). The next two columns describe effects. Effect 1 normally describes the immediate effect of the experimental manipulation on the resource or bird abundance (often implying what competition was for), while effect 2 describes the effect on the species or species group for which the existence of competition was tested. Column 4 describes what kind of interaction could be inferred (exploitation, interference, or diffuse competition, but also facilitation in some cases), and what resource was involved. The reference is given in column 5.

I have grouped the experiments in relation to what manipulation was performed and what outcome was measured. In the first four sections researchers manipulated availability and/or type of cavities thereby (usually) influencing breeding density, availability of roosts, or the availability of cavities in which to store food (larders). In Section 8.1 the manipulation influenced the density of one or two putative competitors and effects were measured on some component of fitness of one target species. In Section 8.2 the outcome of the experimental manipulation was measured as a change in population size. In the two next sections entire communities were manipulated and effects measured either on all species that use cavities (sometimes including mammals) or on overall bird community composition including non-cavity nesters. One can think of such experiments as comparable to 'whole lake experiments' in limnology (see also Section 8.6.3).

In Section 8.5 I move to non-cavity nesters, and discuss the effects of removal experiments whose objective was to determine if interspecific competition occurred between the removed species and another. In most of these experiments the effects of removals were measured as a change in territory size or in breeding numbers of the subordinate species. In some of these studies reciprocal removal experiments were carried out. In the more involved experiment by Martin and Martin (2001) effects of removals on nest predation and feeding rates were measured.

In Section 8.6 I group experiments testing if competition can be proved between birds and animals from another taxon, such as insects or fish, and in Section 8.7 I enumerate experiments whose specific objective was the recovery of endangered species: all examples concern small petrel species.

In Section 8.8 I explore the role of heterospecific aggression, and in Section 8.9, finally, I discuss experiments testing the existence and effects of heterospecific attraction. Conclusions are formulated in Section 8.10.

8.1 Effect of manipulation of cavities available on reproductive or foraging success of presumed competitors (Table 8.1)

Many authors have taken advantage of the fact that cavities may limit breeding population size and measured the possible effects of an increase or decrease in the breeding density of putative competitors on reproductive success. Of the seven experiments in this group, Göran Högstedt's stands out for its elegance and comprehensive approach (Högstedt 1980a). First he demonstrated experimentally that magpies are food limited: magpies provided with additional food laid earlier, suffered less nestling mortality and hence fledged more young (Högstedt 1981). Magpies are thus subject to intraspecific competition for food during the breeding season. Then he documented considerable overlap in the food used by magpies and jackdaws during the breeding season (Högstedt 1980b). Finally he tested if single magpie pairs breeding in small pine woodlots in a matrix of agricultural landscape in southern Sweden competed with jackdaws during the breeding season by putting up nest-boxes for jackdaws in some but not in other woodlots. In half of 20 woodlots Göran Högstedt put up three nest-boxes for jackdaws. He kept the other half of the woodlots as controls without jackdaws. In the second year of the experiment he reversed control and treatment sites whereby each woodlot was used as its own control. His experiment was thus both replicated and repeated between years. There was no difference between experimental and control birds in laydate nor in clutch size, but control birds, with no jackdaws in the woodlot, fledged 1.68 young, which is about 5 times more than the 0.33 young fledged by experimental birds. Furthermore, young from control broods were significantly heavier than young from experimental broods, implying they should survive better. Interspecific competition was primarily for food, causing starvation in magpie nestlings. Additionally magpies lost nestlings through predation by another corvid, the hooded crow. Högstedt proposed that this predation was an indirect effect of interspecific competition by jackdaws. In the woodlots with jackdaws food depletion may have caused the magpies to forage farther from their nest. As a result of this they were less able to defend their nests against predators (Högstedt 1980a). He did not measure a possible reverse effect of magpies on jackdaws, nor did he determine that the size of the magpie breeding populations was reduced by the reduced number of young fledged. It is not obvious that this would be the case because magpies typically have a surplus of non-breeding birds (Baeyens 1979).

Most other similar experiments used tits or flycatchers (Table 8.1). After finding that great tit nestling weight in Wytham Woods, near Oxford (UK), decreased with increasing blue tit density, Ed Minot tested for the existence of competition for food during the breeding season between great and blue tit by removing blue tit nestlings from part of the study site and adding them to another (Minot 1981). Great tit nestlings were lightest where blue tit nestlings had been added, heaviest where no blue tit nestlings remained, and intermediate in the control site, confirming results from correlational data. He thus proved that interspecific competition for food between great and blue tit during the breeding season in Wytham Woods influenced great tit fledgling mass. Given that great tit survival increases with fledgling mass (Adriaensen *et al.* 1998; Dhondt 1971; Perrins 1965) interspecific competition for food affects great tit fitness.

Table 8.1 Field experiments in which resource abundance or competitor density were manipulated and its effect on demographic parameters

Manipulation	Effect 1	Effect 2	Type of competition	Reference
Add boxes for jackdaw	Increase jackdaw numbers	Decrease magpie reproductive success	Exploitation (food)	(Högstedt 1980a)
Remove/add blue tit nestlings	Manipulate food required by blue tits	Increase/decrease great tit nestling mass	Exploitation (food)	(Minot 1981)
Manipulate nest box availability	Manipulate flycatcher and tit density	Flycatcher reproduction, recruitment and survival affected by tit density	Exploitation (food), interference	(Gustafsson 1987)
Provide additional boxes	Reduce takeover of red-cockaded woodpecker holes	Improve red-cockaded woodpecker breeding success	Interference for roost and nest cavities	(Loeb and Hooper 1997)
Manipulate nest box type	Create allopatric and sympatric populations great and blue tit	Shift in prey selection towards smaller prey in sympatry; reduction nestling mass great tit, not blue tit	Exploitation (food), interference	(Török and Tóth 1999)
Provide boxes for pygmy owl close/far from boxes for Tengmalm's owl	Manipulate potential competition with Tengmalm's owl	Impact winter larder size of pygmy owl	Exploitation	(Suhonen et al. 2007)
Remove red-bellied woodpeckers	Increase in number of immigrants of red-bellied woodpeckers	Reduction of red-cockaded woodpecker group size	interference competition (interspecific aggression)	(Walters and James 2010)

Another study that tested for the existence of interspecific competition in the breeding season between great and blue tit was performed by János Török and László Tóth in Hungary. They created study plots where only great tits, only blue tits, or both species together were nesting. Blue tits alone bred when small-holed (26 mm) boxes were offered; both species bred when large-holed (32 mm) boxes were available; and great tits only bred in a site with large-holed boxes in which the authors removed blue tits that attempted to nest (Török and Tóth 1999). They showed that there existed interspecific competition for food by studying what nestlings were fed. Prey size brought to the nestlings of either species was affected by the presence of the other. In allopatry both great and blue tit brought larger prey to the nest than when nesting side by side (Török 1993). This shift in prey selection had a potential effect on fitness in sympatry but only during the nestling stage and only for great tits. Great tit nestlings were lighter in sympatry than in allopatry in four out of five years, while there was no difference for blue tits. Although generating interesting results, this experiment has two problems. First, although the experiment continued over a five-year period, it was not replicated in space. It would have been better if over the five-year study period treatments had been switched between study sites. Second, the blue tit sympatry/allopatry treatments were not equivalent. Since in the blue tit allopatry treatment only small-holed boxes were offered blue tit density must have been high and great tit density very low as compared to the blue tit sympatry treatment where only large-holed boxes were available. In this latter plot blue tit density must have been low but great tit density high (Dhondt *et al.* 1991). There were thus two differences between the blue tit plots: one as regards the intensity of interspecific competition (intended) and another as regards the intensity of intraspecific competition (not intended). I will report on similar but more extensive experiments in the next chapter. Nevertheless Török's experiment does strongly suggest the existence of asymmetric interspecific competition for food that has an effect on great but not on blue tit nestlings mass, as suggested by myself from correlative data (Dhondt 1977).

Lars Gustafsson performed a more comprehensive experiment on the Swedish island of Gotland. He manipulated densities of collared flycatchers and tits independently in two pairs of study plots over a two-year period. In one pair of plots he kept the density of both collared flycatchers and tits low by providing a low density of nest-boxes or by providing a low density of boxes and removing tit nests. In the other study plot pair he provided a high density of nest-boxes, but in one of the plots he made the boxes unsuitable for tits so that only flycatchers used them. This double set of paired plots made it possible to study the effect of tits on collared flycatcher reproduction at high and at low flycatcher density. In both configurations he found effects of tit density on collared flycatcher reproduction (nestling mortality, survival, and recruitment). Because the extra-limital population of collared flycatchers on Gotland is isolated and highly philopatric, and because flycatchers were studied in a large number of neighbouring woodlots, most breeding adults were trapped, and all nestlings banded, a very high proportion of recruits were caught. This made it possible to also determine the effect of competition on fitness by combining juvenile recruitment and adult survival. Flycatcher fitness was about double in plots with low tit density compared to plots with high tit density (Fig. 8.1), indicating a strong effect of interspecific competition. A comparison the author did not make was that between the study plots in which tit density was low but collared flycatcher high or low. This comparison makes it possible to measure effects of intraspecific competition. Calculated fitness differed not or only weakly between plots in which collared flycatchers density was low or high (Gustafsson 1987) implying that intraspecific

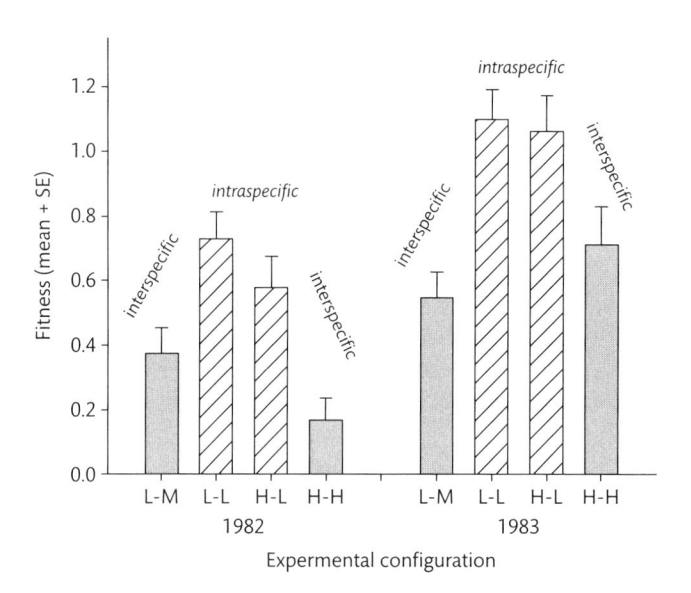

Fig. 8.1 Fitness of collared flycatchers exposed to different experimental levels of intra- and interspecific competition in 1982 and 1983 as indicated along the x-axis. The first letter (L= low or H= high) indicates densities of collared flycatchers. The second letter (L = low, M= medium, H = high) indicates densities of tits. By comparing plots in which either flycatcher or tit density varied, but the other species' density remained unchanged, it is possible to evaluate effects of either intra- or of interspecific competition as indicated on the graph. The comparison of the hatched bars (low tit density but either low or high collared flycatcher density) shows no intraspecific competition in 1983 and weak intraspecific competition in 1982. The comparison of the left (low flycatcher density, but low or medium tit density) or right (high flycatcher density but either low or high tit density) pair of bars shows that in both years interspecific competition had a large effect on collared flycatcher fitness. Note also that 1982 was a 'bad' year, while 1983 was a 'good' year. From Table 5 in Gustafsson (1987).

competition had no (in 1983) or only a weak effect (in 1982) while interspecific competition had a strong effect in both years. This is a surprising result because according to theory two species can only coexist if intraspecific competition is stronger than interspecific competition. This may be possible though if the populations are largely regulated when the populations live apart outside the breeding season. Note that the collared flycatcher is a long-distance migrant.

Using woodpecker species Susan Loeb and Robert Hooper carried out an experiment similar to Högstedt's but with a more applied objective (Loeb and Hooper 1997). The red-cockaded woodpecker is an endangered species that excavates cavities in living pine in south-eastern USA. Excavation of one hole may take over a year and population size is assumed to be limited by number of cavities. This woodpecker often loses conflicts over its cavities against other species, including starling and red-bellied woodpecker (Ingold 1994a, 1994b), so that they cannot themselves breed or breed successfully in them. To test if and to what extent red-cockaded woodpecker reproduction was affected by other species, Loeb and Hooper added three nest-boxes to each of 62 woodpecker cavity clusters (the red-cockaded woodpeckers is a species that breeds socially and cavities of one family are clustered in neighbouring trees) in an area greatly affected by a hurricane. They compared red-cockaded woodpecker cavity occupancy and nest success during one pre-treatment control year and two experimental years. The most common species occupying woodpecker cavities were eastern bluebird and flying squirrel. These species preferred nest-boxes over woodpecker cavities. In the experimental clusters a higher proportion of family groups nested, and those that nested raised more young than in both the control sites and the pre-treatment control year. Furthermore both in control and in experimental clusters red-cockaded woodpeckers were less likely to nest if another species was present in the cluster.

Eric Walters and Fran James tried a different approach to help red-cockaded woodpeckers in northern Florida by removing the competitively superior red-bellied woodpeckers. To their surprise this had a negative effect on red-cockaded woodpeckers: where red-bellied woodpeckers had been removed red-cockaded woodpecker group sizes became smaller, the proportion of their nests that successfully fledged young declined, and/or the proportion that remained on their territory declined. They discovered that the removal caused an increase in the number of red-bellied woodpecker immigrants who filled up vacancies. These immigrants were especially aggressive to female red-cockaded woodpeckers who were then displaced and emigrated at twice the rate of control sites where red-bellied woodpeckers had not been removed. Overall this resulted in a decline of red-cockaded woodpeckers (Walters and James 2010). The 'traditional' approach of removing individuals of a dominant competitor did not have the hoped for effect because removal induced immigration, and newly established birds were more aggressive than established birds. Their experiment nevertheless confirmed the existence of interspecific competition between these two woodpecker species.

The final experiment I describe here is a very different experiment. Jukka Suhonen and colleagues in Finland used nest-boxes in an innovative way to test for the existence of interspecific competition between Tengmalm's owl and pygmy owl during winter (Suhonen *et al.* 2007). Both species use cavities, including nest-boxes, as larders in which they store birds and small mammals during winter. In two winters and in two different sites the authors provided small-holed nest-boxes (entrance hole of 45 mm), that could be used by the smaller pygmy owl only, close to or far away (>2 km) from large-holed boxes (entrance holes of 80 mm) that were often used by Tengmalm's owls. Pygmy owl larders in the plots near large-holed boxes were much smaller than in plots further away. As Tengmalm's owls hunt in dense vole patches and are able to depress vole densities this may lower the hunting success of pygmy owls which would cause larders to be smaller because of food competition with Tengmalm's owls.

All manipulations of the density of potential competitors through selective provisioning of nest-boxes described in this section show at least some effect of interspecific competition on a demographic parameter: nestling mass, reproductive success, winter survival, recruitment, or fitness.

8.2 Effect of resource manipulation on population size of presumed competitors: effects on single species (Table 8.2)

Cavity-nesting species are a particularly convenient group to test for possible effects of interspecific competition on the population size of presumed competitors. All experiments described here concern great or blue tit, and pied or collared flycatcher, species known to prefer nest-boxes for nesting.

Hans Löhrl (Löhrl 1977) followed by Dhondt and Eyckerman (1980) showed that providing small-holed nest-boxes that gave access to blue but not to great tits increased blue tit breeding density. Dhondt *et al.* (1991) proposed that this was the result of competition for roost sites in winter rather than for nest sites during the breeding season. In an aviary experiment Bart Kempenaers and I confirmed this. When during winter a single blue tit was placed in an aviary and given a choice between several large- and several small-holed nest-boxes, it chose to sleep in a large-holed box in each experiment (Dhondt and Eyckerman 1980; Kempenaers and Dhondt 1991). When a great tit was added to the aviary in which we kept a blue tit, in more than half the cases the blue tit moved to a small-holed box for the night (Kempenaers and Dhondt 1991).

The migratory pied and collared flycatchers arrive on the breeding grounds after the mostly resident tits have started nesting. Tore Slagsvold showed experimentally that pied flycatchers are subordinate to great tits when fighting over a cavity and hence interspecific competition may limit their breeding population size when cavities are in short supply (Slagsvold 1978). He made every box in which tits were not nesting uninhabitable, thereby forcing the pied flycatcher to compete with nesting tits if they wanted to acquire a cavity for nesting. Although the flycatchers were quite aggressive towards the tits they were able to take over the tit nest in only 2 out of 40 cases (one of which involved the smaller coal tit). This behaviour is generally dangerous for the flycatcher, as shown both in Slagsvold's own study and in Löhrl's long-term study of collared flycatchers in Germany. The latter reported that *'death caused by fighting between great tits and collared flycatchers is a major source of mortality of the flycatcher'* (Löhrl 1957). A similar conclusion was also reached by Merilä and Wiggins on Gotland. They observed that in plots where nest-boxes were rare the number of collared flycatchers found dead was higher than in plots with a large numbers of nest-boxes (Merilä and Wiggins 1995). In one plot, 17% of breeding flycatchers were killed. This indicates that interspecific competition for nest sites may constitute a significant source of adult mortality both in pied and collared flycatcher populations.

Although not planned as an experiment, Ed Minot and Christopher Perrins took advantage of the fact that in Wytham Woods, Oxford, the nest-box density provided by the researchers varied between different parts of the wood (Minot and Perrins 1986). At low nest-box densities blue tit numbers breeding in nest-boxes were limited by competition with great tits, as shown by the inverse variation in great and blue tit box breeding pairs and by the increase in relative density of blue tits with increasing nest-box numbers.

The final experiment I discuss in this section was performed on the island of Gotland by Lars Gustafsson (Gustafsson 1988b). In two experimental plots and in two different years he doubled nest-box density just before the breeding season, while leaving box density unchanged in three control plots. In a separate experiment he removed tits as they began nesting. Both manipulations caused an increase in collared flycatcher density. In a final experiment he reduced the floor area of all nest-boxes in a plot by half, causing the tits to abandon that plot altogether. In this latter plot collared flycatcher numbers increased 3–9 fold compared to the control sites. Lars Gustafsson concluded that collared flycatcher breeding density was primarily limited through interspecific competition for nest sites.

In all experiments described in this section the breeding density of a socially subordinate species increased when the number of holes available to them for nesting or for roosting was increased. This indicates that suitable cavities are limiting (see also Chapter 5) resulting in a severe effect of interspecific competition for this resource on breeding density of these secondary cavity nesters.

Table 8.2 Field experiments in which resource density was manipulated and its effect on population size

Manipulation	Effect 1	Effect 2	Type of competition	Reference
Reduce entrance of nest-boxes to exclude great tits	Decrease great tit density	Blue tit density increases	Interference? (unreplicated, no control)	(Löhrl 1977)
Remove nest-boxes not occupied by tits	No change in great tit density	Reduction in pied flycatcher breeding density; some mortality	Interference for nest cavities	(Slagsvold 1978)
Reduce entrance of nest-boxes to exclude great tits	Decrease great tit density	Increase blue tit density	Interference for winter roost sites	(Dhondt and Eyckerman 1980)
Vary large-holed nest-box density	High great tit density in all plots	Blue tit density increases with increasing nest-box density	Interference for nest sites	(Minot and Perrins 1986)
Manipulate nest-box density and number of tits breeding	Reduction tit numbers; increase boxes available	Increase collared flycatcher numbers breeding	Interference for nest cavities	(Gustafsson 1988b)
Offer choice of large- and small-holed boxes in aviary with/without great tits in winter		Blue tits move from large-holed to small-holed boxes for roosting	Interference for cavities used for roosting	(Kempenaers and Dhondt 1991)

8.3 Studies of communities of cavity nesters: experiments in which natural cavities were blocked or nest-boxes added generated a diversity of results (Table 8.3)

In a handful of experiments the scientists identified all natural cavities and either blocked them or added nest-boxes and measured the effect of the experimental manipulation on the abundance of cavity nesters. Such a community wide manipulation tested for possible intraspecific competition through limitation of cavities on cavity nester density. If not all species were affected in the same way this would also document the existence of interspecific competition for cavities as a limiting resource. The results varied strongly between experiments going from no effect, over effects on some species, to strong effects.

The first such study was carried out in The Netherlands in parallel with long-term nest-box studies of great tits. In the course of their detailed study of birds nesting in natural cavities in a mixture of old deciduous trees (oak, beech) surrounding young coniferous plantations Hans Van Balen and colleagues performed experiments in which they manipulated access to natural or artificial cavities for starlings (Van Balen *et al.* 1982). In one site most large entrances of natural cavities were narrowed in March to a diameter of around 30 mm, which excluded starlings, and the occupation of the cavities was compared to that in two previous years and to that in an unmanipulated control site. In the area and years in which starlings had access to cavities only 28% of the territorial great tits were able to nest, and of these only 40% were successful. Failure was caused by starling take-over. In the experimental site most great tits could breed and most were successful. To further document the role of starlings a follow up experiment was carried out. They hung twenty large nest-boxes, half of which had a 50 mm entrance diameter (accessible to starlings) while the other half had a 30 mm entrance (not accessible to starlings). In early May, when breeding attempts by starlings and great tits were well under way, the entrance holes of the occupied boxes were changed: the 50 mm entrances were exchanged for 30 mm entrances and vice versa causing the starlings to abandon their nest attempts and exposing great tit nests to potential usurpation by starlings. All boxes occupied by great tits were taken over by starlings, most within 1–2 days, showing that starlings are fierce competitors for cavities and able to expel smaller species even if these have already started nesting.

In the next three studies some or all natural cavities were blocked. In riparian habitat in Arizona, USA, Timothy Brush blocked natural cavities before the breeding season in one plot, while keeping a second plot as a control, and adding nest-boxes to a third (Brush 1983). Compared to the previous year this caused a decrease in the number of pairs of cavity nesters nesting in the plot in which the cavities had been blocked (13.5 to 8) while numbers remained stable in the unmanipulated plot (20 to 24) and increased in the a plot where nest-boxes had been provided (3 to 6). He concluded that natural cavities were limiting, but that interspecific competition for cavities was not important. There are several problems with the experiment. First, it was not replicated as there was one plot only for each treatment. Second, sample sizes were small, limiting the power of the analysis. None of the changes between treatments were statistically significant. Third, Brush combined all species in his analysis, although different cavity-nesting species reacted very differently to the manipulations. Thus ash-throated flycatchers decreased from five to zero pairs after cavity blocking, and increased from zero to three after nest-boxes had been provided, a significant difference (Fisher-exact two-tailed test P=0.018). It could be that natural cavities were limiting in general and that ash-throated flycatchers were excluded from holes because of interspecific competition. Further, different secondary cavity nesters preferred different types of holes as Lucy's warblers, for example, did not use nest-boxes, while ash-throated flycatchers did. Gila woodpeckers displaced by starlings simply excavated a new cavity and did breed, but later in the season. Breeding numbers of the species thus did not change, although their success might have. This non-replicated experiment, therefore, is inconclusive as regards the role of interspecific competition.

After having censused the breeding population of secondary cavity nesters in California oak-pine woodland, Jeffrey Waters and colleagues (1990) blocked a high proportion of natural cavities in two successive years. They blocked cavities just before the breeding season and reopened them after nesting was over. They concluded that population size of secondary cavity nesting species in general did not differ between the treatment and the control plot and that therefore cavities were not limiting in their study site. Nevertheless they reported that '*only ash-throated flycatcher density declined more on the treatment plot than in the control plot*' in the second year of their experiment when they blocked a larger number of natural cavities than in the first year.

Although neither this experiment nor that described above were replicated, and hence had limited power of inference by themselves, combining these results suggests that ash-throated flycatchers may be sensitive to interspecific competition when cavities are in short supply. Cardiff and Dittman (2002) suggested this was because ash-throated flycatchers arrive from migration after many other cavity nesters have begun nesting.

Another experiment in which natural cavities were blocked was a truly large-scale, long-term, well-replicated study. Kathryn Aitken and Kathy Martin studied birds and mammals that use natural cavities in aspen groves in a grassland matrix in British Columbia, Canada (Aitken and Martin 2008). They followed nest abundance on seven treatment and thirteen control sites over a six-year period. They compared cavity use during two pre-treatment years, two treatment years during which high quality cavities were blocked in experimental sites but

Table 8.3 Cavity nesting birds in natural cavities—effects of blocking cavities

Manipulation	Effect 1	Effect 2	Type of competition	Reference
Reduce size of entrance hole of natural cavities	Reduction in starling density	Increased proportion of territorial great tits nesting	Interference competition for cavities	(Van Balen et al. 1982)
Nest holes blocked/boxes added		Secondary cavity nesters decrease/increase	Weak Interference competition for cavities	(Brush 1983)
Nest holes blocked		No reduction in secondary cavity nesters	No effect	(Waters et al. 1990)
Nest holes blocked	Fewer quality cavities available	Starling numbers decrease; Mountain bluebirds increase	Interference competition for cavities	(Aitken and Martin 2008)
Some nest holes protected	Indirect effect on large secondary cavity nesters	Red-cockaded woodpecker increase; Northern flicker switches to other cavities	Interference competition for cavities	(Blanc and Walters 2008)
Block parakeet nest cavities	Take-over of nuthatch nests by parakeets	Reduction in nuthatch breeding density	Interference competition for cavities	(Strubbe and Matthysen 2009)
Add nest boxes	Triple availability of nest sites	Increase breeding density of birds and mammals	Interference for cavities	(Aitken and Martin 2011)

not in control sites, and two post-treatment years after cavities had been reopened. Nest numbers declined by 50% on treatment sites following cavity blocking of high-quality cavities and returned to pre-treatment levels following cavity reopening (see Fig. 8.2). As in previous studies, different species responded in different, sometimes surprising ways, to the experiment. Tree swallows were unaffected by cavity blocking while starlings declined by 89%. Surprisingly, starling numbers did not recover after cavities had been reopened. Although fewer cavities were available in the experimental sites, because cavities had been blocked, the number of mountain bluebird nests increased relative to the control sites and their numbers remained high for two years after reopening. The authors conclude that the subordinate swallows and bluebirds are more plastic in their choice of a nest cavity. Bluebirds especially would be limited by starlings rather than by cavities, while starlings are limited by high-quality cavities. The decrease in starlings would then have allowed an increase in bluebird numbers.

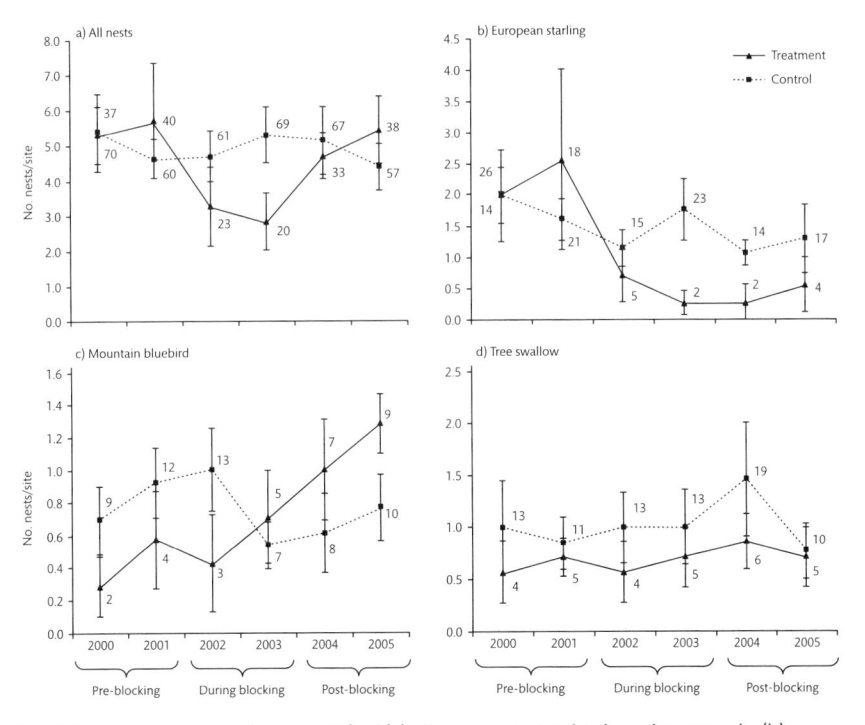

Fig. 8.2 Nest abundance (mean ± SE) of (a) all cavity-nesting birds and mammals, (b) European starlings, (c) mountain bluebirds, and (d) tree swallows on 7 treatment sites (cavity blocking; solid line) and 13 control sites (dotted line) at Riske Creek, British Columbia, Canada. Sample sizes shown beside points indicate the total sample of nests. From Aitken and Martin (2008) with permission from the Ecological Society of America.

In a parallel experiment, Aitken and Martin conducted an 11-year before-after/control-impact study in a mature (> 100 years) mixed conifer forest in British Columbia, Canada using three treatment and four control sites ranging between 7 and 20 ha (Aitken and Martin 2011). During six pre-treatment years, two experimental years, and three post-treatment years they monitored previously active and newly excavated cavities, recording both birds and mammals. During the two treatment years 30 nest-boxes of two sizes were available in each treatment plot. The two nest-box sizes reflected the size of natural cavities used by small (chickadee, nuthatch) and large (flicker, sapsucker) cavity nesters. These nest-boxes were added in the July of the last pre-treatment year and were removed in late July of the second treatment year after which the study continued for another three years. Nest-box addition tripled the nest site availability from a mean of 1.2 cavities ha^{-1} before boxes were added to an average of 3.5 cavities ha^{-1} with boxes. The observation that in the pre-treatment years only about 10% of the existing cavities were occupied suggested that either cavities were not limiting

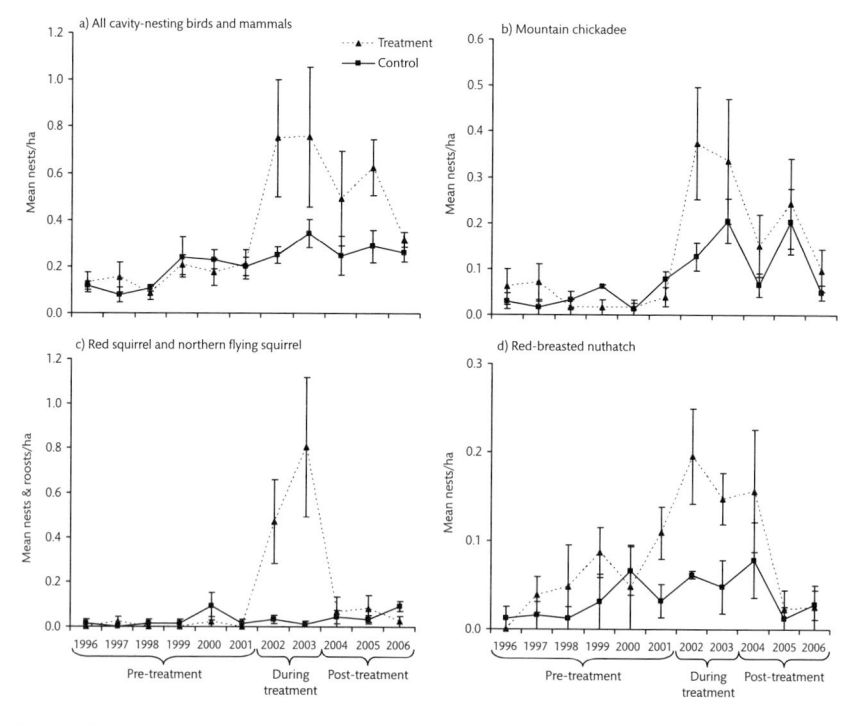

Fig. 8.3 Density of cavities and boxes occupied (mean ± SE) by a) all cavity-nesting bird and mammal species; b) mountain chickadees; c) red squirrels and northern flying squirrels (nests and roosts); and d) red-breasted nuthatches on three treatment (boxes added) and four control sites, in interior British Columbia, Canada. From Aitken and Martin (2011) with permission from John Wiley and Sons.

or that the quality of unoccupied holes was inadequate (Lohmus and Remm 2005; Remm *et al.* 2008). If the first hypothesis was correct the addition of nest-boxes would not cause an increase in cavity nesters, while if the second hypothesis was true adding boxes should increase the density of at least some species.

The results were quite clear (Fig. 8.3), but as usual in well done experiments contained some surprises. Density of mountain chickadee, red squirrel, and northern flying squirrel increased during treatment years and declined following box removal, with no such changes on the control plots. The surprising result was that red-breasted nuthatch density increased on treatment sites and not on control sites, although only one pair bred in a nest-box. The authors suggest that red-breasted nuthatches, when settling in the autumn, evaluate habitat quality, and that the presence of additional holes (through the provision of nest-boxes) increases apparent habitat quality by reducing interspecific competition for cavities. This competition could perhaps also be for winter roost sites, as found for blue tits (Dhondt *et al.* 1991). Aitken and Martin conclude that the addition of nest-boxes may release subordinate species from nest site limitation, and that even excavators may benefit from box addition if it reduces overall nest site competition.

The results of this experiment are particularly important as it is the only long-term, replicated nest-box addition experiment with controls and before/after treatment data for cavity nesters in mature mixed conifer forests. The results clearly demonstrate that natural cavities limit breeding density of cavity-nesting birds and mammals and cause interspecific competition.

Rather than block natural cavities completely Lori Blanc and Jeff Walters, who studied hole-nesting bird communities in longleaf pine habitat in Florida, USA, manipulated access to the cavities of red-cockaded woodpeckers by using metal restrictor plates (Blanc and Walters 2008). They placed these plates in eight experimental plots, which kept northern flickers out of red-cockaded woodpecker cavities or made it impossible for the flickers to enlarge the cavity entrances. They left cavities unchanged in eight unmanipulated control plots. To determine a possible effect of the experimental manipulation they compared data between one pre-treatment year, one treatment year, and one post-treatment year. Flickers responded to their reduced access to red-cockaded woodpecker cavities by simply switching to other cavities in snags. This in turn caused a reduction in the nesting abundance of large secondary cavity nesters (American kestrel and eastern screech owl), documenting an indirect effect of the experimental manipulation (see also Section 5.4).

In a more targeted variant of this approach, Diederik Strubbe and Erik Matthysen tested the effect of blocking nesting cavities used by the introduced and expanding ring-necked parakeet on native cavity nesters in Belgium (Strubbe and Matthysen 2009). When they blocked a high proportion of cavities previously occupied by parakeets, nuthatch breeding density decreased significantly through usurpation of their holes by parakeets. The nuthatches' habit of reducing the size of the entrance hole of their cavity by plastering it with mud does not work against the parakeets, because the latter start breeding six weeks before the nuthatches, and thus preemptively occupy the best nest holes. What is particularly interesting about this study is that cavity quality for parakeets and for nuthatches was calculated (and was similar), and that it was therefore possible to test the hypothesis that a large number of unoccupied cavities does not mean there is a surplus of potential holes if the unoccupied holes are of insufficient quality (Fig. 8.4). Thus, after blocking cavities, there still remained 15 and 11.8 cavities ha^{-1} in the two study sites, which is a high density considering that nuthatch breeding density was not higher than 0.3 pairs ha^{-1} in either the pre-experimental years or in the control sites. Their results clearly show that in the year when parakeets had to find new cavities for nesting they specifically targeted high-suitability sites. The observation that in experimental years and sites nuthatch density declined by about half suggests that, through interspecific competition, not enough cavities of sufficient quality remained for nuthatches to remain and breed.

The sometimes subtle and indirect effects of experimental reduction or addition of natural cavities shows essentially two things. First, in many instances cavities are limiting for some secondary cavity nesters but not for others. Dominant

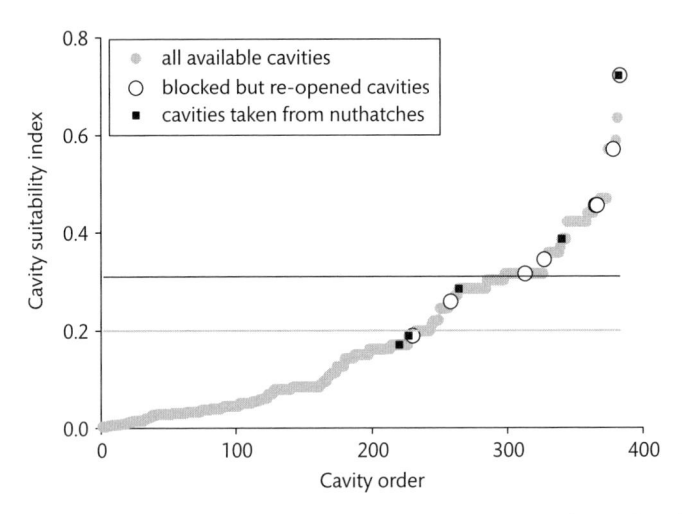

Fig. 8.4 Quality of cavities located in the experimental areas based on the predicted values of a logistic model consisting of the cavity characteristics influencing parakeet cavity choice. Cavities are ordered from lowest to highest index of suitability. Horizontal black line shows mean parakeet cavity quality in the 2006 pre-blocking period, grey line shows mean cavity quality after blocking. Note that parakeets preferentially usurp high quality cavities occupied by nuthatches. From Strubbe and Matthysen (2009), with permission of Elsevier.

species can either preemptively occupy the best sites, or take-over occupied sites by expelling the subordinate species so that interspecific competition for cavities does impact the number of successfully breeding individuals of some species. Second, only detailed, long-term, and replicated field experiments can detect such effects. Finally, more attention needs to be given to effects of cavity suitability in studies of nest site limitation of cavity nesters, and how cavity suitability varies between species.

8.4 Interactions between cavity and open nesters: does adding nest-boxes influence the density of open-nesting species? (Table 8.4)

Given that in a large number of studies adding nest-boxes increases breeding density of secondary cavity nesters considerably (see Section 5.2, 8.3), such an increase might have an impact on other bird species in the community especially through competition for food. All experiments discussed in this section documented that adding or removing nest-boxes resulted in an increase/decrease in secondary cavity nesters. In two of the eight experiments this coincided with a decrease in density of open-nesting species, in one with an increase in density of open- nesting species (see also Section 8.9), while in four there was no effect (Table 8.4.)

I will start by discussing these negative experiments, which were carried out by multiple authors in diverse habitats inhabited by different bird communities. The first study was a well replicated eight-year study (two pre-treatment years in five plots; six treatment years with nest-boxes in two plots, and no boxes in three unmanipulated control plots) in rich subalpine birch forest in Swedish Lapland. In the plots where nest-boxes had been added pied flycatcher density increased about twenty-fold while their numbers remained very low in the control plots (Enemar and Sjöstrand 1972). This caused the overall numbers of birds in the study plot to almost double. Nevertheless, the density of other passerines did not decrease in the experimental plots as compared to the control plots, a pretty unambiguous result. The second study ran four years (two pre-treatment; two treatment years with nest-boxes in half the plots) in coniferous habitat in Colorado, USA, where a variety of cavity-nesting species (mountain chickadee, pygmy nuthatch, house wren) increased, but the numbers of open-nesting species did not change (Bock and Fleck 1995). The third study was both correlative and experimental and was carried out in four plots in coniferous habitat in Arizona. In the experimental part two plots were provided with nest-boxes while two other plots were kept as controls. Again, numbers of open nesters did not decrease as cavity nester abundance rose, leading the authors to conclude that '*Interspecific competition for food during the breeding season appears to be unimportant in ponderosa pine bird communities*' (Brawn *et al.* 1987). In a similar study Marion East and Christopher Perrins removed nest-boxes from a semi-mature broadleaved woodland near Oxford, UK, while keeping the nest-boxes in an adjacent woodlot. This experiment was, therefore, unreplicated in space. Nest-boxes that had been present since at least 1962 were rendered unsuitable by removing the front panel in the fall of 1981, and maintained like that for three breeding seasons. In those years tits bred only in natural cavities. This caused great tit density to decrease, although blue tit numbers did not decline significantly. No change in density of open nesters coincided with the change in tit density.

Sara Sanchez and colleagues performed an experiment similar to that of East and Perrins (1988), but not only compared breeding numbers but also winter abundance. They studied a series of pine and beech plots, half of which received nest-boxes (Sanchez *et al.* 2007). The treatment had no effect on abundance of non-cavity nesters during the breeding season, but did result in an increase in abundance of non-cavity nesters during the non-breeding season in pine forest but not in beech. They underlined that cavities can be important during the non-breeding season as roosts not only for cavity but also for some non-cavity nesters such as dunnocks, although a detailed look at their data does not make clear which species would be affected in their study.

In stark contrast to these experiments that clearly showed the absence of interspecific competition between cavity-nesting and open-nesting species during the breeding season, Olav Hogstad's experiment in two coniferous study plots in Norway showed the opposite outcome (Hogstad 1975b). In a control plot of 110 ha, in which he did not provide nest-boxes, Hogstad simply plotted the territories of all breeding birds. In the experimental plot of 9 ha he alternated periods without and with nest-boxes during an eight-year study. Overall he had data for four years with and four years without nest-boxes. Hanging nest-boxes increased the density of cavity nesters about twenty-fold, primarily because large numbers of pied flycatchers settled, although some great tits also bred in them. The total number of open nesters was reduced by about half in years with nest-boxes compared to the years without them. After comparing annual variations in control and experimental plots he concluded that three open nesting species were adversely affected by the increase in cavity nesters: chiffchaff, willow warbler, and chaffinch. Chiffchaffs were absent in two of the four years with nest-boxes and willow warblers were absent in all four years, suggesting strong effects of interference competition. Note that this experiment was also unreplicated in space, but the immediate changes in open nester breeding densities in response to changes in nest-box availability in alternate years does suggest that interspecific competition did influence breeding numbers. Given that the effect was immediate, interference competition rather than exploitation competition for food played an important role in this study, which is interesting, because the studies I described above were designed to test for the existence of interspecific competition for food (exploitation competition).

A second experiment that resulted in a decrease in the densities of open nesters was carried out with a very different bird community by Carl Bock and colleagues in riparian woodland in Arizona, a very species-rich habitat that contained 17 cavity nester species and 53 species that build open nests (Bock *et al.* 1992). They started by counting birds for three years in a series of small circular plots and then added nest-boxes in 50 plots while keeping 49 plots as controls. Over the following three years they found that, as cavity nesters increased in the plots with nest-boxes compared to control plots, the density of open nesters decreased significantly in these relative to the control plots. The authors concluded that diffuse competition was important in this very rich habitat with very high bird densities, and that interspecific interactions influenced the composition of the avian assemblage. They argued that this was the outcome of interspecific competition for limiting resources. Three years later Carl Bock repeated a similar experiment in a very different montane habitat in Colorado, and found no effect of the addition of nest-boxes on the abundance of open nesters (see above; (Bock and Fleck 1995)). They write that they '*do not know why assemblage-wide (diffuse) interspecific competition might be more important among southwestern riparian birds than among species of montane and/or higher latitude forests*' but propose that '*secondary consumers such as avian insectivores are more likely to limit their prey (and therefore to be limited by it) in relatively productive ecosystems*' such as the riparian woodland (Bock *et al.* 1992).

A third, somewhat unintuitive, outcome of increasing the density of cavity nesters by providing nest-boxes is an *increase* in the numbers of open-nesting species. As a forester, Gustav Wellenstein was interested in controlling insects that damage pine forest plantations. The multiple approaches used included the introduction of nests of different ant species (*Formica rufa*, *F. polyctena*) and the provision of nest-boxes. Both ants and birds reduced insect numbers (Wellenstein 1968). What is relevant in the context of this book is that he laid out 8 one ha plots with four different nest box densities each replicated twice. He reported how many successful broods fledged in each season from the nest-boxes, and also censused territorial open-nesting birds on the plots, making it possible to test if the variation in cavity nest abundance (caused by different numbers of nest-boxes) resulted in variation in the number of open nesters (Wellenstein 1968). The number of cavity nesters (mostly pied flycatchers and tits) correlated strongly with nest-box density but also with the numbers of open nesters (Fig. 8.5), a result that has mostly been overlooked in the ornithological literature. He suggested this was caused by a 'social effect'. A few remarks concerning this experiment: first, the nest-box densities used were extremely high, going all the way to 36 boxes ha^{-1}. Second, as the objective of Wellenstein's study was to find economical ways to reduce numbers of forest insect pests, these bird results were

Table 8.4 Experiments testing for the effect of interspecific competition between cavity and open nesters at community level

Manipulation	Effect 1	Effect 2	Type of competition	Reference
Provide nest-boxes at different densities	Increase cavity nesters in proportion to box density	Increase number open nesters	Facilitation *	(Wellenstein 1968)
Provide nest-boxes	Increase pied flycatchers	No effect on open nesters	None	(Enemar and Sjöstrand 1972)
Provide nest-boxes	Increase cavity nesters	Decrease number open nesters	Interference (space)	(Hogstad 1975b)
Provide nest-boxes	Increase cavity nesters	No effect on open nesters	None	(Brawn et al. 1987)
Remove nest-boxes	Decrease cavity nesters	No effect on open nesters	None	(East and Perrins 1988)
Provide nest-boxes	Increase cavity nesters	Decrease number open nesters	Diffuse competition (food?)	(Bock et al. 1992)
Provide nest-boxes	Increase cavity nesters	No effect on open nesters	None	(Bock and Fleck 1995)
Provide nest-boxes	Increase cavity nesters	No change in breeding numbers; increase in winter abundance in pine, not beech	Interference; (cavities, especially in winter)	(Sanchez et al. 2007)

* Heterospecific attraction: see Section 8.9.

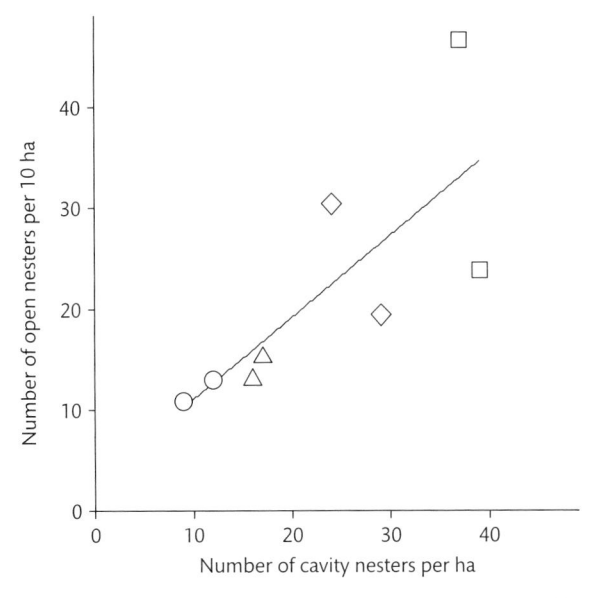

Fig. 8.5 Number of birds nesting on replicate 1 ha plots in mixed pine and spruce forest in western Germany between which nest-box density varied. Circles: 9 boxes ha^{-1}; triangles: 16 boxes ha^{-1}; diamonds: 25 boxes ha^{-1}; squares: 36 boxes ha^{-1}. The results show that as the number of cavity nesters increases, open nester densities also increase (r^2=0.58; P=0.028), which the author attributed to an unspecified social effect. After Table 3 in Wellenstein (1968).

only discussed in the context of how cost effective the approach was (the most cost effective treatment would be 5 nest-boxes ha^{-1}), and not how it was possible that the increase in cavity nesters coincided with an increase in open nesters (Pfeifer 1955; Pfeifer and Keil 1960; Weinzierl 1957).

Since the groundbreaking work by Mikko Mönkkönen and colleagues this effect has been called heterospecific attraction, and I will discuss it in Section 8.9 (Mönkkönen *et al.* 1990).

8.5 Effects of direct removals on habitat use and population size of subordinate species (Table 8.5)

Most species defend territories against conspecifics thereby, in many cases, limiting the number of individuals that can breed, or at least breed in optimal habitat. Removal experiments to test for the existence of intraspecific competition are numerous and some have already been discussed in Chapter 3. A small number of such experiments have also been performed to explore whether

Table 8.5 Removal experiments to test for interspecific competition

Manipulation	Effect 1	Effect 2	Type of competition	Reference
Removal of single individuals				
Remove chaffinch on island	Empty part of habitat	Expansion of great tit territory	Interspecific territoriality; interference	(Reed 1982)
Remove chiffchaff or willow warbler	Vacate individual territories	Replaced by same or other species	Interspecific territoriality; interference	(Saether 1983)
Remove brambling or willow warbler males	Vacate individual territories	Replaced by same species only	limitation by intraspecific territorial behaviour; No interspecific competition	(Fonstad 1984)
Remove male variable or Jambandu indigobird	Vacate individual territories	Replaced by same (rarely other) species	Intraspecific sexual selection	(Payne and Groschupf 1984)
Removal of birds from an entire area				
Remove junco or golden-crowned sparrow in winter	Empty part of habitat	Juncos replaced by juncos; golden-crowned sparrow first replaced by GCSP, later by juncos	Interference for preferred winter habitat	(Davis 1973)
Remove red-headed woodpecker in winter	Remove dominant species	Home range of red-cockaded woodpecker expands; number of white-breasted nuthatches increases	Interference; exploitation?	(Williams and Batzli 1979a)
Remove bell miner	Empty habitat	Numbers of 3 species increase; 6 additional species settle	Interspecific territoriality; interference	(Loyn 1985; Loyn et al. 1983)
Remove blackcap	Empty part of habitat	Garden warbler numbers increase	Interspecific territoriality; interference	(Garcia 1983)
Remove least flycatcher	Empty part of habitat	American redstart numbers increase	Interference for preferred breeding habitat	(Sherry and Holmes 1988)
Reciprocal removals of orange-crowned and Virginia warbler	Remove one species	Reduced nest predation on both species; Virginia warbler nest site shift; increased feeding rate	Asymmetric interference; exploitation; apparent competition through nest predation	(Martin and Martin 2001)
Introduction house finch, decline through disease	House finch abundance increases, decreases	House sparrow abundance decreases, increases	Exploitation?; interference?	(Cooper et al. 2007)

interspecific interactions could be detected and if species differences in habitat use could be the result of one species excluding the other from habitat preferred by both (Table 8.5).

When different bird species have the same habitat preference this may lead to displacement at the level of the territory or at the level of the habitat (see Subsection 3.4.2). To test the former one can remove individuals of one species and observe if they are replaced by a conspecific or by a heterospecific. To test the latter one needs to remove all males in a plot. Both types of experiments have been carried out. Replacements by birds of the same species only indicates the existence of intraspecific competition for space, or that one species was consistently dominant to the other, while replacement by either species provides evidence for interspecific competition for space.

8.5.1 Removals of single territory holders in species that defend interspecific territories

Four studies tested whether removal of a territory holder of one species resulted in expansion or occupation of the vacated site by the same or by different species. Normally great tit and chaffinch have overlapping territories, although on Eigg, a small island off the Scottish coast, Timothy Reed found that they defended interspecific territories and responded aggressively to each other's song, but not to willow warbler control song (Reed 1982). On a mainland site only 9 km away this was not the case. To test if habitat separation between chaffinch and great tit was the result of interspecific aggression Reed removed nine chaffinches from one site on Eigg, while not removing them on the control site. Removal resulted in great tits expanding their territory but not in more great tits defending territories. Nothing changed in the control sites where chaffinches were allowed to remain. On Eigg chaffinch populations seem to be limited by intraspecific competition, as unringed chaffinches appeared after removal. Great tit population size is limited by interspecific competition for space, as territory holders expanded and shifted their territories after chaffinch removal. In the absence of chaffinches great tit numbers would be able to increase over time.

Berndt-Erik Saether carried out a similar experiment in Norway with willow warbler and chiffchaff although he also tested for reciprocal effects between the species. He found that removed individuals of a species were replaced by a male of either species and suggested that the earlier arriving chiffchaff did not saturate the habitat so that space remained for willow warblers to occupy interspecific territories (Saether 1983). Given that removed birds were replaced, this experiment suggests that both species are limited by intra- and interspecific competition.

In the third such experiment, Tore Fonstad tested a hypothesis proposed by Olav Hogstad (Hogstad 1975a) that the limited overlap between brambling and willow warbler was the result of interspecific competition. The experiment permitted differentiation between the hypotheses that limited territory overlap between territories of the two species was simply caused by differences in habitat preference or was rather the result of interspecific aggression (Fonstad 1984). Playback and removal experiments clearly showed the absence of interspecific interactions, but as most removed males were replaced by conspecifics (but never by heterospecifics), the experiments proved that population size was limited by intraspecific territorial behaviour.

The fourth experiment in this section concerns male variable and Jambandu indigobirds in Cameroon. Because these species are brood parasites they do not need to sequester resources to rear their offspring and their interspecific call-site territories only serve for mating. The authors tested the relative importance of interspecific competition and of intraspecific sexual selection by removing individual males. If male call-site dispersion resulted from interspecies competition experimental replacements should be by either species. If, on the other hand, male dispersion resulted from intraspecific interactions in the context of sexual selection, then removed males should be replaced only by another male of the same species (Payne and Groschupf 1984). Although males responded aggressively to experimental presentations of interspecific stimuli, nine out of ten removed males were replaced by other males of the same species, leading the authors to conclude that the distribution of replacements was consistent with the hypothesis of sexual selection but not with one of interspecific competition.

8.5.2 Removals of all birds in an area

In contrast to the previous experiments in which only a few territorial males were removed, in another seven experiments all birds were removed from a larger area to test whether removal resulted in reoccupation by individuals of the same or of another species.

Of the three experiments in this group that were made during the non-breeding season, the first was carried out in the Hastings reservation, California, USA, by John Davis. He investigated if differences in winter habitat use by juncos and by golden-crowned sparrows, along an open field with willow thickets at one end, was the result of differences in habitat preference or of interspecific competition (Davis 1973). The proportion of golden-crowned sparrows trapped close to the willow thickets was much higher than elsewhere on the study plot, implying a preference for that habitat. John Davis removed each of the species from different zones in the study plot. Removed juncos were replaced by other juncos while removed golden-crowned sparrows were initially replaced by small numbers of sparrows but later by juncos too. When Davis released the golden-crowned sparrows in March they returned to the willow thickets they had occupied previously, thereby displacing the juncos again. The larger golden-crowned sparrows were able to keep juncos out of their preferred habitat although very little overt aggression between the two species was observed. Habitat use of golden-crowned sparrows in winter was limited by intraspecific competition, while juncos were affected both by intra- and interspecific competition.

The second winter-removal experiment was based on observations of between-species differences in habitat use of the dominant red-headed and the subordinate red-bellied woodpecker. Joseph Williams and George Batzli believed competition between the woodpeckers caused the exclusion of the subordinate species from certain parts of their study site (Williams and Batzli 1979b). To test this hypothesis they removed the dominant woodpecker during winter (Williams and Batzli 1979a). Red-bellied woodpeckers expanded their home range by using the area vacated by the red-headed woodpecker. At the same time white-breasted nuthatches became more numerous. The authors concluded that release from interspecific competition allowed the subordinate species to expand its foraging and habitat niche, but did not conclude that their experiment generated an increase in population size of the subordinate species.

Some experiments that provide insight on the existence of interspecific competition were actually carried out to test very different hypotheses. Bell miners in Australia breed in colonies and feed almost exclusively on dome-like coverings made by plant-feeding psyllid bugs. Nevertheless, when an area is occupied by a colony of bell miners the trees are in poor condition because bell miners remove an insufficient number of psyllids and trees suffer from them. Richard Loyn and colleagues wanted to determine if the infestation of a Eucalyptus forest by psyllids could be reduced by other insectivorous birds that are excluded by the interspecifically territorial bell miner. They therefore removed all bell miners from a 3 ha plot, while keeping another psyllid infested site and a non-infested site as controls. In the experimental plot the three previously present species increased eight-fold, and six additional bird species invaded the area, while in the control sites bird numbers changed little (Loyn et al. 1983; Loyn 1985). Although this was not the objective of the experiment, it did document a powerful effect of interspecific territoriality. Once bell miners no longer inhibited other species from exploiting the psyllids as food, the increased numbers of insectivores caused psyllids numbers to rapidly decline in the experimental area, with no similar change in the control sites.

The other four experiments were carried out during the breeding season, and therefore tested if interspecific competition reduced breeding population size.

Ernest Garcia tested the hypothesis that blackcaps and garden warblers compete interspecifically for space during the breeding season by removing the blackcaps in one site during one year of his three year study (Garcia 1983). The removal of blackcaps allowed garden warblers to settle in areas they did not occupy in the control years, showing that the more aggressive and earlier arriving blackcap can exclude garden warblers from suitable breeding habitat. Garden warbler breeding population size is limited because of intra- and interspecific competition.

Tom Sherry and Dick Holmes carried out a similar experiment in Hubbard Brook Experimental Forest, New Hampshire, USA, testing if the subordinate American redstart was excluded from preferred habitat by the dominant least flycatcher (Sherry and Holmes 1988). Theirs was a longer term experiment with more experimental replications and a larger number of control sites than that of Garcia. Wherever least flycatchers disappeared naturally or were removed experimentally American redstarts settled or increased in numbers. In one plot that least flycatchers recolonized after experimental removals had stopped, redstart numbers gradually declined indicating that interspecific competition between these species limits redstart habitat use and hence population size.

The final experiment in this group was particularly well carried out (reciprocal removals in numerous experimental plots over a three-year period with multiple control plots) and, because the effect of removals on reproduction was also studied, yielded particularly interesting insights. Peter and Tom Martin studied the consequences of coexistence for orange-crowned warblers and Virginia's warblers. In Arizona these two small (8–9 g) migratory warblers have overlapping breeding territories and could potentially compete. For both species the removal of the other species led to a considerable reduction of nest predation, so that the breeding pairs produced more offspring (Martin and Martin 2001). This is a nice example of apparent competition (Holt 1977). The removal of the smaller Virginia warbler had no effect on orange-crowned warblers. On plots where orange-crowned warblers were removed, however, Virginia's warblers shifted nest sites to locations very similar to orange-crowned warbler nest sites implying interspecific interference competition for high-quality nest sites that are therefore in limited supply (Fig. 5.2). Males also brought more food to the incubating female and nestlings were provided with more food by both parents. The reduced nest predation in removal plots increased reproductive success for both species about 2.5 times, although fecundity did not differ between treatments. The substantial fitness cost of coexistence is thus the result of the functional response by predators to more abundant prey (see also Section 5.1). Interspecific competition for nest sites forces Virginia warblers to nest in less favoured sites that they leave when orange-crowned warblers are removed. The authors concluded: '*This pattern of indirect interactions underlying costs to both the behaviourally dominant and subordinate competitors, and direct interactions causing costs to only the subordinate competitor, may prove to be a more general pattern in nature, although few intensive studies are available to test this association*' (Martin and Martin 2001).

I also place within this section the results of the 'natural experiment' reported by Caren Cooper and colleagues, because of its very large scale (Cooper *et al.* 2007). They showed that, after the introduction of house finches in eastern North America, house sparrow numbers decreased, and that when house finch numbers declined because of a large-scale epidemic the downward trend in house sparrow numbers was reversed (see Fig .2.1). While the initial decrease in house sparrows, as house finch populations grew, could be a mere coincidence, the reverse house sparrow population trajectory, as house finches declined because of a disease epidemic, provides more convincing evidence of interspecific competition between these two granivores.

Most of the experiments reported in this section document the exclusion of one or more individuals of a subordinate species from suitable habitat occupied by a dominant competitor, indicating that interspecific territorial behaviour or heterospecific aggression can limit numbers both during the breeding season and in winter. An increase in numbers of the subordinate species when the dominant species was removed implies that intraspecific competition among birds of the subordinate species also occurs and shows that some individuals are excluded from breeding. In all of these experiments in which interspecific competition was established (except that reported by Cooper *et al.* 2007), competition was for space. Competition can be both asymmetric and indirect. The fact that not all experiments that tested the putative competition between species actually supported this hypothesis (Fonstad 1984) underlines the importance of testing conclusions from observational field observations experimentally.

8.6 Competitive interactions between birds and species of a different class

Some of the most famous experimental studies of interspecific competition concern interactions between organisms of different classes. James Brown and James Munger experimentally demonstrated a complex set of interactions in the desert, including interspecific competition, among desert rodents and between these and ants (Brown and Munger 1985; Munger and Brown 1981). Given the preponderance of ants in most ecosystems of the world, it is not surprising that some scientist have experimentally tested the possible existence of interspecific competition between birds and ants. Other studies that I will discuss concern interactions between hummingbirds and insects, between birds and fish, and between birds and mammals.

8.6.1 Competition between birds and ants (Table 8.6)

After observing a low density of breeding birds in an area with a huge supercolony of *Formica lugubris* (1,200 nests on 70 ha of spruce forest), Catzeflis hypothesized that because birds and wood ants eat the same food and wood ants form large colonies in boreal forest, ants and birds might compete (Catzeflis 1979). In northern Sweden Paul Haemig set out to experimentally test this hypothesis of exploitation competition between birds and wood ants (Haemig 1992, 1994). He selected pairs of trees close to an ant colony and applied a chemical repellent to the lower trunk of one of the two trees while keeping the other tree as a control. Arthropod biomass was lower in the trees with ants (*Formica aquilonia*) than in trees ants could not access. In both study years foliage and trunk gleaners (tits, chaffinches, warblers, woodpeckers, and treecreepers) visited the trees without ants more frequently and for longer time periods than the trees with ants. Pied flycatchers, which used the trees solely as perches for sallying to catch flying insects, visited the trees with ants more frequently. In the year in which cone-foraging species were abundant these did not show a preference for experimental or control trees. Ants thus partially excluded birds that use similar resources by means of exploitation competition but did not have an adverse effect on birds using a different resource. The effect on pied flycatchers might be an example of facilitation, foraging ants causing insects to fly thereby making them more available for the flycatchers.

To test the idea that competition between ants and birds was solely exploitation competition for food or that the mere presence of ants might also affect birds directly, Haemig constructed two artificial 'trees' in the forest, provided them with equal amounts of food for ants and for birds, and placed them close to each other at equal distances from a wood ant nest (Haemig 1996). Wood ants were excluded from one tree using the repellent, but allowed to forage on the other. After 25 great tit visits, Haemig switched the treatment between the two trees to correct for possible location effects. In all replicates great tits visited the tree without ants more frequently, and for longer periods of time, than the tree with ants. Furthermore great tit foraging bouts on the trees with ants were shorter when ants were more active, thus demonstrating that interference competition from ants also exists and influences bird foraging behaviour. One mechanism of this interference competition could be formic acid deposited by the ants, since at high concentrations it is toxic to birds (Bennett *et al.* 1996).

In a final experiment to study interactions between birds and ants, Haemig put up nest-boxes for tits and flycatchers in trees with and without ants (Haemig 1999). He found that during the breeding season interactions of wood ants with nesting birds differed with level of predation risk. Under low predation risk (forest interior, low number of predators) only 8% of tits nested in trees with ants. Along the forest edge, where he counted about four times more bird and mammal predators than in the forest interior, 45% of tits nested in trees with ants, a significant difference. For tits this was adaptive as along the forest edge reproductive success was about double in the nests on ant trees than on non-ant trees. The interaction between nesting birds and ants, therefore, was opposite at high and low predation risk. How ants interacted with predators was not clear.

A Finnish team from the University of Jyväskylä studied treecreeper foraging behaviour and nest success in territories that varied naturally or experimentally in ant abundance. Both ants and birds significantly reduced arthropod abundance

Table 8.6 Competition between ants and birds

Manipulation	Effect 1	Effect 2	Type of competition	Reference
Remove ants	Increase in arthropods	Foliage gleaners increase use of ant-free trees; pied flycatchers increase use of trees with ants; no effect on cone feeders	Exploitation; facilitation	(Haemig 1992, 1994)
Create 'trees' with equal food for birds with/without ants	Tree choice based on ants, not food	Birds prefer artificial trees without ants	Interference	(Haemig 1996)
Provide nest-boxes on trees with/without ants with low/high predation risk		Birds prefer nest sites without ants at low predation risk and reverse at high predation risk	Interference; facilitation	(Haemig 1999)
Remove ants	Increase in arthropods	Increase in treecreeper foraging success	Exploitation	(Aho et al. 1997)
Remove birds from army ants	Effect on birds not measured	Ant foraging success increases	Parasitism of birds on ants; exploitation competition	(Wrege et al. 2005)

(Jantti *et al.* 2001). Birds foraged more often and for longer periods on trees with no ants, confirming the existence of interspecific competition between treecreepers and ants (Aho *et al.* 1997). Non-experimental data further showed that food abundance was higher in territories without wood ants and that treecreepers in these territories started breeding earlier, suggesting that interspecific competition for food delayed laying. In addition to hatching later, nestlings in territories with ants achieved lower body mass near fledging and suffered higher mortality than nestlings in territories without ants. Furthermore in territories with no ants second clutches were larger than first clutches, while the reverse was true in territories with ants. Consequently, double-brooded treecreeper pairs produced an average of 2.3 more fledglings of higher quality in territories without ants than in territories with ants. The results suggest that wood ants reduce territory quality of Eurasian treecreepers by means of food depletion and have negative effects on the breeding success of individual birds. (Aho *et al.* 1999; Jantti *et al.* 2007).

The Finnish and Swedish studies demonstrate how interspecific competition between ants and birds impacts bird fitness (and although not measured, perhaps ant fitness as well) both by exploitation and by interference competition. All experiments had controls and were replicated.

A very different kind of interaction between birds and ants concerns the army ants in the neotropics which are attended by a community of birds, insects, some lizards, and even mammals. Fifty species of birds regularly follow army ants as 'professionals' (Willis and Oniki 1978) and benefit from the association by taking arthropods flushed by the swarming ants. Willis and Oniki suggested that the presence of attending birds influenced arthropod behaviour by flushing prey back into the oncoming raid and that this effect was necessary for swarm-raiding in the Ecitonini to evolve. They thus implied that a mutualistic relationship existed between birds and ants. More recently Peter Wrege and colleagues tested this hypothesis experimentally by studying ant foraging success in the natural presence/experimental absence of birds by flushing the birds (Wrege *et al.* 2005). The presence of antbirds decreased ant foraging success considerably and Peter Wrege concluded that birds were parasites of ants, because the birds benefitted from the relationship while the ants suffered a cost. One could argue, though, that since birds and ants use the same food the interaction could also be a competitive one, but this still needs experimental proof.

8.6.2 Competition among hummingbirds, and between hummingbirds and insects (Table 8.7)

Flowers continually produce nectar that is exploited by various birds and a great variety of insects. This system is particularly well suited for experimental manipulations because foraging observations at flowers can be made easily, interactions

Table 8.7 Competition among hummingbirds and between hummingbirds and insects

Manipulation	Effect 1	Effect 2	Type of competition	Reference
Vary time and location where feeders available	Manipulate resource predictability	Competition increases when predictability increases	Exploitation; interference	(Pimm 1978)
Vary food quality in feeders; manipulate bird density	Manipulate resource quality and intraspecific competition	As dominant density increases subordinates feed more on poor quality feeders	Exploitation; interference	(Pimm et al. 1985)
Cage flowers	Exclude hummingbirds	Insect foraging on flowers increases	Exploitation; interference	(Carpenter 1979)
Manipulate flowers	Vary flower access for birds, insects or both	Residual nectar in flowers visited by either taxon much reduced compared to flowers inaccessible to both	Exploitation	(Gill et al. 1982)
Build artificial flowers that can be used by both birds and bees, or only birds	Vary food access to either birds or insects	Hummingbird feeding success reduced by bee aggression	interference	(Gill et al. 1982)
Remove one of two bee species; exclude hummingbirds	Vary abundance of putative competitors	Increase numbers of 2nd bee species; broaden foraging niche of bees	exploitation	(Laverty and Plowright 1985)

between foragers can be observed directly, variations in resource abundance can be measured precisely, feeders with sugar solutions can be provided experimentally, and exclusions of various competitors are possible. Nevertheless only a relatively small number of short experiments have tried to unravel interphyletic interactions. In this section I include all experiments involving hummingbirds, even if they did not address interactions with insects.

In systems in which flower nectar was used both by hummingbirds and insects, three different studies manipulated access to nectar in order to determine if competition could be documented. All found that the volume of nectar was reduced by either taxon. None of the studies lasted more than a few days.

In the Sierra Nevada, CA, USA, Lynn Carpenter excluded hummingbirds from flowers by placing small-meshed cages over them (Carpenter 1979). Bumblebees increased both the number and duration of visits to plants inaccessible to hummingbirds compared to control plants, implying the existence of exploitation competition between the taxa. Natural variation in hummingbird numbers at sites where moths fed showed that hummingbird visits led to aggressive chases, leading Carpenter to conclude that interference competition also existed between birds and moths.

Frank Gill and colleagues studied foraging of hummingbirds and stingless bees at *Heleconia* flowers during a week-long study in Costa Rica. Very ingenious reciprocal exclusion experiments showed that both birds and bees reduced food resources, indicating the existence of reciprocal exploitation competition. The stingless bees, however, were quite aggressive toward the long-tailed hermits, leading the authors to conclude that there is also interference competition of bees towards hummingbirds (Gill *et al.* 1982). To test this latter idea they constructed artificial flowers that were accessible to either both bees and hummingbirds, or to hummingbirds only. The results were unambiguous: in two experiments hummingbirds foraged successfully in 85% and 87% of visits to flowers that bees could not use, while they were chased away by the bees in 58% and 88% of visits to flowers that bees used.

In a final experiment ruby-throated hummingbirds on Amherst Island, Ontario, Canada were excluded from patches of jewelweed flowers by hanging cotton screening from trees, and the long-tongued golden Northern bumble bees were removed through capture (the bees were released again two days later) (Laverty and Plowright 1985). A second bumble bee species, the half-black bumble bee, which has a shorter tongue, was not manipulated. While hummingbirds can access all nectar in jewelweed flowers, bumble bees can only reach halfway into the spur of the flower. When all three species were present, hummingbirds used flowers on the outside of the flower patches, while bumble bees foraged from less exposed flowers. When hummingbirds were excluded (or were naturally rare) bumble bees rapidly increased in numbers. Coincident with this increase, the foraging niche of the long-tongued bumble bee expanded as they also used the flowers on the outside of the patches, an example of competitive release. When the long-tongued golden Northern bumble bee was removed, the number of half-black bumble bees increased gradually. This response was reversed when the captive bumble bees were released. When present, hummingbirds depleted the nectar in the outer flowers by 8 a.m. so that bumble bees could no longer use them. This was not the case in the absence of hummingbirds, which made the expansion of the foraging niche of the long-tongued bumble bee possible. Resource use by the bumble bees was clearly affected by the foraging of the hummingbirds, providing evidence for exploitation competition. The study, however, did not show any effect of bumble bees on hummingbirds, nor any evidence of interference competition between any of the three species.

These experiments, like those of Stuart Pimm (Pimm 1978; Pimm *et al.* 1985) document a rapid exhaustion of nectar by foraging of either hummingbirds or insects, demonstrating the existence of exploitation competition affecting both birds and insects. In some species and depending on what insects were present, interference competition, resulting from aggressive defence of the resource, was also observed. However, as Lynn Carpenter pointed out, such short-term experiments, even if they document a strong interaction, are not really evidence for the existence of interspecific competition because an impact on fitness cannot be measured (Carpenter 1979).

8.6.3 Competition between birds and fish (Table 8.8)

Interactions between fish and water birds are complicated in that (1) benthic feeding fish and water birds—such as diving ducks—use similar food resources and hence may compete for zooplankton and benthic invertebrates; (2) piscivorous fish and water birds prey on fish and hence may compete for fish; (3) benthic feeding and herbivore fish species alter their environment and have a major impact on the presence of macrophytes (Wright and Phillips 1992), which affects invertebrates living in the ponds (Haas *et al.* 2007).

When fish are added to or removed from ponds the effects on bird numbers can be both direct (less food for birds because fish and birds eat the same food), and indirect (less food because fishes alter habitat and water quality, thereby affecting food availability for birds).

Some of the experiments studied only effects on single species (Eriksson 1979; Pehrsson 1984), while others studied entire systems (Haas *et al.* 2007). Of the eight experiments in which fish abundance has been experimentally manipulated, most linked fish abundance to a change in bird abundance, although a few studied an impact on growth or survival of ducklings.

In Sweden Mats Eriksson found that goldeneyes, a species of diving duck that feeds primarily on aquatic invertebrates, preferred lakes without fish over those with fish. In 14 small lakes that contained fish the number of fledged goldeneyes per lake was about four times smaller than in 8 lakes without fish (Eriksson 1979). To test for cause and effect he carried out two experiments. In the first he divided one of the lakes without fish (and with goldeneyes) into two halves with a net barrier and introduced perch into one of the halves. Over the course of five years fish were present in one half or in both halves of the lake before being eventually all removed. Although he found some effect of fish presence on invertebrate abundance, there was no measurable effect on bird numbers. In a second experiment he chose two neighbouring lakes that contained equal amounts of fish and no goldeneyes. After he killed all the fish using rotenone in one of the lakes invertebrates increased compared to the control lake. Three to five years later goldeneye fledged about four times more young in the lake from which the fish had been removed compared to the lake in which fish had remained. He concluded that fishes and birds competed for invertebrates. He ascribed the lack of an effect in the split-lake experiment to insufficient predation pressure of the introduced fish, since he did not observe a clear decline in invertebrates in the lake half into which fish had been introduced, so that food did not become limiting.

Starting from a similar observation as Eriksson, Olof Pehrsson observed that mallard density in the breeding season was higher in lakes lacking fish and in which there were more and larger invertebrates, compared to lakes with fish (Pehrsson 1984). He did not manipulate fish abundance, so in that sense this study is not strictly experimental, but used existing variation to test if fish presence and the concomitant reduced food supply could have an effect on mallard reproductive success. He measured feeding efficiency of ducklings he brought to the lakes. In fishless lakes mallard ducklings had a higher feeding efficiency leading to the conclusion that competition for food between fish and mallards did affect mallard reproduction. A similar result was obtained by Wright and Phillips (1990) in a before/after comparison when fish were removed from a gravel pit lake.

Since then there have been a series of whole pond or whole lake experiments in which fish were either added and/or removed and effect on bird numbers measured (Table 8.8). Fish removal always showed a positive effect on birds, and fish addition always a negative one. In most studies effects of fish on water vegetation and on invertebrates was also reported. Effects of fish on birds could either be direct (fish eating invertebrates and competing for food with birds) or indirect (fish changing vegetation, water turbidity, and quality and hence affecting invertebrate density and abundance). The most detailed, replicated, and compelling experiment was carried out over a nine-year period by Karin Haas and colleagues in a series of ponds of about 7 ha each located side by side along the Speichersee, a shallow artificial lake in Bavaria, southern Germany (Haas *et al.* 2007). In four years (1997, 2003–05) three ponds each were stocked in spring with a low density of carp, while in 1997 three ponds were also stocked with a high density of carp. Three to six ponds were kept unstocked as controls. In the intervening years all ponds were treated in the same way and stocked or not stocked with fish. Water birds were censused June–September, and ponds were sampled for macroinvertebrates, plants, and phytoplankton. Macroinvertebrate densities and plant biomass were higher in control ponds than in ponds with carp, especially late in the season. As expected if competition influenced bird numbers, both herbivore, omnivore, and carnivore birds (that mostly did not feed on fish) were more abundant in ponds with no fish. Somewhat surprisingly, there were also more piscivorous birds in ponds not stocked with fish. The authors concluded that:

'our replicated whole-system experiment clearly supports the hypothesis that benthivorous fish and migratory water birds compete for limited macroinvertebrate and plant resources and that competitive effects of fish are density dependent. Competition appeared to be asymmetric in favor of fish and resulted in behavioral avoidance of fish ponds by water birds in late summer' (Haas et al. 2007).

Changing fish stocking in ponds does not always have the same effect on all water birds. Thus when roach were introduced into a large eutrophic lake, the subsequent increase in the abundance of this cyprinid was accompanied by a decline in tufted duck numbers but by an increase in over-wintering piscivorous great crested grebes. A reduction in the roach population during the mid-1980s was accompanied by a recovery of tufted ducks and a decline of grebes, implying a cause and effect between fish and bird abundance as in the previous example (Winfield *et al.* 1992). The authors later confirmed that diet overlap between tufted duck and roach was generally high due to the common consumption of chironomid larvae and molluscs (Winfield and Winfield 1994).

In conclusion, all experiments testing for a possible effect of interspecific competition between fish and water birds showed an effect of fish on birds. In many cases competition was both direct, because birds and fish preyed on the same food, and indirect, because fish altered the habitat and hence water quality and food diversity and abundance. In most cases the effects were probably asymmetric, the fish having an impact on bird numbers or reproduction while an inverse effect was not documented although usually also not tested. This conclusion, that fish introductions adversely impact birds, can be applied for conservation without the need to perform conclusive field experiments in each case. This is illustrated by work in Patagonia. A few years before Laguna Blanca in Patagonia, Argentina, was designated a RAMSAR site, four fish species were introduced into the lake. Within 20 years one of the endemic frogs living in the lake had gone extinct, and the numbers of black-necked swans and coots had drastically declined (Ortubay *et al.* 2006). Although these results of changes in bird and frog abundance do not prove the existence of competition beyond any reasonable doubt,

Table 8.8 Fish and birds

Manipulation	Effect 1	Effect 2	Type of competition	Reference
Fish removal/addition	Increase/decrease invertebrate prey	Increased goldeneye use, nesting success when fish absent	Exploitation	(Eriksson 1979)
Natural variation in fish abundance	More insects when less fish	Mallard ducklings grow more rapidly where no fish	Exploitation	(Pehrsson 1984)
Variation in fish abundance	Use existing variation in invertebrate prey	Mallard ducklings grow more rapidly where no fish	Exploitation	(Wright and Phillips 1990)
Fish removal	Weed growth; increase in invertebrates	Increase herbivore birds; increase ducks feeding on invertebrates	Exploitation	(Phillips 1992)
Fish removal	Increase in macrophytes, chironomids, gastropods	Increase nesting by tufted duck, pochard, shoveler; increase wintering waterfowl numbers	Exploitation	(Giles 1994)
Fish removal	Increase in macroinvertebrates; increase in macrophytes	Increase in diving duck numbers	Exploitation	(Hanson and Butler 1994)
Fish introduction, removal		Decrease, increase tufted duck; increase, decrease GC grebe	Exploitation	(Winfield et al. 1992)
Variation in carp abundance	More fish, less invertebrates, algae and macrophytes	More fish, fewer birds; quantitative response	Asymmetric exploitation	(Haas et al. 2007)

the existing evidence of general adverse effects of fish on birds and macroinvertebrates strongly supports the interpretation that competition between fish and birds, on the one hand, and declines in endemic fauna elements on the other, are causally related and that therefore fish removal can be recommended as a conservation measure.

8.6.4 Birds and mammals

As already discussed in Section 8.3 of this chapter, red-cockaded woodpeckers suffer from the presence of other bird species that take-over the holes they laboriously excavate. They may also compete with southern flying squirrels that also use their holes. Two studies explored the extent to which the presence of southern flying squirrels has an adverse impact on red-cockaded woodpeckers. Interestingly, although the experimental approaches were similar (removal of flying squirrels), and both studies were carried out correctly and had sufficient replication and controls, their results were not the same.

In South Carolina (USA), Kevin Laves and Susan Loeb removed a large number of southern flying squirrels over a two year period, both when using the cavities at night and through Sherman trapping. In both years and in all four experimental areas, red-cockaded woodpeckers produced significantly more fledglings in experimental than in control clusters (on average about 50% more). The authors conclude that their results showed:

'that southern flying squirrel use of red-cockaded woodpecker cavities during the breeding season has a significant impact on red-cockaded woodpeckers and that management of red-cockaded woodpecker populations should include activities that either minimize southern flying squirrels populations in red-cockaded woodpeckers clusters or limit access of southern flying squirrels to red-cockaded woodpeckers cavities' (Laves and Loeb 1999).

More recently John Kappes and John Davis explicitly approached the question of interactions between flying squirrels and red-cockaded woodpeckers in Florida in the context of a multi-species community where the possible existence of both direct and indirect effects between species were postulated (Kappes and Davis 2008). They hypothesized that flying squirrels, on top of having direct negative effects, could also have a positive indirect effect on red-cockaded woodpeckers if squirrels suppressed red-bellied woodpeckers. After monitoring the use of red-cockaded woodpecker cavities during a pre-experimental year, they regularly removed flying squirrels from red-cockaded woodpecker cavities for one year in half of the clusters, while keeping the other half of the clusters as controls. During the removal phase flying squirrel occupancy decreased by 53% on the removal territories relative to controls, but red-bellied woodpecker occupancy increased by 46%. They suggested that the reason why red-cockaded woodpeckers failed to respond positively to the removals was because red-bellied woodpeckers responded more rapidly to the removal of the flying squirrels than the red-cockaded woodpeckers by preemptively occupying the vacated holes. They therefore concluded that southern flying squirrels indirectly benefit red-cockaded woodpeckers by suppressing red-bellied woodpeckers, although the direct negative effects and the indirect positive effects of southern flying squirrels on red-cockaded woodpeckers more or less balanced out in the Florida study where red-bellied woodpecker densities are relatively high.

Both experiments document interspecific competition between a mammal (the southern flying squirrel) and a bird (the red-cockaded woodpecker), although the actual impact on the woodpeckers varies between sites in relation to the abundance of other competitors. One can expect that in tropical forests nest webs should include more such indirect interactions between a diverse set of cavity users.

8.7 Competition between burrow-nesting seabirds can have a severe impact on numbers: application of our understanding of interspecific competition for conservation (Table 8.9)

In earlier sections I have illustrated that intra- and interspecific competition can impact breeding populations of cavity-nesting forest birds. Among the procellariform seabirds, competition for burrows in which they nest can also be severe, whereby typically the larger species, such as a shearwater, causes problems for a smaller species, such as a small petrel. As so many of these seabirds are declining or threatened, reducing competition for nesting burrows has been used to support declining populations of some of the smaller petrel species. Various authors have reduced access to nest sites by the larger species or provided nest-boxes or smaller burrows.

One of the more spectacular such efforts is that by David Wingate who brought the endemic Bermuda petrel or cahow back from near extinction. Cahows had been driven to extinction on all larger islands by the introduction of pigs, cats, and rats. Cahows only survived on tiny offshore islands free from human disturbance and human-introduced mammals. On these islands, however, there was insufficient soil in which to excavate burrows and cahows were forced to breed in natural erosion crevices of the aeolian limestone sea-cliffs, which are the optimal breeding site for the larger white-tailed tropicbirds. Cahows breed earlier than tropicbirds so that, when tropicbirds arrived at a nest site, cahow chicks that had just hatched were killed, because cahows leave the nestlings unattended during day time when they forage at sea. In 1954 a 'baffler' was developed that had a hole of sufficient size for the petrel to pass through, but kept the tropicbird out. Later Wingate constructed elaborate burrows with 'plunge-holes' that mimicked natural petrel burrows and were never used by tropicbirds. This increased cahow nesting density and breeding performance considerably (Wingate 1977). The Bermuda petrel thus almost went extinct first by predation of introduced mammals on the larger islands, and then by interference competition with the larger tropicbirds for adequate nest sites.

In a mixed colony of shearwaters and petrels in the Azores, nest sites for the larger Cory's shearwater (840 g) were limiting (Ramos et al. 1997). When 28 artificial burrows were provided for shearwaters these were rapidly occupied, indicating intraspecific competition for nest sites. Providing also 20 smaller artificial burrows helped the smaller species, although several of these were excavated by Cory's shearwater. Some little shearwater (172 g) and band-rumped storm petrels (49 g) were found dead at the entrance of their burrows when excavated by Cory's shearwater. The authors conclude that interference competition for nest cavities between Cory's shearwater and small petrels, and within small petrels, is severe. In a later study, also from the University of the Azores, nest-boxes were designed to exclude larger species. This resulted in a rapid increase of the breeding populations of the Madeiran storm petrels (synonymous with band-rumped storm petrel): 12% increase in the first year and 28% over the original colony size in the second year (Bolton et al. 2004). Furthermore, the breeding success of storm petrels nesting in boxes averaged 2.9 times greater than that of birds at natural sites. Providing nest-boxes rather than burrows makes it impossible for larger species such as Cory's and little shearwaters to excavate the nests of the storm petrel and helped the population of the smaller species to recover.

As in cavity nesters in forests, adequate burrows can be limiting for seabirds, causing intra- and interspecific competition that limits breeding population size and reduces breeding success. Given the large size differences between them the use of selective nest-boxes can have great conservation value.

Table 8.9 Competition among burrow-nesting sea birds

Manipulation	Effect 1	Effect 2	Type of competition	Reference
Provide small-holed cavities	Reduce access larger species	Improve nest success and breeding numbers Madeiran storm petrel	Interference	(Bolton et al. 2004)
Provide small-holed cavities	Reduce access larger species	Improve nest success cahow	Interference	(Wingate 1977)
Provide cavities of various sizes	Reduce access larger species	Improve nest success and breeding numbers	Interference	(Ramos et al. 1997)

8.8 Heterospecific aggression and interspecific territories

In Chapter 3 I showed that intraspecific competition for space is common, at least among various tit species, and that many studies find that suitable habitat is partitioned in defended territories that cover the entire area. As many bird species are territorial, one can assume that this also applies to many other species. In most studies of bird communities the territories of many different species overlap and heterospecific aggression is rare. But there exist a number of studies, both in temperate and tropical regions, that document not only that closely related species in sympatry are aggressive towards one another but that this can also lead to their territories being exclusive from one another.

Numerous examples of interspecific territories have been described since the seminal paper by Gordon Orians and Mary Willson (Orians and Willson 1964). For heterospecific territories to be the outcome of interspecific competition it needs to be established that the species involved respond to one another aggressively, one (or both) thereby excluding the other. This can be done using song playback experiments and/or removal experiments (see, for example, the interaction between chaffinch and great tit in Scottish islands described in Chapter 3). Two recent reviews have shown that heterospecific aggression occurs in a wide range of taxa and can impact community composition (Peiman and Robinson 2010) or cause character displacement (Grether *et al.* 2009).

In this section I will limit myself to song playback experiments. Bird songs typically convey species-specific information so that in most cases birds do not respond to the songs of other species (Emlen 1972). Therefore, when closely related species, using similar resources, respond aggressively not only to playback of conspecific song but also to that of heterospecific song this suggests that interspecific competition could be involved. It is difficult to quantify the importance of this behaviour from studies of single species pairs that typically would only report positive results. I therefore limit myself here to two studies that are particularly important because they attempt to explore how general, and in what conditions, heterospecific aggression occurs and leads (or does not lead) to territorial exclusion. (Leisler 1988; Robinson and Terborgh 1995). A more complete list can be found in Peiman and Robinson (2010).

Berndt Leisler summarized experiments on six species of European *Acrocephalus* warblers that all breed in or in association with marshes. Using the strength of responses to conspecific and heterospecific song playback he found three forms of interspecific relationships: interspecific territoriality (aggression and exclusive use); partial exclusion (aggression and no exclusive use); and tolerance (no aggression and no exclusive use). The outcome of an interaction depended on spatio-temporal overlap between the species. Thus, if species differed in breeding phenology, they did not respond to heterospecific song (the short distance migrating moustached warbler breeds six weeks before the long-distance migrating reed warbler and neither responds to heterospecific song playback). If species differed in habitat use but had overlapping breeding phenology (sedge and aquatic warbler) the response to heterospecific song playback was five to seven times weaker than that to conspecific song. If species overlapped in breeding phenology and in habitat use, but did not co-occur at a site, they did not respond to heterospecific song (sedge, reed, and marsh warbler). These same species, however, when occurring at the same location, did respond to heterospecific song, whereby the earlier arriving species was aggressive towards the later arriving species (sedge aggressive to reed and marsh warbler; reed aggressive to marsh warbler only). When the later arriving species invaded the already occupied territory of the earlier species, it was attacked. Overlaps of these three warbler species territories were small (5–17%) with the exception of that between sedge and marsh warbler (51%). Finally, the smaller reed warbler (12 g) and the larger great reed warbler (31 g) were segregated by habitat because of frequent aggressive interactions. Great reed warblers nest in deeper water with *Phragmites* reeds, while reed warblers defend territories in shallower areas, both occupying exclusive interspecific territories. In interactions, reed warblers were subordinate to great reed warblers. The former, though, responded more strongly to playback of heterospecific song than the latter. That interspecific territories were the result of interspecific aggression can be further inferred from the observation that, after the great reed warblers left the area at the end of June, the subordinate reed warblers expanded their home range and colonized the areas vacated by the dominant species (Leisler 1988). What is interesting about this study is that it illustrates that a combination of factors determine how closely related species coexist. Given that ecological separation between some species is still incomplete this has lead to various scenarios of their coexistence.

The second study that looked in detail at the possible role of heterospecific aggression leading to interspecific territories on a large scale was that of Scott Robinson and John Terborgh in the Peruvian Amazon (Robinson and Terborgh 1995). They mapped territories of more than 330 bird species across a complete successional habitat gradient and found species pairs in over 20 genera that showed contiguous, not overlapping, territories, while other species pairs had partially or completely overlapping territories, a result very much reminding one of the situation among *Acrocephalus* warblers reported above. They then used reciprocal heterospecific playback experiments among selected species pairs that had one of the three types of territorial overlap: none, partial, or complete. Their results were particularly interesting. Most species pairs showing contiguous, non-overlapping territories, showed evidence of interspecific aggression (10 of 12 tested), whereby the response was asymmetric in most cases (7 out of 10). Most species pairs with completely overlapping or partially overlapping territories showed no evidence of heterospecific aggression (two out of nine, and one out of six tested showed heterospecific aggression). In the majority of cases, response to conspecific playback was stronger than that to heterospecific playback. The heavier species was consistently the aggressor.

They concluded that '*patterns of habitat selection, species richness and abundance may be strongly affected by interspecific aggression*'. Based on their extensive data they concluded that in this part of the Amazon:

'there are 85 species out of a total of 435 species in terrestrial habitats whose habitat occupancy may be limited by the presence of a closely related species in an adjacent habitat. On the elevational gradient of the eastern Andes, the distributions of roughly one third of all bird species abut those of close relatives and display a release in a locality where the relative is missing. We thus conclude that interspecific aggression is an important mechanism in species turnover along both successional and elevational gradients (Robinson and Terborgh 1995).

Both Leisler and Robinson and Terborgh addressed Murray's hypothesis that interspecific aggression is the result of mistaken identity (Murray 1976, 1981, 1988) and were able to reject that hypothesis conclusively.

8.9 Heterospecific attraction

In Fig. 8.5 I presented results showing that increased densities of nest-boxes not only resulted in more cavity nesters breeding, but could also caused an increase in the number of open-nesting species observed on the experimental plots (Wellenstein 1968). This result, clearly, was the opposite to predictions made on the assumption that interspecific competition structured bird communities (see Section 8.4) but has been largely overlooked in the ornithological literature having been cited only a handful of times. It suggests that species that use similar resources, and are therefore assumed to compete with one another, are nevertheless attracted to each other when settling, implying the existence of facilitation rather than competition. This reminds one of the concept of conspecific attraction, a phenomenon already reported in the 1930s: territorial birds often occur in same-species clusters, leaving apparently equally suitable habitat unoccupied. This suggests that when settling individuals actively seek out already established birds because they benefit from their proximity (Stamps 1988). That this occurs can be demonstrated experimentally by broadcasting song playback (e.g. Hahn and Silverman 2007) or by displaying bird models (Kress 1983; Kress and Nettleship 1988).

In 1988–89 Mikko Mönkkönen and colleagues performed a two-year experiment on six small islands in a lake in central Finland to investigate if interspecific competition between resident tits and migrant passerines influenced avian community composition and feeding ecology. On three islands they experimentally increased the density of resident tits (by providing nest-boxes and food during winter and by releasing birds from the mainland), while on three other islands they reduced tit density by removing them. In the second year of the experiment they reversed the treatment between the islands, thereby controlling for possible effects of island quality. They found that the experimental increase in resident cavity nesters caused an *increase* in migratory open nesters while they did not find evidence for interspecific competition for food between migrants and residents (Mönkkönen *et al.* 1990). This result places the discussion about possible effects of interspecific competition in a broader context. Interactions between species with similar resource needs can have both negative (competition) and positive (facilitation) effects. For heterospecific attraction to have evolved, the costs of breeding close to a potential competitor must be smaller than its benefits (Mönkkönen and Forsman 2002). In this section I will explore heterospecific attraction in more detail.

Mikko Mönkkönen's experiment started a suite of studies on 'heterospecific attraction' whereby birds base settlement decisions on 'inadvertent social information' (density or presence of others) or 'public information' (the performance of others) in order to make vital decisions, especially related to habitat choice, the exact site where to settle (Danchin *et al.* 2004; Seppanen *et al.* 2007), and how much to invest in reproduction (Forsman *et al.* 2007). I summarize the experiments regarding heterospecific attraction in this section, although experiments testing for effects of density manipulation on reproductive success could have been listed in Section 8.1, and those testing for effects of density manipulations on population size in Section 8.4. (See table 8.10.)

The Heterospecific Attraction Hypothesis predicts that in a heterogeneous landscape migrants use the presence of settled residents as a cue for profitable breeding sites. Two factors contribute to this behaviour. The first is related to the fact that in northern forests migrants need to make very rapid decisions as to where to settle, as every day by which reproduction is delayed incurs a fitness cost (see for example Enemar 1987; Forsman *et al.* 2007). The second is related to the fact that northern resident bird populations tend to be primarily winter limited (see Section 3.3.2 and Oksanen 1987). The survivors are therefore expected to occupy the best breeding sites (Fretwell and Lucas 1969), thereby providing honest information about habitat quality for guild members. However, because they do not fully fill the habitat, energy will remain available for migrants (Forsman and Mönkkönen 2003; Mönkkönen *et al.* 2006) so that settling close to a potential competitor does not incur a foraging cost but can provide a foraging benefit (Mönkkönen *et al.* 1996) or an anti-predator benefit (Thomson *et al.* 2006).

After the initial 1990 paper, Mikko Mönkkönen and collaborators carried out two more similar experiments in northern Finland to determine to what extent increasing or decreasing the numbers of resident cavity nesters influenced the number of migrants that settled on experimental plots. All experiments were carried out with a before/after crossover design with replicates, meaning that the experiments were repeated in successive years and experimental treatment of plots

Table 8.10 Experiments testing for effects of facilitation through heterospecific attraction

Manipulation	Effect 1	Effect 2	Type of interaction	Reference
Add nest-boxes, food and tits; remove tits	Increase/decrease tit numbers	Increase/decrease migrant numbers	Facilitation	(Mönkkönen et al. 1990)
Introduce mallards on lakes	Increase mallard density	No response of teal measured by numbers breeding	None	(Elmberg et al. 1997)
Feed, remove cavity nesters	Increase/decrease resident cavity nester density	Increase/decrease numbers of arboreal insectivore migrants	Facilitation	(Mönkkönen et al. 1997)
Remove, introduce cavity nesters	Increase/decrease resident cavity nester density	Increase/decrease migrant numbers	Facilitation	(Forsman et al. 1998)
Reduce, increase tit density	vary resident cavity nester density	Pied flycatcher density co-varies with tit density; flycatchers settle earlier and accrue fitness benefits on high tit density sites	Facilitation	(Forsman et al. 2002)
Increase, decrease, tit density	Increase/decrease resident cavity nester density	Increase/decrease migrant foliage gleaning guild numbers	Facilitation	(Thomson et al. 2003)
Broadcast song least flycatcher, American redstart	Provide social information	Flycatchers attracted to both songs; redstarts attracted to conspecific song, avoid heterospecific song	Facilitation; interference competition	(Fletcher 2007)
Manipulate cavities to influence distance between nesting great tit and pied flycatcher	Influence potential effects of interspecific competition	No effect on flycatcher reproduction; reduction in great tit reproductive success close to flycatcher	Facilitation; exploitation competition	(Forsman et al. 2007)
Randomly manipulate tit density using boxes, and tit removals and additions	Create density gradient of resident tits	Collared flycatchers settle earliest at intermediate tit densities where they have a larger clutch and more male offspring than at low/high tit densities.	Facilitation; exploitation competition	(Forsman et al. 2008)

Add food larders on and outside great grey shrike territories	Separate effect of presence of larders from that of presence of great grey shrikes and larders	Red-backed shrikes breed at higher densities where larders were created and earlier in presence of natural or experimental larders; no effect of great grey shrike per se	Facilitation; use of social information	Hromada et al. 2008
Randomly manipulate tit density using boxes, and tit removals and additions	Create density gradient of resident tits	Migrants (species number and density) increase monotonically with tit density	Facilitation	(Forsman et al. 2009)
Manipulate blue tit brood size	Separate number and quality of fledglings as public information	No effect on great tit juvenile dispersal	none	(Parejo et al. 2008)

was switched between years. All experiments showed an increase in migrant density on the plots in which tit density had been increased compared to plots in which tit densities had been reduced, but which migratory species responded varied between experiments: in the first experiment mainly chaffinches and willow warblers increased (Mönkkönen *et al.* 1990); in the second experiment redwings especially increased (Forsman *et al.* 1998), while in the third experiment chaffinches but not willow warblers increased (Thomson *et al.* 2003). Mönkkönen also carried out a similar experiment on seven islands in Minnesota, USA where the bird community was much more diverse. The response of arboreal insectivores was similar to that in Finland but weaker, with the generalist red-eyed vireo showing the strongest response (Mönkkönen *et al.* 1997). The authors concluded that especially habitat generalists that used food resources similar to the tits (foliage gleaners) should respond to the density manipulations of tits in a positive way.

Inspired by Mönkkönen's ideas and by observational presence/absence data revealing a positive, not a negative, association between the migratory mallard and migratory teal, Elmberg *et al.* (1997) tested if heterospecific attraction would influence settlement decisions of teals. The introduction of wing-clipped mallards on oligotrophic lakes in northern Sweden, however, did not cause an increase in teal numbers.

In North America Fletcher (2007) tested the possible reciprocal effect of least flycatcher and American redstart (both migrants) on each other by broadcasting song of one or the other species on a series of small plots. He found that, compared to the previous year, the change in flycatcher numbers was larger where he broadcast song of either species (though more strongly on plots with conspecific song than on plots with heterospecific song) compared to control plots (with no song broadcast) implying an effect of both conspecific and heterospecific attraction. American redstarts, though, avoided plots where least flycatcher song was broadcast, indicating an effect of interference competition, confirming effects of conspecific and heterospecific song playback of these two species by Martin *et al.* (1996). Hromada and colleagues tested for the possible use of public information to explain heterospecific attraction of the migratory red-backed shrike to resident great grey shrikes (Hromada *et al.* 2008). To determine if red-backed shrikes were attracted to the larders on great grey shrike territories, or the presence of the latter, they compared laying dates and densities of red-backed shrikes on sites where great grey shrikes were present (with natural larders) and sites with experimentally created larders only. Red-backed shrikes settled at higher densities and laid earlier on sites with larders (regardless of the presence of great grey shrikes), documenting the use of heterospecific larder size in settlement decisions or simply responding to the extra food.

Once the positive effect of resident density on migrant habitat choice had been established, the next question was to determine if this behaviour was adaptive, and how decisions were made. This was done by measuring reproductive success of pied flycatchers that had been lured to breed in plots with high tit density or close to nesting tits. A first experiment showed that flycatchers reproduced more successfully close to tits than further away (Forsman *et al.* 2002), while a second experiment found no effect on flycatcher reproduction (Forsman *et al.* 2007). Surprisingly, this latter experiment showed that great tits suffered from pied flycatcher proximity through exploitation competition for food.

In his first paper Mönkkönen had already suggested that heterospecific attraction should only work until tit population density reached a certain level after which the benefits for the migrants of settling close to resident tits would be less than its costs. All the papers from Mönkkönen's group in which heterospecific attraction had a positive effect on migrant settlement decisions mentioned that the breeding densities of the resident tits were too low for the adverse effects of interspecific competition to be larger than its positive effects, but in Finland and in Minnesota density of resident cavity nesters could not be increased sufficiently to test this idea. They thus moved 1,000 km south to the Swedish island of Gotland using Lars Gustafsson's collared flycatcher study for the next experiment. On Gotland they exposed collared flycatchers to experimental tit densities varying by an order of magnitude in 13 plots, density in one plot reaching almost 4 pairs ha^{-1} (Forsman *et al.* 2008). Although flycatcher density did not correlate with tit density as would have been expected under the 'pure' Heterospecific Attraction Hypothesis, they did settle first in plots of intermediate tit density, demonstrating a preference for these. In plots with intermediate tit densities, flycatcher females invested more in reproduction. Clutch sizes and proportions of eggs that gave rise to male offspring were higher in plots with intermediate tit density than in plots with low or high tit densities (Fig. 8.6). Given that tit density had been manipulated experimentally across plots, and therefore did not really reflect natural plot resource abundance, it was not surprising that collared flycatcher reproductive success (number of fledglings, nestling body mass) decreased monotonically with tit density (Fig. 8.7), confirming Gustafsson's earlier experimental evidence that interspecific competition with tits adversely affected flycatcher reproduction (Gustafsson 1987). The same experiment also showed that the density and diversity of migrants increased monotonically with tit density, a result not expected from theory (Forsman *et al.* 2009).

Expanding the idea that resident birds also use heterospecific public information to make settlement decisions, Parejo and colleagues manipulated blue tit fledgling success in a number of study plots in France. They did not find that great tit juvenile dispersal differed between plots in which the number of blue tit fledglings had been increased or decreased (Parejo *et al.* 2008).

All the work regarding heterospecific attraction underlines the complexity of between-species interactions. One experimental (Fletcher 2007) and one correlational study (Sebastian-Gonzalez *et al.* 2010) showed that settlement decisions can be influenced both by the presence of conspecifics and of heterospecifics.

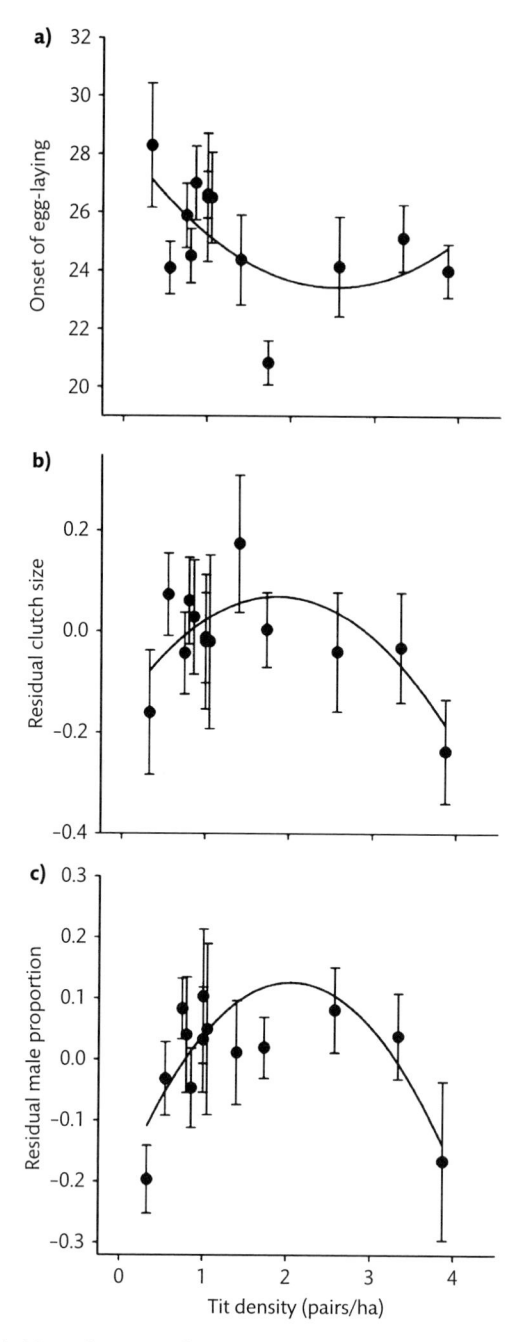

Fig. 8.6 Breeding habitat selection preference and offspring investment of collared flycatchers as a function of tit density (pairs ha^{-1}). Tit density and squared tit density explained flycatchers (A) habitat preference as indicated by the onset of egg laying; offspring investment as indicated by (B) clutch size, and (C) proportion of male offspring. In panels B and C, data points represent residual study plot means (± SE), after excluding effects of other factors on the response variable. From Forsman et al. (2008), with permission from Oxford University Press.

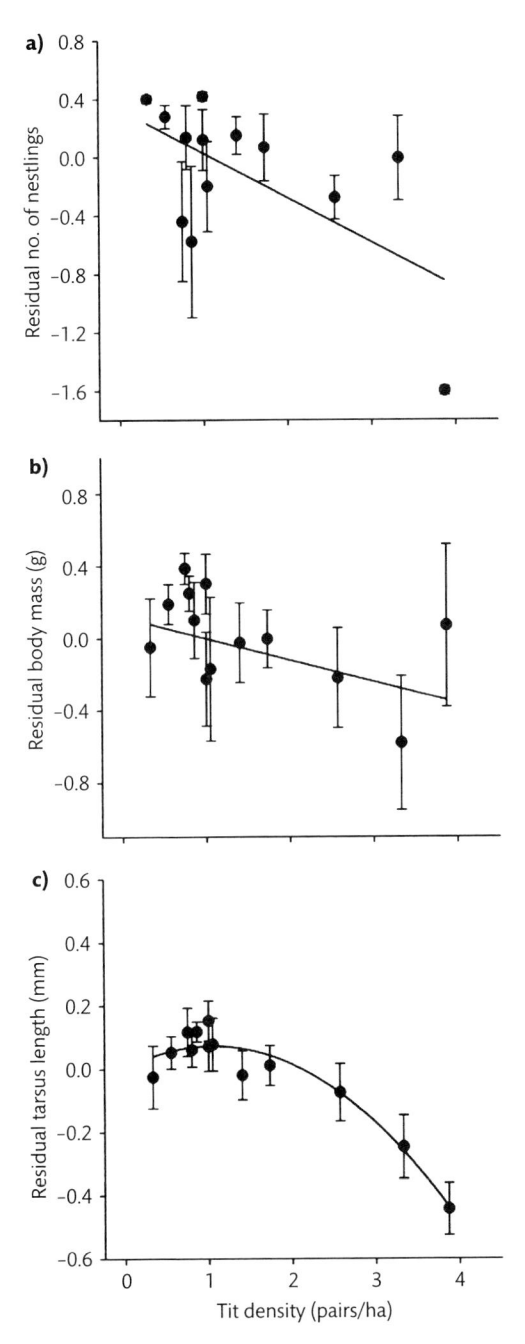

Fig. 8.7 Collared flycatchers breeding success expressed through (A) nestling number, (B) nestling body mass, and (C) nestling tarsus length as a function of tit density (pairs ha^{-1}). Data points represent residual study plot means (± SE), after excluding effects of other factors on the response variable. From Forsman et al. (2008), with permission from Oxford University Press.

Depending on conditions (location, year), interactions can have an adverse effect on one species, on the other species, on both or on neither species, but can also have positive effects. In the boreal forest it appears that it is advantageous for pied flycatchers to settle preferentially where tits are established, although tits can suffer a cost. Technically this would mean that flycatchers are parasites (a +/- interaction) rather than competitors (a -/- or a -/0 interaction) of tits. It would be interesting to test when and where pied or collared flycatchers and tits compete with each other (meaning that both suffer a cost), where/when neither suffers a cost, and where/when one but not the other suffers a fitness cost of breeding together.

When discussing the effect of biotic interactions on community composition it will be important to take this into account. More experimental work on the possible effects of between-species interactions in tropical avifaunas could be very enlightening.

8.10 Conclusions

Table 8.11 An overview of the field experiments testing for the existence of interspecific competition

Section	Number of experiments	Exploitation competition	Interference competition	Both	Neither	Facilitation	Remarks
8.1	7*	3	2	2	0	0	*Indirect effect through predation
8.2	6	0	6	0	0	0	
8.3	7	0	6*	0	1	0	*Indirect effect
8.4	8	1 (diffuse)	2	0	4	1	cf. 8.8
8.5	11	0	7	2	2	0	
8.6.1	5	3*	2*			2*,*	
8.6.2	6	2	1	3	0	0	
8.6.3	8	8	0	0	0	0	
8.6.4	2	0	2*	0	0	0	*one with indirect effect
8.7	3	0	3	0	0	0	
8.8 **	27	0ᵃ	13	0	13	0	ᵃ:not tested
8.9	12	2*	1*	0	2	10*	

Notes: * is placed either when a comment on one of the experiments is added (8.1, 8.3), or represents a double effect of an experiment that reflected both facilitation and competition (8.6.1; 8.9).

**: data on *Acrocephalus* experiments not included; (see text for detail)

8.10.1 The quality of field experiments

Comparing the list of experiments discussed in this chapter to that reported by John Wiens 20 years earlier shows that the quantity, quality, and duration of the more recent experiments has increased considerably. Among the 35 studies Wiens listed, only eight were replicated and had adequate controls and very few lasted for more than one year (Wiens 1989). Most of the more recent experiments were both replicated and had controls and some lasted for up to 11 years. As a result of the high quality of many experiments, inferences are stronger. Furthermore experiments manipulating entire communities, while logistically more demanding, show the complex and subtle effects of interactions. Replication of experiments in different geographic areas show that results can vary depending on specific conditions (see Section 8.6.4). One can speculate that the increased quality of recently published field experiments reflects that either field ecologists are more aware of the need to carry out experiments 'correctly', and that this is reflected in the publications discussed, and/or that journal editors and referees have become more demanding and that therefore experiments that are not executed according 'to the book' no longer get accepted for publication.

8.10.2 What is interspecific competition for?

The majority of the 102 experiments listed in this chapter and summarized in Table 8.11 documented the existence of interspecific competition either between different bird species, or between birds and species belonging to other taxa. Thirteen showed the existence of facilitation.

Among the 71 experiments documenting an effect of interspecific competition the majority (63%) found interference competition, about a quarter (27%) showed exploitation competition, and 10% found evidence for both exploitation and interference competition. I would, however, not conclude that among birds interference competition is generally more important than exploitation competition. The high proportion of experiments finding interference competition simply reflects the high proportion of experiments testing competition for cavities or involving cavity-nesting birds (Sections 8.1, 8.2, 8.3, 8.4, 8.6.4, and 8.7), and the relatively large number of experiments testing the existence of interspecific territoriality (Section 8.8). All experiments testing for interspecific territoriality (competition for space), and most experiments using cavity nesting birds (competition for sites to nest or roost in) involve interspecific aggression and hence interference competition.

What we can conclude is that interspecific competition is currently ongoing and affects abundance or demographic variables in many bird populations. Competition can be for space, cavities, food, or access to food. What the relative importance is of each of these is, at present, unclear.

8.10.3 What is often lacking in these experiments?

In Connell's 1983 review of field experiments testing the importance of interspecific competition he only listed two bird studies that could be used to compare the relative strengths of intra- and interspecific competition (Connell 1983). With the exception of song playback experiments, surprisingly few experiments were designed to measure effects of both intra- and interspecific competition simultaneously, suggesting that authors often assumed the existence of intraspecific competition and therefore did not feel the need to explicitly make the comparison, although the data were available (see for example Fig. 8.1). It would be useful to routinely document the existence of intraspecific competition when testing for the existence of interspecific competition.

This review of field experiments to test for the existence of interspecific competition in birds underlines that such experiments require a major effort and that it is very challenging to carry them out. This is why very few experiments, if any, have attempted to discover which demographic processes cause changes in population size if it was documented that population processes were adversely affected by interspecific competition, a remark already made by Ian Newton (Newton 1998). When mechanisms that caused changes in numbers were identified, they were usually implicitly related to behaviour, specifically immigration.

Furthermore, very few tested for reciprocal effects in both potentially competing species. Song playback experiments and some removal experiments showed that in most cases heterospecific aggression was asymmetric, leading to individuals of the dominant species excluding those of the subordinate species.

Nevertheless we can conclude from this chapter that interspecific competition between individuals of different bird species, or between birds and other taxa with overlapping resource needs, is widespread. In the next chapter I will review my own long-term field experiments on interspecific competition between great and blue tit, in which I tried to identify at least some of the population processes causing changes in blue tit numbers and in which I measured both effects of intra- and interspecific competition.

9
Long-term experiments on competition between great and blue tit

In Chapter 8 I reviewed field experiments that tested for the existence and ecological effects of interspecific competition. One of the observations that is unavoidable is that no such experiments tested simultaneously for possible effects of interspecific competition on population size and on demographic processes that generate these effects on population size, a remark made by Ian Newton (1998) and still valid. Furthermore very few tested for reciprocal effects in both potentially competing species and only a handful tested for the existence of both intra- and interspecific competition. The probable reasons for this absence are the very demanding logistics, long-term funding, and intense field work needed for such extended manipulative studies that study both variation in population size and collect sufficiently detailed observations on individuals to measure not only reproduction but also survival and dispersal. This latter variable especially is very difficult to measure since it requires multiple study sites in which all birds are individually marked and recaptured, or ring recoveries reported by members of the public.

In this chapter I shall summarize my own long-term experiments that have explored intra- and interspecific competition in tits. I shall begin by explaining why I thought that interspecific competition between great and blue tit might occur, although until the 1970s it was generally accepted that these two species coexisted without competing with one another. The initial correlational analyses led to a first set of short-term experiments in Ghent study plots. These confirmed the existence of competition between great and blue tits. This result (Dhondt and Eyckerman 1980) plus the observation of rapid directional change in great tit body size over a ten year period (Dhondt *et al.* 1979) formed the basis for the long-term field experiments in the Antwerp and Ghent sites. The questions asked in the research proposal given to the Belgian Fund for Scientific Research in 1978 were:

- Can we document intraspecific competition in blue tits, and identify what resource(s) is limiting? (I assumed that intraspecific competition in great tits was well documented.)

- What is interspecific competition between great and blue tit for?
- Which demographic variables are impacted by intra- and interspecific competition?
- Can differences in the intensity of interspecific competition cause changes in selection pressures and lead to evolutionary changes?

This initial four-year project was extended repeatedly and led to most of the results presented in this chapter, most of which are published. I also present a few novel analyses and results. The detailed results are presented in Appendix 3.

9.1 Interspecific competition in tits: the origin of the idea

In the 1970s, the work of Kluyver, Van Balen, Lack, Perrins, and my own on population dynamics of the great tit had already established, to a large extent, the importance of density-dependence in regulating numbers. Given that food niches of various tit species overlap, and given that intraspecific competition occurred, I thought that interspecific competition might also occur. So, in 1977, I used the long-term data on great and blue tits from the Ghent study sites to test whether

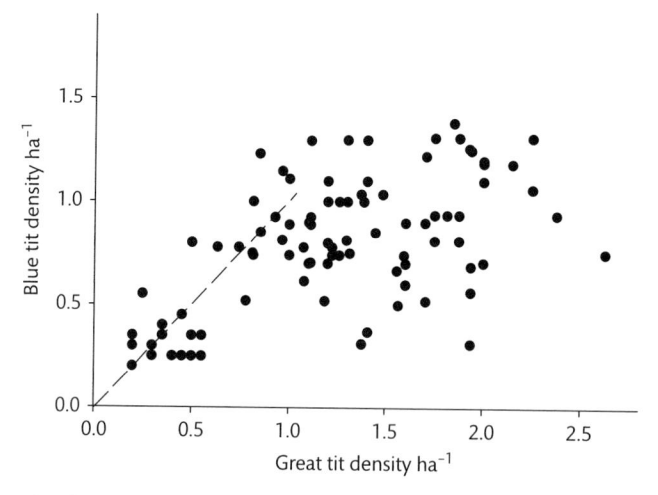

Fig. 9.1 Density of great and blue tit in relation to one another in Ghent (Belgium); four study plots with large-holed nest-boxes only. Up to about 1 pairs ha^{-1} densities of the two species vary in parallel, as shown by the line drawn at a 45° angle between 0 and 1 pairs ha^{-1}. Great tit densities increased to about 2.5 pairs ha^{-1}, while blue tit densities levelled off below 1.5 pairs ha^{-1}. This suggests that either intra- or interspecific competition limit blue tit breeding populations. Ghent study sites: Hutsepot (HP) 1964-98; Maaltepark (MA) 1959-77; Zevergem (ZE) 1960-84; Citadelpark (CI) 1959-78. After Dhondt and Eyckerman (1980).

significant correlations could be found, not only between demographic variables and conspecific density, but also between these and heterospecific density (Dhondt 1977). This initial analysis showed a clear effect of blue tit density on great tit reproduction but only a weak effect of great tits on blue tit reproduction (Appendix 3, Table 1).

In the late 1970s we knew that, where they coexisted, great and blue tits fed their young to a large extent on the same arthropod food, mainly defoliating caterpillars. Lack (1966) had shown that the breeding densities of the two species tended to vary in parallel across large regions. This suggested that the dynamics of the two species were influenced by similar environmental factors. In the hypothesis of no competition the densities of the two species should either fluctuate independently or be linearly related. In Fig. 9.1 the annual breeding densities of blue and great tit in four of the Ghent sites, with large-holed nest-boxes only, are plotted against one another. The graph shows that, up to a density of about 1 pairs ha^{-1}, the two species tend to fluctuate in parallel. When great tit densities further increase to about 2.5 pairs ha^{-1} blue tit densities did not further increase, but levelled off and remained below 1.5 pairs ha^{-1}. There are two possible explanations for this observation. One is that blue tits defend larger breeding territories than great tits, causing blue tit numbers to be limited by intraspecific competition for territories at a lower density than great tits. The other explanation is that great tits limit blue tit breeding density through interspecific competition. If we could experimentally reduce great tit density and thereby cause an increase in blue tit breeding density above the previous plateau level of 1.5 pairs ha^{-1}, we would have shown that blue tit breeding density was not limited by intraspecific competition for territories at that density, but was instead influenced by interspecific competition with great tits.

We decided, therefore, to experimentally test the second hypothesis first, that blue tit numbers are influenced by great tit numbers. (We would later show that the first explanation—that blue tit numbers were limited by territorial behaviour—was not correct, and that at low blue tit density a lot of unclaimed space remained between the defended territories; at high blue tit density, though, the entire plot was occupied and defended by blue tits (Dhondt *et al.* 1982)). Since during the breeding season blue tits adversely influenced great tit reproduction, but the reverse did not occur (Dhondt 1977) and great tits could apparently limit blue tit abundance (Fig. 9.1), we postulated that any adverse effects of interspecific competition of great on blue tits should occur primarily outside the breeding season. What resource was involved in this presumed interspecific competition was unclear, but we believed that food would be important because John Krebs had shown an increase in blue tit (not in great tit) breeding density in a plot near Oxford where he had massively fed the birds throughout winter, whereas he found no such increase in an unfed control plot (Krebs 1971). Hans Källander, too, had documented interspecific competition for food during winter between these spe-

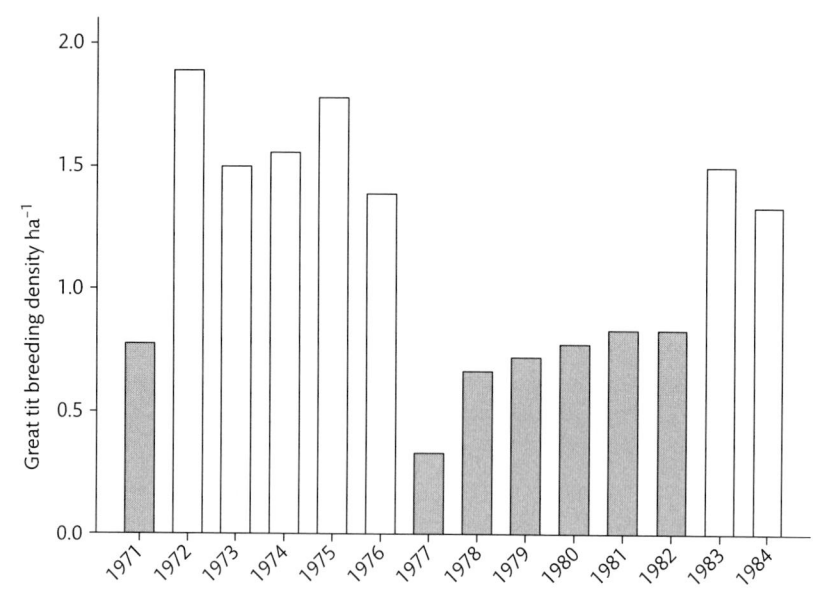

Fig. 9.2 Great tit breeding density in pairs ha⁻¹ in plot GT. White bars represent years in which large-holed boxes were available and density was high; grey bars represent years without nest-boxes (1971) or with small-holed boxes only during winter (1977–82).

cies (Källander 1981). Furthermore, the fact that cold winters severely affect blue tit numbers suggests that food might be limiting during winter (Chapter 4).

To manipulate great tit winter density we decided not to provide extra food (as Krebs and Källander had done), but to reduce the size of the entrance holes of the nest-boxes, thereby making them inaccessible to great tits. In our study sites, great tits use nest-boxes as winter roosting sites, and we believed that fewer first-winter great tits would settle in a plot in which roosting sites were in short supply, and that therefore great tit winter density would be reduced. If in winter the two species competed for food, a reduced great tit winter density should result in a decrease in the intensity of interspecific competition for food allowing an increase in blue tit density. In spring we mapped great tit territories and re-enlarged the entrance hole of one nest-box in the centre of each great tit territory. Great tits that had stayed in the plot could thus breed in a nest-box, allowing us to monitor their reproduction. Based on mapped spring territories, great tit breeding densities in the absence of large-holed nest-boxes during winter in Plot GT were lower than when such boxes were present and similar to the density before nest-boxes had been provided (Fig. 9.2).

We replicated this experimental manipulation in a second plot in a different year and obtained the same results: in both study plots blue tit density increased to a level never previously observed in the plot before (Fig. 9.3). The result was as predicted on the assumption of interspecific competition in winter. It confirmed a similar result obtained by Hans Löhrl in an unreplicated field experiment with no controls that we had not known about when carrying out our own experiment (Löhrl 1977).

As is the case with most field experiments, we were in for a few surprises. At that time we had become interested in testosterone levels in free-living birds (De Laet *et al.* 1985) and needed winter testosterone data from great tits living in a high-quality woodlot at low density. Plot GT was an obvious choice since it was a high-quality plot, and we had experimentally reduced great tit density by excluding them from nest-boxes. During 1977–78, the second winter of the competition experiment, we therefore enlarged the entrance hole of every tenth nest-box in Plot GT to trap a sufficient number of male great tits that roosted in them at night. Rather than only check the boxes with enlarged entrance holes, we checked all the nest-boxes and were completely surprised when we found large numbers of blue tits roosting in the small-holed nest-boxes (Dhondt and

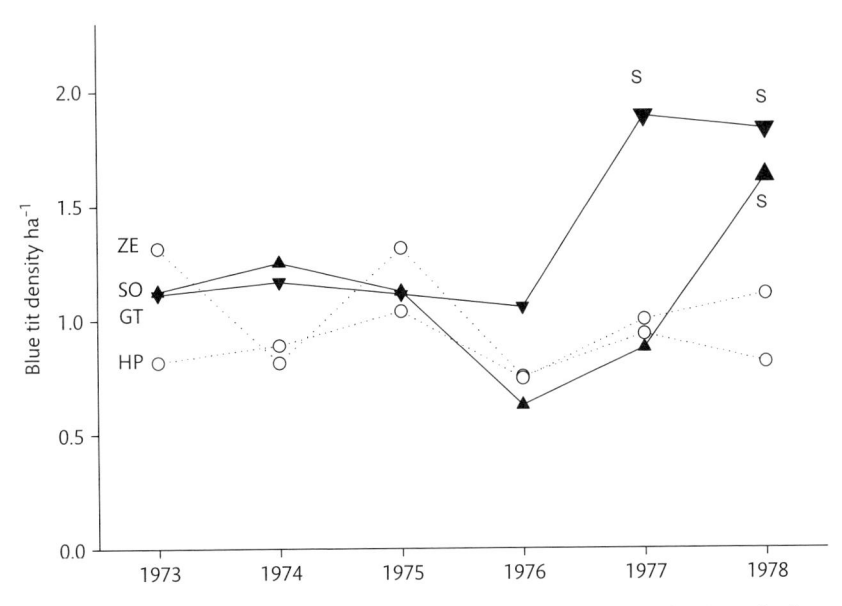

Fig. 9.3 Effect of providing small-holed nest-boxes on blue tit breeding density. In both experimental sites (dark triangles) in years in which small-holed boxes were available (large symbols, S) density increased above any level observed in years with only large-holed boxes. No such increase was observed in the control plots in which nest-box configuration remained unchanged (large-holed nest-boxes only; open symbols). Experimental sites: GT (Gontrode), SO (Soenen); control sites ZE (Zevergem); HP (Hutsepot). After Dhondt and Eyckerman (1980).

Eyckerman 1980). As Fernando Nottebohm once said during a talk on how he discovered the ongoing replacement of some kinds of neurons in the adult brain of birds '*We had entered the right room through the wrong door*'. This result, which suggested that in our plots winter competition between great and blue tits was primarily for nest-boxes as roosting sites, was unexpected.

A final remark about the initial experiment: in hindsight it was poorly designed, because we changed two conditions in the experimental plots simultaneously: great tit numbers and the availability of roosting sites for blue tits. At that time, however, we did not know that blue tits would use small-holed nest boxes for roosting during winter.

Having realized that the original experimental set-up was poor I added a third nest-box configuration when I began further experiments in 1979: I created plots with large-holed boxes only, small-holed boxes only, but also a configuration in which both nest-box types were available in surplus. That way great tit density was manipulated independently of the manipulation providing small-holed boxes. I expected high great tit densities when only large-holed boxes were available and when large- and small-holed boxes were both provided. The initial experiment in Plot GT had suggested that when small-holed nest-boxes were present a larger number of local-born blue tits were recruited into the breeding population than in earlier years (Dhondt and Eyckerman 1980). To confirm this possible result I planned to continue the new experiment for a longer time period to obtain detailed information on possible differences in demographic variables between treatments. By comparing plots with different sets of nest-box configurations it would be possible to determine effects of intra- and interspecific competition on demographic variables independently, and also to determine if interspecific competition during winter was for roosting sites only, for food only, or for both resources simultaneously (Table 9.1).

9.2 Is winter competition between great and blue tit for roosting sites only, for food only, or for both resources?

In all treatments nest-box densities were very high. When large-holed boxes are available either by themselves (L) or in combination with small-holed boxes (LS) great tit density should be high because high-quality winter roosting sites are available for great tits; this will intensify any competition for food with blue tits. When small-holed boxes are available (S, LS) protected roosting sites are available for blue tits (Table 9.1). In the configuration S, great tit density was expected to be low (because of limited breeding and roosting sites), while in the configuration LS great tit density was expected to be high. If winter competition is only for food, we expected blue tit breeding densities to be reduced at high great tit density and

Table 9.1 Nest-box configurations used to manipulate winter resources for blue tits, expected great tit breeding densities in the treatments, and expected blue tit breeding densities under different hypotheses as to what resource is limiting during winter. L = large-holed nest-box (diameter of opening 32 mm); S = small-holed nest-box (diameter of opening 26 mm); L+S= both nest-box types present. Food: interspecific competition during winter is for food only; roosting sites: interspecific competition during winter is for roosting sites only; food and roosting sites: interspecific competition during winter is both for food and for roosting sites. If competition was for neither resource, blue tit breeding densities should be the same in all treatments.

Nest-box configuration	Expected great tit density	Protected roosting sites for blue tit?	Expected blue tit breeding density under different hypotheses as to which resource is limiting during winter		
			Food is limiting	Roosting sites are limiting	Both food and roosting sites are limiting
L	high	no	low	low	low
L+S	high	yes	low	high	intermediate
S	low	yes	high	high	high

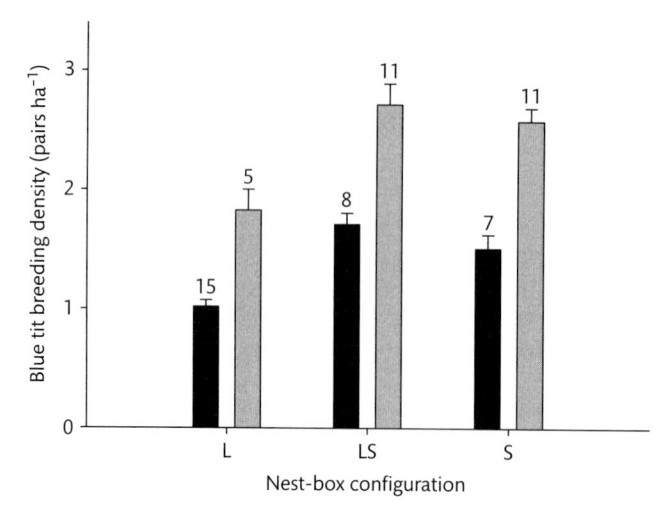

Fig. 9.4 Blue tit breeding density (mean ± SE) in plots with different nest-box configurations. Dark columns: Ghent sites, light columns: Antwerp sites; numbers above columns give number of years mean values are based on. Note that when small-holed (26 mm diameter entrance hole diameter; not accessible to great tits) nest-boxes are available, blue tit breeding density is significantly higher than when large-holed (32 mm diameter; accessible to great and blue tit) boxes only are offered.
L: large-holed boxes only; S: small-holed boxes only; LS: large- and small-holed boxes both present in plot. Note also that densities at Antwerp are higher than at Ghent in any configuration, reflecting the higher habitat quality at Antwerp. After Dhondt et al. (1991).

expected we would thus observe lower blue tit densities in configurations L and LS than in configuration S (Table 9.1). If competition is for roosting sites only, we had reduced winter competition for this resource by providing small-holed nest boxes and expected to observe higher blue tit densities in configurations S and LS than in configuration L. If winter competition is both for food and for roosting sites, then we expected blue tit density to be low in configuration L (competition for food and roosting sites), high in configuration S (reduced competition for food and for roosting sites), and intermediate in configuration LS (competition for food but not for roosting sites).

The results were unambiguous (Fig. 9.4): blue tit densities did not differ between the plots with only small-holed boxes and plots in which both small-holed and large-holed boxes had been provided (Dhondt *et al.* 1991). This result indicated intense competition for roosting sites, but no effect of high great tit densities during winter on blue tit breeding density, suggesting no winter competition for food. This last result was surprising given results discussed in Chapter 4, Section 4.2. Winter competition in Belgium between great and blue tits is therefore for roosting sites and not for food.

9.3 Experimental manipulations to vary the intensity of intra- and of interspecific competition

Once I had found that great and blue tits competed for food in the breeding season (Dhondt 1977) and for cavities used for roosting outside the breeding season (Dhondt and Eyckerman 1980) I started manipulative studies to try to identify what demographic variables differed between blue tit and great tit populations subjected to different levels of intra- and interspecific competition. While interspecific competition for roosting sites in winter impacted blue tit breeding density, interspecific competition during the breeding season, while having a strong negative effect on great tit reproductive success, did not affect great tit breeding population size in my study populations. Furthermore these experiments would make it possible to test the hypothesis that by increasing the intensity of both intra- and interspecific competition, great and blue tit morphometric or life-history traits might change through natural selection, because in tits, as in Darwin's finches (Grant and Grant 2006), rapid micro-evolution can occur (Dhondt *et al.* 1979). I will revisit this matter in Chapter 10.

By varying the nest-box types offered it is possible to manipulate the intensity of intra- and interspecific competition for each species separately as indicated in Table 9.2.

A comparison of tit populations between a plot with large-holed boxes only, and one with large- and small-holed boxes, measures the effect of intraspecific competition among blue tits and of interspecific competition for great tits. In this pair of plots great tit density is 'high' in both plots, while blue tit density is 'low' in the plot with only large-holed boxes and 'high' in the plot with both box types. Table 9.3 lists the between-plot comparisons that can be made.

The only effect that cannot be easily measured using this approach is that for intraspecific competition in great tits. This comparison is between a plot with

Table 9.2 Nest-box configurations to vary the intensity of intra- and interspecific competition in great and blue tit. In our study plots great tit are normally not found nesting in natural cavities, while only a low proportion of blue tits use natural cavities for reproduction (Dhondt et al. 1982)

	Great tit		Blue tit	
	Intraspecific competition	Interspecific competition	Intraspecific competition	Interspecific competition
Large-holed boxes only (L)	high	low	low	high
Both box types (LS)	high	high	high	high
Small-holed boxes only (S)	low	high	high	low

both large- and small-holed boxes and a plot with small-holed boxes only. In both plots blue tit density is high, while great tit density is high in the plot with both box types, but low in the plot with small-holed boxes only. For this reason, in the latter plots I changed the entrance hole of a small number of boxes to 32 mm, so that a limited number of great tits could breed in these boxes. The problem, however, with this approach is that (1) intense intraspecific competition between great tits for the limited number of cavities could result in a non-random set of great tits using the boxes for nesting; and (2) except for the initial years in plot GT in which small-holed boxes only were available, I did not census great tits in plots with small-holed boxes only. I did not, therefore, know how many (if any) great tits bred in natural holes when no large-holed boxes were available. Given that the effects of intraspecific competition among great tits were well documented I decided that this was a minor handicap in the approach.

In this long-term experiment I provided three study plots near Ghent and three plots near Antwerp, each with one of the three nest-box configurations. In the surroundings of Ghent, where Jan Hublé started nest-box studies in 1959 with large-holed nest-boxes only, we selected five of the study sites in which we could do a before–after comparison: three were kept as controls with large-holed boxes only, in one we provided small-holed boxes with a small number of large-holed boxes, and in one we added small-holed boxes to obtain the treatment with both large- and small-holed boxes. In Antwerp I started three new study plots, each having one of the nest-box configurations (later I added a second plot with large-holed boxes only). When, after five years, it appeared that demographic traits differed between treatments, I changed the nest-box configuration from large-holed to small-holed in the Antwerp plot T (with a small number of large-holed boxes in some years). Five years later I changed it again, now to the treatment with both large- and small-holed boxes. In the other Antwerp plots the nest-box configuration remained unchanged throughout the 15 years of the study so as to measure possible between-year changes. The two sets of plots were about 60 km apart. The Ghent plots were all isolated fragments in a rural or suburban matrix, and, as later became clear, were of much poorer quality than the Antwerp plots (see Fig. 9.4). This had the advantage though that it was possible to determine if the intensity and importance of competition varied between habitats that differed in overall quality and resources. Three Antwerp plots were part of the *Peerdsbos*, a large forested estate, while one was part of a large fragment about 2.5 km south of the *Peerdsbos*.

In all nest-boxes routine data were collected during the breeding season (species, laydate, clutch size, number of unhatched eggs, number of fledglings, and identity of adults) and all nestlings were ringed. In Antwerp we also measured nestling tarsus length and body mass at 15 days of age for all nests of both species. As all nestlings and breeding adults were ringed and sexed on plumage, (cf. Dhondt

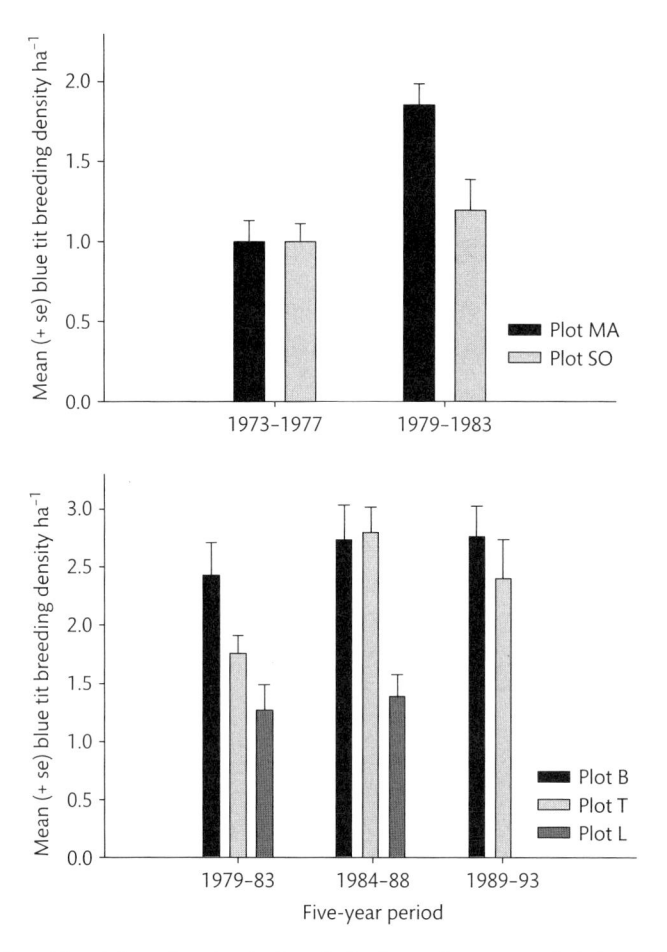

Fig. 9.5 Mean blue tit breeding density in pairs ha^{-1} over five-year periods in various study plots. In each pair of plots the nest-box configuration was changed in the experimental plot, while it was kept unchanged in the control plot. In the top panel small-holed boxes were added to the large-holed boxes present in Plot MA, while in plot SO only large-holed boxes were present throughout. The period*plot interaction was significant (P<0.01) indicating that density changed with treatment. In the lower panel both box types were present throughout in plot B, large-holed boxes only were present throughout in plot L. In plot T the nest-box configuration was large-holed boxes only in the first five-year period, small-holed boxes only in the second five-year period, and both box types in the third period. The period*plot terms: comparison T and L: P=0.01; T and B (period1979–83 versus 1984–88: P=0.04); T and B (period 1984–88 versus 1989–93): ns. In all cases when small-holed boxes were added blue tit density increased compared to the control sites. After Dhondt and Adriaensen (1999).

1970) it was possible to study dispersal and immigration by sex. Any unringed bird was considered an immigrant. In Appendix 3 (Table 2) the nest-box configurations are listed for all study plots.

9.4 Effects of intra- and interspecific competition on blue tit density and demographic variables

The effect of nest-box configuration on blue tit density was very robust: in all plots in which a high density of small-holed nest-boxes was provided, blue tit density was higher than in comparable plots with large-holed boxes only (Fig. 9.4, Fig. 9.5) while great tit densities were not significantly affected by treatment (see Appendix 3, Table 7). This made it possible to compare the demographic parameters between sets of plots exposed to different treatments. By comparing mean values of demographic variables over two five-year periods with different treatments, it was possible to measure effects of intra- and interspecific competition on blue tit demographic variables (see Table 9.3). In each comparison between two plots there was a five-year period in which the treatment in both plots was the same, and a five-year period in which in one of the plots the treatment had been changed.

9.4.1 Effects of competition during the breeding season

9.4.1.1 Intra- and interspecific competition in blue tits—experimental results

To determine effects of intraspecific competition on blue tit reproduction we compared plots SO (control plot) and MA (experimental plot) in Ghent, and plots B (control) and T (experimental plot) in Antwerp. In both comparisons great tit density was high, while blue tit density was changed from low to high in the experimental plots making it possible to test for effects of intraspecific competition among blue tits (Table 9.3).

A number of effects of intraspecific competition became apparent, although there were some differences between the regions (Appendix 3, Table 3).

In both regions blue tit laydate was delayed at higher density. In Ghent intraspecific competition also affected the proportion of eggs that produced fledglings (nest success), but this did not translate into a significant effect on number of young fledged per female per season (reproductive rate). In Antwerp intraspecific competition did not affect nest success, but a significant effect of intraspecific competition for food could still be documented on 15-day nestling weights. The increase in intraspecific density caused nestling mass to decrease by 2.6% in the Antwerp treatment plot compared to the control site. Perhaps more

Table 9.3 Comparisons of different nest-box configurations that can be used to study effects of intra- or interspecific competition between great and blue tit. (L: plot with large-holed nest-boxes; S: plot with small-holed nest-boxes; L+S: plot with both nest-box types at high density)

Comparison	Species with variable density	Species with constant density	For great tit	For blue tit
L and L+S	Blue tit low or high	Great tit–high	Effect of *interspecific* competition	Effect of *intraspecific* competition
L+S and S	Great tit high or low	Blue tit–high	No or limited data for great tit	Effect of *interspecific* competition
L+S and S (few L)	Great tit high or low	Blue tit–high	Effect of *intraspecific* competition	Effect of *interspecific* competition

surprising was the effect on adult body mass. Intraspecific competition did not affect male body mass, but did impact female body mass.

Taken together these results implied the existence of intraspecific competition among blue tits before laying (a delay in laydate) and while raising young. This latter effect was more pronounced in the poorer Ghent plots than in the Antwerp plots, because at Ghent nest success decreased when density was increased, while at Antwerp intraspecific competition only affected body mass. The limiting resource was presumably food, with pressure points occurring when females were preparing to lay and when young were being raised.

As intraspecific competition had been established it was now meaningful to test for the presence of interspecific competition. This was done by comparing plots in which boxes for blue tits remained unchanged while those for great tits were manipulated (Table 9.3).

Except for the lack of an effect of interspecific competition on laydate, we found the same effects of interspecific competition as found for intraspecific competition: interspecific competition affects 15-day nestling mass and adult female mass. Curiously, the effects were about the same for intra- and for interspecific competition. Furthermore the results suggested a possible effect of interspecific competition by great tits on clutch size of blue tits (P=0.07), although no effect of intraspecific competition was found. These results indicate that great and blue tit compete for food during the breeding season, but that blue tits are affected by interspecific competition over a shorter time period than by intraspecific competition, given that both influenced nestling mass and female condition, and that intraspecific but not interspecific competition influenced laydate (Appendix 3, Table 4).

9.4.1.2 Intra- and interspecific competition in blue tits—correlational analyses

The correlational analysis confirmed a strong effect of intraspecific competition on nest success, but also showed that reproductive rate and even clutch size decreased with increasing blue tit density, an effect not found in the 10-year experiments in which nest-box configurations were manipulated (Dhondt 2010b). Interspecific competition had a weak effect only on blue tit reproductive rate, but no effect on clutch size nor nest success. Possible effects on mass were not tested (Fig. 9.6).

The intensity of intraspecific competition was strongest in the poorest plots (as measured by density and nestling mortality). This aligns well with the experimental results, because in the poor plot pair in Ghent there was an effect of intraspecific competition on nest success, while this was not found in the high-quality plots in Antwerp. When cavities were not limiting, food during the breeding season was therefore more limiting in plots in which breeding density was lower.

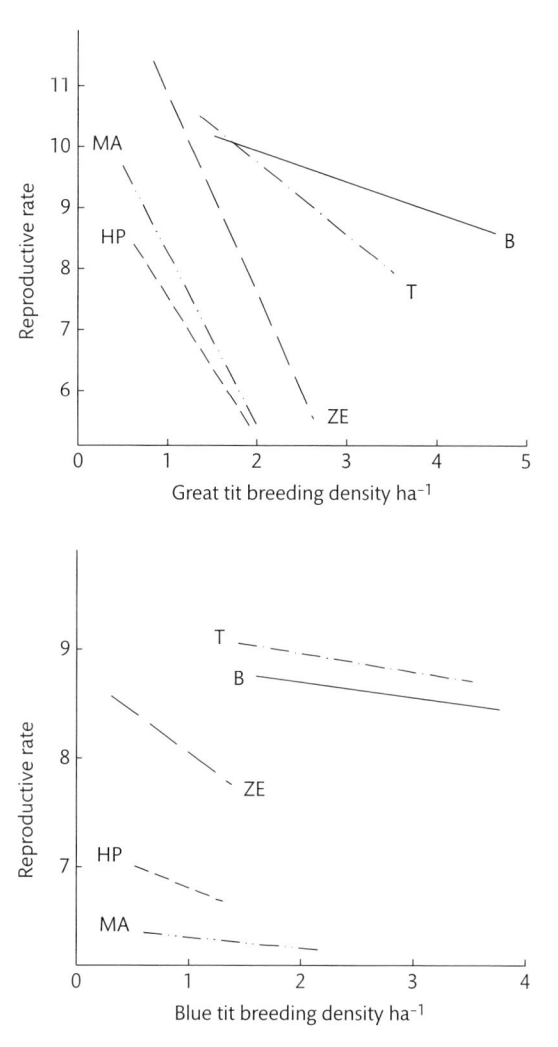

Fig. 9.6 Intensity of intraspecific competition for great tit and blue tit reproductive rate. The regression lines are drawn between the lowest and highest density observed in each study plot for the species. Since in the best model the interaction term between conspecific density and plot is present, the slopes, which represent the intensity of competition, differ significantly between plots. The slopes of the regression lines are steeper in plots in which habitat quality and density is lower than in plots in which habitat quality and density is higher. The study plots are identified next to each line. B and T are Antwerp plots; ZE, MA, and HP are Ghent plots. From Dhondt (2010b), with permission from John Wiley and Sons.

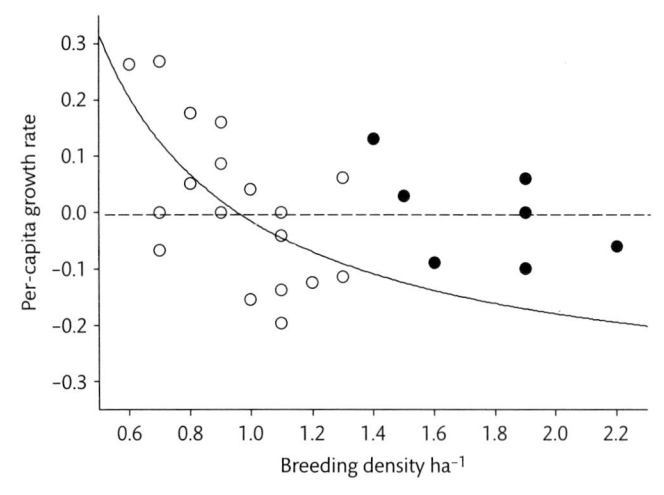

Fig. 9.7 Blue tit per-capita growth rate of the population in the Ghent Plot MA (Maaltepark). Open symbols, line, represent the per-capita growth rate values when large-holed boxes only were present and blue tit population density fluctuated around a mean of 0.96 pairs ha^{-1}. Filled symbols represent per-capita growth rate values when small-holed boxes had been added and blue tit density fluctuated around a mean value of 1.71 pairs ha^{-1}. The question we need to answer is which of the demographic variables (adult survival, juvenile recruitment, emigration, or immigration) changed when small-holed boxes were provided and per-capita growth at a given density increased. Dhondt, unpublished.

9.4.2 Effects of intra- and interspecific competition on blue tit per-capita growth rate

Given that several demographic variables are affected by both intra- and interspecific competition it is interesting to test to what extent the per-capita growth rate (r) is also dependent on conspecific and heterospecific density, as r represents the algebraic sum of all demographic variables. Although the use of time series data to detect conspecific density-dependence is often criticized (Lebreton 2009) it might be interesting to use time series of the complete counts of the birds breeding in the nest-boxes in our study plots to evaluate possible effects of density on r. For great tits, the birds found in the boxes represent all or nearly all breeding pairs (when large-holed boxes are available in surplus), although for blue that may not always be the case.

The problem with calculating r-values as differences between the log numbers in successive years, and plotting these values against the numbers of the initial year, is that the values on the two axes are not independent. Several authors have proposed techniques to address this problem but all have some issues (reviewed in Pollard *et al.* 1987). Pollard *et al.* (1987) developed a distribution-free test that calculates the probability that a significant density-dependent relationship between r and log(N) is caused by chance. The correlation between per-capita growth rate and log(N) was significant to highly significant in all plots (Appendix 3, Table 5). With the exception of Plot L, for which we have data for only nine breeding seasons, none of these correlations were caused by chance according to Pollard's test (Appendix 3, Table 5). We can thus conclude that all populations studied are regulated through density-dependence. Note that in many populations a non-linear relationship is a better fit than a linear one (illustrated in Fig. 9.7 for the blue tit population in Plot MA).

For one of the plots we can directly compare r values at high and low blue tit population density (Fig. 9.7). In the period 1959–78 only large-holed nest-boxes were available in the Ghent plot MA, and blue tit breeding density fluctuated between 0.6 and 1.3 pairs ha^{-1} (mean 0.96 ± SE 0.048). Between 1979 and 1986 small-holed nest-boxes were added, thus relieving blue tits from competition with great tits for roosting sites (and possibly also for breeding sites), but without reducing great tit numbers. Blue tit density then fluctuated between 1.3 and 2.2 pairs ha^{-1} (mean 1.71 ± SE 0.109), a mean value that was significantly higher than in the earlier period (t_{25} = 7.38, P<0.0001). Values of r for a given density thus increased when small-holed boxes were present (Fig. 9.7).

Knowing that per-capita growth rate decreases as conspecific density increases and that interspecific competition has also been documented, it is logical to test to what extent heterospecific density has an additional adverse effect on per-capita growth rate. The results of a multiple regression analysis (using PROC MIXED in SAS) for the four study sites with the longest time series confirm significant effects of conspecific density, and show that in plots HP in Ghent and in plot B in Antwerp heterospecific density does appear to influence blue tit per-capita growth rate (Appendix 3, Table 6). In Plot HP per-capita growth rate was also affected by the interaction between great and blue tit density. This may be related to the fact that over the study period in Plot HP great tit density gradually increased and blue tit density gradually decreased.

9.4.3 What causes the increase in blue tit breeding density when small-holed nest-boxes are available?

The next question that needs to be addressed is why blue tit breeding density increases when small-holed boxes are provided: where do the additional birds come from? This increase can either be caused by an improved adult survival rate, an increased local recruitment rate (because of reduced offspring dispersal or increased offspring local survival), or an increased immigration rate. The former does not seem likely as female mass decreased with increasing density of blue or great tit (see above) and is confirmed by a capture-mark-recapture analysis. Dhondt and Adriaensen (1999) found that although there was significant between-year variation in blue tit adult survival rates in both plots B and T, these variations occurred in parallel between males and females and between the two plots. The lower female mass in breeding seasons with increased density did not have a detectable effect on their subsequent survival probability.

The only remaining demographic variable that can explain the change in blue tit breeding density when small-holed boxes are provided is a change in recruitment rate, that is the number of new recruits joining the breeding population per same-sex adult in the previous breeding season. The question that needs to be asked, however, concerns the origin of these recruits: do they derive from local born young (Local Recruitment Rate), immigrant young (Immigrant Recruitment Rate), or both groups? There are two ways to explore a possible effect on recruitment. One is to calculate the proportion of first-time breeders that were born locally or immigrated from outside the plot. The second is to calculate the proportion of young fledged that recruit locally.

Dhondt and Adriaensen (1999) found in the initial five-year period that local recruitment of blue tits (calculated as the proportion of fledglings recovered breeding in their plot of birth) was much higher in plot B (small-holed boxes; high blue tit density treatment) than in plot T (no small-holed boxes; low blue tit density treatment). The causal link between treatment and local recruitment was shown after changing the experimental treatment in plot T. When in plot T small-holed boxes were provided and blue tit breeding density increased, a higher proportion of fledgling males stayed in their birth plot to breed as compared to the five-year period in which large-holed boxes only were available and blue tit breeding density was low. The local recruitment for females was low under both treatments in both plots and did not change significantly between treatments (Fig. 9.8). Dispersal of male fledglings was thus inversely density-dependent, but most females dispersed outside the plots regardless of density (see also Matthysen *et al.* 2001). The most extreme difference occurred in the first breeding season following the change in the nest-box configuration. Coincident with this increase, the proportion of local born birds in the breeding population also increased and, therefore, the proportion and number of birds immigrating to the population from elsewhere decreased.

The reduction of winter competition between great and blue tits for roosting cavities caused by providing small-holed nest-boxes has an effect on dispersal, particularly of young males: a higher proportion of fledglings remained in their birth plot, thereby increasing the proportion of local born birds in the breeding population while also increasing breeding population size. This change could be either the result of the change in habitat quality (resulting from the availability of small-holed nest-boxes for roosting), an example of conspecific attraction (caused by an increase in blue tit density), or be caused by an increase in local survival of fledglings. Since the increase in local recruitment was already apparent in the first breeding season after small-holed boxes had been provided (see Fig. 9.8, period 2) it must be the presence of protected roosting sites that caused the reduction in dispersal. We can thus conclude that interspecific competition for roosting sites impacts the availability of a critical winter resource,

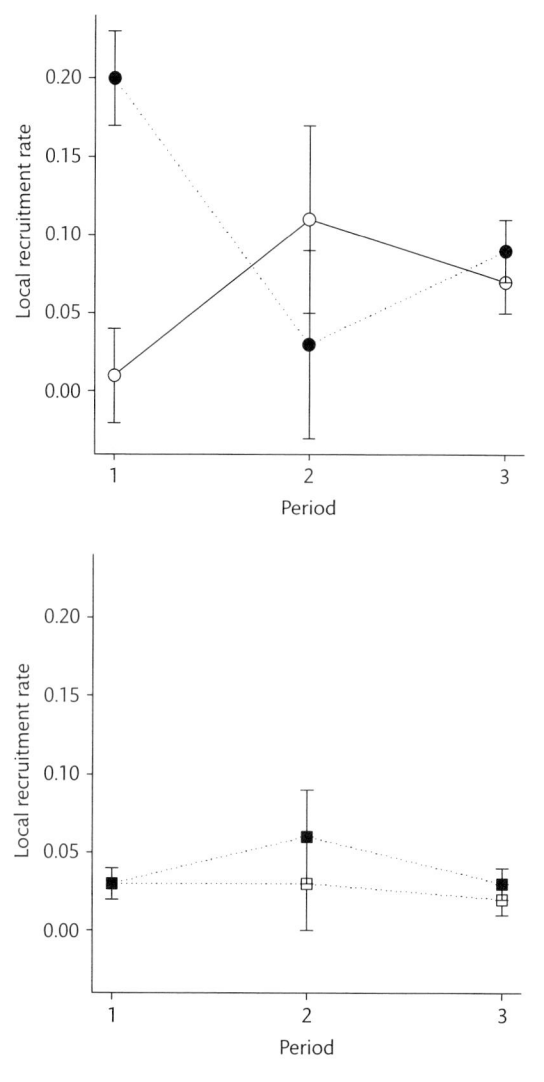

Fig. 9.8 Blue tit local recruitment rate (mean ± SE) in the Antwerp study plots T (experimental site; open symbols) and B (control site; closed symbols) Period 1: pre-experimental years during which blue tit density was low (1980–83); Period 2: first breeding season in Plot T after providing small-holed nest-boxes (1984), and thus first breeding season with a high blue tit density; Period 3: subsequent high density years in Plot T (1985–93). In Plot B Blue Tit density was high throughout because both large- and small holed boxes were present. Upper panel males; lower panel: females. Note that female local recruitment rate was consistently very low and did not change between periods, nor did it differ between plots. In males, however, the first breeding season after providing small-holed nest-boxes before winter, caused a major increase in local recruitment rate compared to the sudden decrease in the control plot B (this decrease is most likely because of the extremely cold winter 1984–85). In later years the recruitment rates in the two plots were the same. After Dhondt and Adriaensen (1999).

influencing dispersal decisions and/or juvenile survival, and leading to an increase in breeding population size.

9.5 Effect of intra- and interspecific competition on great tit density and demographic variables

Neither in Ghent nor in Antwerp did great tit density change in the experimental plots after blue tit density was experimentally increased by providing small-holed nest-boxes. Great tit breeding density, in our sites, was thus limited by intraspecific competition with no additional effect of interspecific competition (Appendix 3, Table 7).

Two manipulations make it possible to explore how great tit demographic variables responded to an experimental increase in blue tit density. These are discussed in the next sections.

9.5.1 Effects of blue tit density on great tit reproduction– experimental results

To determine effects of interspecific competition of blue on great tits I compared Plots MA and SO in Ghent, and plots B and T at Antwerp, during two 5-year periods (see Table 9.3). At Ghent both great tit clutch size and nest success were affected by interspecific competition but, although the change goes in the expected direction, the effect of the experimental manipulation on total reproductive rate was not significant (Appendix 3, Table 8). The cause of this lack of significance is probably linked to the very high proportion (89%; the second highest proportion over 27 years in plot MA) of breeding pairs that successfully raised second broods in 1979, linked to the very low great tit density (the second lowest over 27 years in this plot), while in the control plot SO more average values were observed.

At Antwerp, although none of the demographic variables were significantly affected by the experimental change in nest-box configuration, there was still evidence for interspecific competition for food during the nestling stage. Both adult and 15-day nestling condition (mass/tarsus length) were impacted by blue tit density (Appendix 3, Table 9). Interestingly both parents were equally affected.

There are two possible explanations why at Ghent we do, and at Antwerp we do not, find strong evidence for interspecific competition. In the Antwerp plots, food conditions during the breeding season are much better for great tits than in the Ghent plots, as illustrated by the much higher nest success values at Antwerp (0.83 to 0.91) compared to Ghent (0.63 to 0.76); as a result effects on clutch size, nest success, and reproductive success do not appear, but the condition of the young fledged is affected by interspecific competition. Effects on other demographic variables would only become apparent when resources are sufficiently scarce, as they are in the Ghent plots. Alternatively it could be that the difference in blue tit density between the control and the treatment period in plot T was not large enough to result in a change in the intensity of interspecific competition.

9.5.2 Effects of intra- and interspecific competition on great tit demographic variables—correlational results

In Subsection 9.4.1.2 I described how I calculated effects of intra- and interspecific competition on blue tit demographic variables, combining in a single correlational analysis the data from five study plots. I performed the same analysis for great tits, and obtained similar results (Fig. 9.6) (Dhondt 2010b). There exists a strong effect of both intra- and interspecific competition on great tit reproductive rate and nest success, but an effect of intraspecific competition only on clutch size. As in blue tits, the intensity of intraspecific competition (expressed at the slope of the regression line relating a demographic variable, such as clutch size or nest success, to breeding density) varied between study sites differing in quality. It was strongest in the poorest plots. For nest success the intensity of interspecific competition also varied between plots, while its effect on reproductive rate was simply additive to that of intraspecific competition. Overall the intensity of intra- and of interspecific competition is stronger for great than for blue tits, meaning that demographic variables are more affected by competition in great than in blue tits (Dhondt 2010b).

9.5.3 Effects of intra- and interspecific competition on great tit per-capita growth rate

In Subection 9.4.2 I explained the approach used to calculate per-capita growth rates from time series data. For great tits, as for blue tits, the analyses show that in the six study sites for which we have long-term data the negative correlation between per-capita growth rate and log(N) is not caused by chance (Appendix 3, Table 5), and as for blue tits the relationship between r and density is non-linear (Fig. 9.9). Contrary to the results found for blue tits though, great tit per-capita growth rate is not significantly influenced by blue tit density (Appendix 3, Table 6). This result is not really unexpected since we did not find an effect of the experimental manipulation of blue tit density on great tit numbers. Although some great tit demographic variables are adversely affected by interspecific competition with blue tits, in the end great tit numbers are not reduced by increased blue tit densities. We can assume, however,

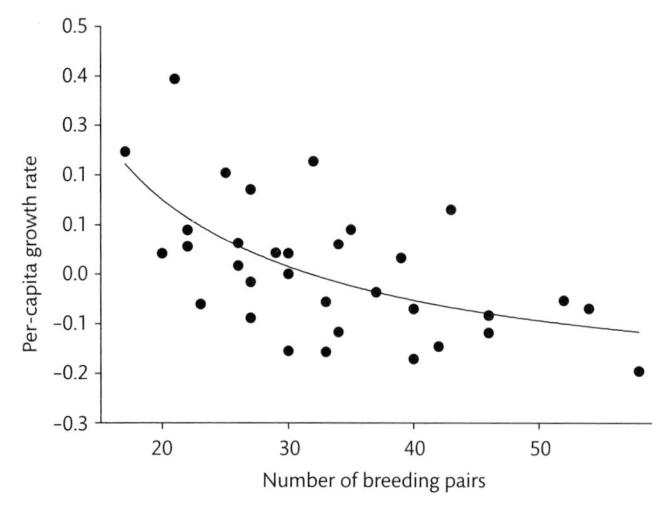

Fig 9.9 Great tit per-capita growth rate in the Ghent plot Hutsepot (HP, 27 ha) for the period 1964–98 plotted against the number of breeding pairs. The fitted line is a first order inverse polynomial, ($r^2 = 0.35$; P=0.0002). Dhondt, unpublished.

that variation in blue tit density could change selection pressures on certain great tit traits. This is explored in Chapter 10.

9.6 How similar are the results of experimental and correlational studies?

It is interesting to compare results obtained from correlational analyses, based on long-term data in many different study plots, to the more limited results stemming from field experiments. For blue tits the experiments provided information on effects of both intra- and interspecific competition, while for great tits the experiments only gave information on effects of interspecific competition (see Table 9.3).

Experimental manipulations showed an effect of interspecific competition on great tit clutch size (weak in high-quality sites), nest success (in poor-quality sites), and no significant effect on overall reproductive rate (number of young fledged per pair per season). The more detailed data from the experimental study in Antwerp, however, did show that great and blue tits compete for food during the breeding season, because both great tit fledgling and adult body condition were lower in the plots/years during which blue tit density had been increased. The correlational data did not show that blue tits impacted great tit clutch size, but did show a significant effect of blue tit numbers on great tit nest success and reproductive rate. This latter result was not obtained in the field experiments. Note that the correlational analysis combined data from 5 study sites that had been studied for up to 28 years. As explained above, between-year variation can be large and sometimes caused by factors out of the control of the experimenter. If such effects are limited to one of the study plots only, a single control site in an experimental study might be insufficient. Long-term series are more likely to buffer such effects, while if such effects occur in relatively short-term studies in small populations these disturbances will inflate variance and reduce the significance of the results.

For blue tits correlational studies detected effects of intraspecific competition on clutch size, nest success, and reproductive rate, and a weak effect of interspecific competition on clutch size and reproductive rate. The field experiments documented a weak interspecific effect on clutch size, an effect of intraspecific competition on nest success in the poor-quality study plots, and an effect of both intra- and interspecific competition on fledgling and adult female body mass.

Although the precise results are not identical, both approaches document the existence of intra- and interspecific competition for food during the breeding season. Both approaches also show that the intensity of competition varies between habitats that differ in richness, being strongest in the poor quality sites.

One of the problems with my experimental approach is that I did not manipulate bird densities (or food) directly, but influenced densities by manipulating nest-box configuration, hence the availability of cavities for roosting and for nesting. Extreme between-year variation in density could thus override the effects of the intended experimental configuration. One result of this is that blue tit density did not always vary as planned. Thus in plot T blue tit density in 1983 reached 2.4 pairs ha^{-1} in a year when only large-holed boxes were present and blue tit density was 'low', but only 1.6 pairs ha^{-1} in 1993, and 1.76 pairs ha^{-1} in 1992 years when both nest-box types were present and hence blue tit density was expected to be 'high'. Control sites should show similar variations, but that is not always the case, and in any study there are limits on how many replicates can be studied.

As will be shown below, the experimental approach, in which treatments varied between plots, could provide insights not obtained by simple longitudinal studies, especially regarding dispersal and selection pressures.

9.7 Density and dispersal

One of the more difficult demographic variables to measure is dispersal. Ideally one would know the total number of surviving juveniles wherever they settled and calculate the proportion of these that either recruited into the breeding population of the wood in which they were born or bred elsewhere (proportion dispersed). Only in very few studies can this be done. Both the Antwerp and Ghent study encompassed a series of neighbouring plots so that a certain proportion of the dispersers were recovered but certainly not all. This made it possible, for example, to show that the proportion of great tit juveniles that were recovered breeding outside the birth plot increased with the number fledged (see Fig. 6.6.), thus documenting that dispersal in great tits is density-dependent. It also made it possible to prove that a higher proportion of second brood than first brood great tits disperse (Dhondt and Hublé 1968). Nevertheless the alternative approach (percentage local born breeders) makes it possible to compare dispersal annually or over short periods, and hence compare dispersal between experimental treatments.

The question I want to address here is whether it is possible to document an effect of interspecific competition or perhaps even of heterospecific attraction from the data collected in the Ghent and Antwerp studies.

9.7.1 Effect of blue tit density on great tit local recruitment

To study the effects of changes in experimental nest-box configuration on recruitment into the breeding population, which were performed every five years, one can only use data from breeding seasons 2–5 in each treatment, because the birds recruited in the first year of the five-year period were born before the nest-box configuration was established. This is why I here compare four-year periods.

In the plots in which nest-box configuration was changed by adding small-holed nest-boxes to large-holed boxes, thus increasing blue tit density, local recruitment of great tit males was much *higher* at high than at low blue tit density. This was true both at Ghent (five-fold difference from 5.3 to 27.6%) and Antwerp (double from 9.8 to 18.9%), while in the same periods there was only a minor change in male recruitment rate in the plots in which nest-box configuration had remained unchanged (31.6 to 31.3% in Antwerp; 16.1 to 17.4% in Ghent) (Appendix 3, Table 10). Local recruitment of female great tits did not vary significantly with blue tit treatment (Appendix 3, Table 10). The results thus indicate that juvenile dispersal of male great tits on the one hand increases with conspecific tit density (an example of intraspecific competition; see Fig. 6.6), but on the other hand is reduced when blue tit density is higher (a possible example of heterospecific attraction). For female great tits there is no effect of blue tit numbers on dispersal.

9.7.2 Effect of blue tit density on great tit adult dispersal

In the Ghent study hardly any adult great or blue tits moved between plots after having bred. In the Antwerp study, where the plots were closer together, 17 adult great tits were recaptured breeding in a different plot in a later season. This number is sufficient to show that the experimental blue tit density manipulation had an effect on great tit post-breeding dispersal. The results are unambiguous (Appendix 3, Table 11). In Plot B, in which blue tit density was high throughout, 3% of the females and no males were recovered breeding out of their breeding plot in either period. When blue tit density was high in Plot T (breeding seasons 1989–92) no birds emigrated, but when blue tit density was low (breeding seasons 1979–82) 14% of the surviving females and 7% of the surviving males were found breeding outside Plot T at a later date. Note that all these birds except one had bred successfully before dispersing, so that failed breeding could not be invoked as the cause of the move. A three-way loglinear analysis indicates that the interaction between period and movement (stay, disperse), corrected for sex, was very significant (G^2= 17.24, 2 df, P=0.002), indicating that at high blue tit density adult great tits are *less* likely to move out of their breeding area, another possible example of heterospecific attraction.

Thus in great tits both post-juvenile and post-breeding dispersal is influenced by the number of blue tits present, but in a non-intuitive way: great tits may use blue tit density as inadvertent social information when selecting breeding habitat. A similar result was obtained from correlational data in a large-scale study in France (Parejo *et al.* 2008). The authors concluded that '*great tits might use interspecific inadvertent social information more in the form of density than actual breeding performance*', and suggested that because local density may be evaluated all year long '*habitat selection based on density is likely to be a more ubiquitous strategy in resident birds such as the great tits as they spend much of the year in the reproductive habitat*'.

9.7.3 Effects of great tit density on blue tit juvenile local recruitment

In the initial experiment in which we provided small-holed nest-boxes we had already observed that more locally-born blue tits settled to breed in their birth area in years with small-holed nest-boxes compared to years without them (Dhondt and Eyckerman 1980). To separate effects of intraspecific and interspecific interactions on blue tit recruitment rates we compared local recruitment rates between treatments with small-holed boxes only and with large-holed boxes only, and between treatments with small-holed boxes only and with both box types present (Table 9.3). Only the Antwerp study provides this information, so this experiment does not have a replicate.

Providing small-holed nest-boxes, whereby blue tit density increased, resulted in an increased local recruitment (Section 9.4.3 and Fig. 9.8). In Plot T 1.8% of breeding males had been born locally at low blue tit density, while 14.7% had been born locally in the period when blue tit density was high, while in the control Plot B there was no difference between periods (Appendix 3, Table 12). Providing small-holed nest-boxes, which were used as roosting sites during winter and resulted in an increase in blue tit breeding density, led a higher proportion of local born birds to remain in their birth plot. Rather than documenting an effect of intraspecific competition the results showed the opposite result. Either we observed conspecific attraction or simply that the presence of high-quality winter roosting sites made the habitat more attractive, inciting the juveniles to remain. Female local recruitment was not affected significantly by this manipulation.

The effect of interspecific competition was obtained by comparing the four-year periods between which blue tit density was similar, but great tit density differed (Table 9.3). The data showed that, under increased great tit density, juvenile male blue tits dispersed more, with no effect on females (Appendix 3, Table 12). This supports the idea that interspecific competition by great tits influences male blue tit dispersal.

9.8 What have we learned about competition between blue and great tit?

The results summarized in this chapter leave little doubt that both great and blue tit populations are subjected to intra- and interspecific competition. What is more interesting, though, is to explore how each species' population dynamics is affected by competition, for what resources they compete, and to determine the extent to which the intensity of competition varies in time and space.

9.8.1 Competition for space

If all other impacts on their populations fail to keep them below the maximum number that an area can support, the sizes of the breeding populations of both great and blue tit are ultimately limited by territorial behaviour, thus by intraspecific competition for space. This, however, may ultimately relate to food availability. Although the two species are often aggressive to one another, interspecific territorial behaviour between them, or even with other species, does not seem to occur. The only example that suggests that great tit numbers can be limited by interspecific territorial behaviour is that described by Reed on a small Scottish island, where it can be assumed that resources were very limiting (see Subsection 3.4.2). In many tit studies intraspecific competition for territories occurs regularly. Surplus individuals are expelled from high-quality habitat so that intraspecific competition for space not only determines breeding density in these habitats, but also the distribution of birds across habitats (see Section 3.2, Buffer

Hypothesis). If birds cannot establish a territory they become floaters, some of which will try to breed as intruders or guest pairs, or occupy sites that become vacant (Krebs 1970; Lambrechts and Dhondt 1988). Some female blue tits prefer to breed polygynously on high-quality territories, rather than monogamously in less productive sites (Dhondt 1987).

9.8.2 Competition during the breeding season

During the breeding season competition can occur for nest cavities. If these are in short supply the size of the breeding population will be reduced. In such situations both intra- and interspecific competition can impact breeding density. Their effects will depend on the distribution of cavities of different quality and the sizes of their entrance holes. Intra- and interspecific competition for food during the breeding season can impact a variety of demographic variables including laydate, clutch size, nest success, nestling mass (which influences post-fledging survival), the probability of undertaking a second brood, and thus the overall reproductive rate. These effects vary between habitats of different quality but also between the two species. Overall the impact of intra- and interspecific competition on reproduction is stronger in poor-quality, low-density habitats than in high-quality, high-density habitats, and in great than in blue tits. Interspecific competition affects largely the same demographic variables as intraspecific competition, suggesting that the same resources are limited in both types of competition. Competition not only influences the number and quality of young produced, but can also have an effect on adult condition, which could (although not documented in my studies) translate to effects on adult survival. Female condition seems to be more impacted by competition than male condition.

9.8.3 Competition outside the breeding season

The 'outside the breeding season' period is quite long and should be partitioned into various periods as done, for example, by Jan Ekman in his willow tit study (Ekman 1984b). The first, and possibly one of the main periods during which future breeding density is determined, is the period soon after juveniles become independent and during which juvenile mortality is very high (Dhondt 1979). Dispersal of juveniles is rapid and explosive (Dhondt 1971, 1979; Goodbody 1952). More females than males disperse in the first weeks after independence, suggesting that intersexual competition influences dispersal, because, on average, males are dominant over females. As mentioned above, fledgling weight is strongly correlated to survival. Given that fledgling weight is influenced by both intra- and interspecific competition, we can consider that the survival of juveniles after independence is indirectly influenced by intra- and interspecific competition for food during the breeding season. While juvenile dispersal and survival are strongly influenced by intraspecific competition in great tits, as shown elegantly by Kluyver's experiments in which he reduced the number of young great tits allowed to fledge (Kluyver 1971), it is not clear if interspecific competition also has an effect on these variables. Thus great tit dispersal increases with increasing conspecific density (intraspecific competition) but is lower at high blue tit densities, suggesting the existence of heterospecific attraction.

Blue tit dispersal is higher at high great tit density, reflecting an effect of interspecific competition on their dispersal, but lower in the presence of small-holed nest-boxes. The underlying mechanisms of these effects on dispersal are probably aggression over space, food, and nest-boxes for roosting, especially in late summer and autumn. The role of intraspecific aggression is well documented in great tits (Drent 1984; Kluyver 1971) but not studied in blue tits.

Jean Clobert and Robin McCleery found that great tit local recruitment rate decreased both with increasing great tit density and with increasing blue tit density (McCleery and Clobert 1990), a result partially opposite to mine. One difference between the studies was that they used natural variation in density of the two species in a study with large-holed boxes only, while I manipulated blue tit density by offering both large- and small-holed nest-boxes. What was more surprising though was that the observed effects depended on the origin of the parents. Immigrants tended to recruit fewer offspring than local born parents. The exact mechanisms that cause variation in the extent of dispersal are not well known, but recent work on the relationship between personality (a heritable trait) and dispersal suggests that the study of dispersal is even more complicated, interesting, and relevant than previously thought (Dingemanse *et al.* 2003). Dispersal seems to be the result of an interplay between individual traits (such as gender, phenotypic quality, fledging date, and even parasite infestation (Heeb *et al.* 1999)) and intra- and interspecific competition.

Once most juveniles have settled down, they compete for food and for roosting cavities, and this both with con- and heterospecifics. Competition for roosting sites can be quite severe as tits sleep alone in a cavity and each individual great and blue tit claims multiple roosting sites (Kluyver 1957; own observations). Furthermore, cavities with certain characteristics (size, depth, size of entrance hole) are preferred over others. In both great and blue tits competition for cavities leads to a strong male bias among birds found roosting in nest-boxes, as first reported by Kluyver for great tits (Kluyver 1957). If all nest cavities are accessible to great tits, most blue tits are excluded from them through interspecific aggression.

During winter, competition for food also influences tit survival and subsequent breeding numbers (see Chapter 4), but the extent to which interspecific competition is important is still unclear.

9.8.4 The effect of competition on breeding density

Intraspecific competition for territorial space can limit breeding density of great, and probably also blue, tits in optimal habitat. One of the unexpected results of my experiments in which I manipulated nest-box types, was that the presence of small-holed nest-boxes during winter made the study plot more attractive to blue tits as a wintering site, reduced blue tit juvenile dispersal (especially of males), and that this resulted in a clear increase in blue tit breeding density. As a result a larger number of locally born birds were present in the breeding population (Fig. 9.8). Although at these increased densities reproductive rates declined through intraspecific competition for food, the increased breeding density was maintained because of the increase in the local recruitment rate.

In great tits the effects of intra- and interspecific competition on reproduction did not seem to change the equilibrium breeding density in high nor in medium quality habitats.

The study of dispersal is thus essential to understand how breeding density is determined.

9.8.5 Variation in the intensity and effects of competition in space and time

So far I have discussed primarily the results of the Belgian studies in only two regions about 50 km apart. Given that the results between these two regions differ one can wonder how effects of competition would vary between sites across Europe. Some of this variation has already been discussed in Chapter 4, but I now want to revisit this.

Recent effects of climate change on populations raises the question to what extent this change could impact intra- and interspecific competition. Recent climate change may lead to increased coexistence of plants (Adler *et al.* 2006; Angert *et al.* 2009), rotifers and water fleas (Hampton 2005), but may also lead to competitive exclusion (for example: snowmelt plants (Heegaard and Vandvik 2004)). In our own historical past Modern Man seems to have driven Neanderthal Man to extinction through interspecific competition during periods of climate change around 40,000 years BP (Banks *et al.* 2008).

It is therefore interesting to evaluate possible effects of climate change in species between which interspecific competition has been well documented. A recent analysis by Stenseth *et al.* (unpublished) of six long-term data sets of great and blue tits across Europe explored (1) if the existence of interspecific competition was found in all studies, and (2) if the competitive interactions between the species were influenced by long-term climate change. In three of the populations no evidence for the existence of interspecific competition was found (Schluechtern, Germany; Liesbos, Netherlands; Fågelsångsdalen, Sweden), while in the other three sites interspecific competition was documented (Plot HP, Ghent, Belgium; Plot B, Peerdsbos, Belgium; Marley Wood, UK). In these three latter sites stable coexistence was mathematically supported (Stenseth *et al.* unpublished ms). The analyses, however, also indicated that the model selected varied with spring temperature in the two Belgian populations, and since spring temperature has increased in both Belgian populations (Visser *et al.* 2003), continued spring warming could eventually jeopardize coexistence. In Plot HP the blue tit, but not the great tit, population showed a threshold interaction with April temperature: when April temperatures were below this threshold, coexistence between great and blue tits was stable with both populations existing at densities of around 1 pairs ha^{-1}. When April temperatures were above the threshold, however, blue tit densities declined while great tit densities increased. Not much additional temperature change would be needed to lead to competitive exclusion of blue tits by great tits. In Plot B, the analysis indicates stable coexistence at temperatures below a threshold of mean spring temperature, but unstable coexistence above that threshold temperature. At Marley Wood (Oxford), populations of the two species were unaffected by the environmental variables modelled and a single model showing stable coexistence fits the data (Stenseth *et al.* unpublished ms).

These results illustrate two points clearly. First, studies of the same species using similar methods show the existence of intra- and interspecific competition in some locations, while in others this is not the case. Second, changes in conditions, such as those resulting from climate change, can affect the possible outcome of interacting species leading to continued coexistence or competitive exclusion.

9.9 Concluding comments

The studies of intra- and interspecific competition using blue and great tits show the complexity of such interactions and the large numbers of factors that influence the existence and outcome of competition. Although I have mostly discussed competition by itself, it is obvious that many other factors influence its existence, intensity, and outcome. When conditions become more constraining, because food becomes more limiting (as for example caused by acid rain), because predator or disease pressures change, or because habitat quality changes for whatever reason, both intra- and interspecific competition could intensify. If this leads to competitive exclusion of some species by others, the composition of entire communities could be impacted.

These studies also show that factors influencing the existence of competition vary in time and space and that the effects, even in two closely related coexisting species with similar ecologies, are often not the same. To understand the role that competition plays, detailed, replicated, long-term studies using individually marked individuals are necessary.

10
Evolutionary effects of interspecific competition

In the introduction I explained that some authors believe that interspecific competition is such a powerful force that when species compete this can only be a transient situation. This was the paradigm in the 1950s as illustrated by a quote from Brown and Wilson (1956):

'However, interspecific competition of the direct conspicuous, unequivocal kind is apparently a relatively evanescent stage in the relationship of animal individuals or species, and therefore it is difficult to catch and record. What we usually see is the result of an actually or potentially competitive contact, in which one competitor has been suppressed or is being forced by some form of aggressive behavior to take second choice, or in which an equilibrium has been established when the potential competitors are specialized to split up the exploitable requisites of the environment' (p. 60).

In this logic competing species will thus either exclude one another or will evolve rapidly so that they can coexist without competing further. This implies that interspecific competition can be a potent evolutionary force, but cannot often be observed. This led critics to reject the idea that interspecific competition was important at all, best phrased perhaps by Joseph Connell in his famous paper on the 'Ghost of Competition Past'. He wrote: *'the notion of coevolutionary shaping of competitors' niches has little support at present'* and lists what kind of field experiments would be required to demonstrate the importance of interspecific competition (Connell 1980).

10.1 Ecological character release and the Niche Variation Hypothesis

There exist multiple examples of an expansion of the ecological niche in species-poor environments (Chapter 7). The Niche Variation Hypothesis predicts that when a population is released from interspecific competition, because of the absence of a putative competitor, its niche becomes wider because of an increase in inter-individual variation (Van Valen 1965). As explained in detail

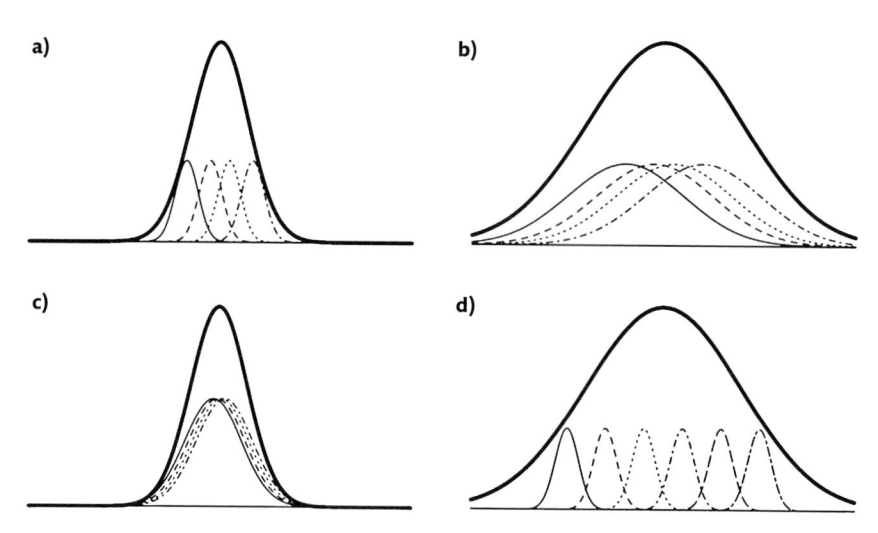

Fig. 10.1 Illustration of three potential forms of ecological character release. The thick curves represent the population niche width, while the thin curves represent individual niche widths. (a) represents the situation of a population in the presence of interspecific competition whereby individual niche width is narrower than population niche width; (b) represents the scenario under which both individual and population niche width increase ('parallel release'); (c) represents the scenario under which individual but not population niche width increases; differences between individuals become smaller ('individual release'); (d) represents the scenario under which population niche width increases although individual niche width remains unchanged ('niche variation hypothesis'). In this scenario among-individual variation increases. From Bolnick et al. (2010) with permission from The Royal Society and the author.

by Daniel Bolnick, and illustrated in Fig. 10.1, there is another way through which niche expansion, at the population level, can come about (Bolnick *et al.* 2010; Bolnick *et al.* 2003). If individual niche widths increase this will also cause an increase in population niche width ('parallel release'). A third possible response to a release from interspecific competition is 'individual release'. In that scenario individual niches become wider, but inter-individual differences become smaller causing population niche width to remain unchanged (Bolnick *et al.* 2010).

In reviewing data from studies of 93 species covering gastropods, insects, and vertebrates Bolnick *et al.* (2003) concluded that individual specialization is a widespread phenomenon, thus providing support for the frequent occurrence of the mechanism underlying the Niche Variation Hypothesis.

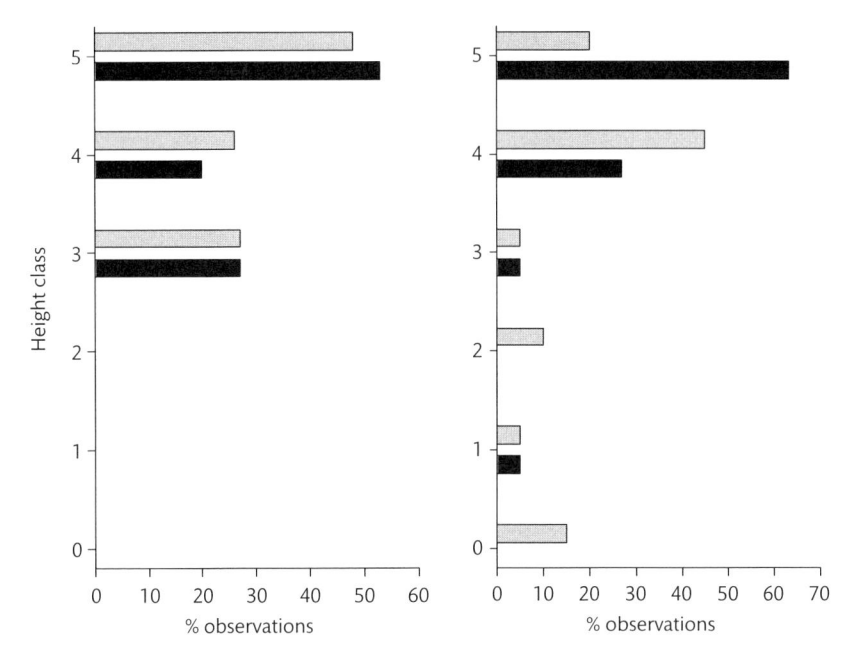

Fig. 10.2 Foraging height of male (black bars) and female (grey bars) coal tits in Wytham Woods, Oxford, England (left graph) and Killarney, Ireland (right graph). In the Irish study site willow tits are absent and the foraging niche of coal tits is wider than in the English site. Sexual dimorphism is also more pronounced. After Gosler and Carruthers (1994).

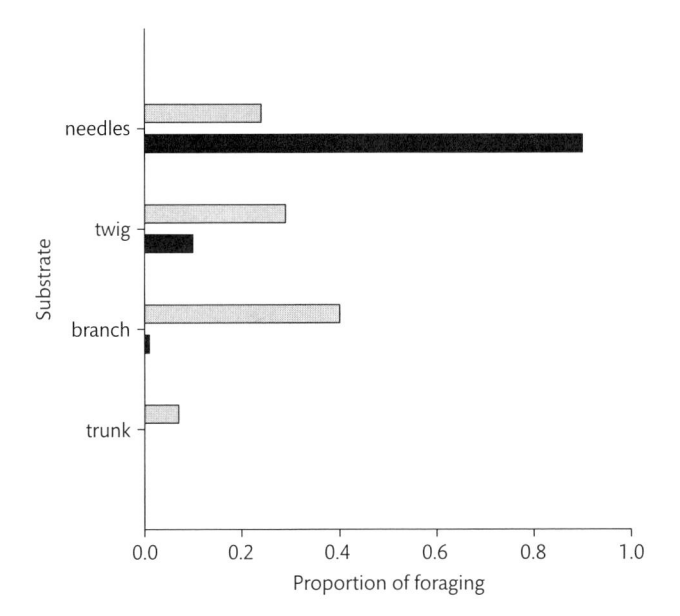

Fig. 10.3 Use of foraging substrates by coal tits in pine forest in mainland Sweden (black bars) and on the island of Gotland (grey bars). Note that in mainland Sweden coal tits forage mostly on needles, while on Gotland coal tits use a greater diversity of substrates and population niche width is therefore much wider. After Gustafsson (1988a).

10.2 Testing the criteria for ecological character release

10.2.1 Coal tits

Coal tits are the smallest tit species in European forests. Across their range, that covers North Africa, Europe, and Asia, they show considerable variation in body size, beak size, and colouration. Beak shape varies with habitat, being significantly more pointed in coniferous habitats than in broad-leaved habitats (Snow 1955). Comparisons of coal tit foraging niches in forests with and without crested and/or willow and marsh tits in Sweden, Ireland, and Denmark, documented an expansion of the coal tit niche towards the space vacated by the absent competitors (Fig 10.2, Fig. 10.3) (Alerstam *et al.* 1974; Gosler and Carruthers 1994; Sørensen 1997b), thus providing indirect evidence for the existence of interspecific competition. Direct evidence for interspecific competition was provided by the experimental work of Rauno Alatalo and colleagues in coniferous forests in Sweden. They showed that coal tits are 'forced' to forage in needles in the presence of the larger tit species: when the larger tit species were removed coal tits

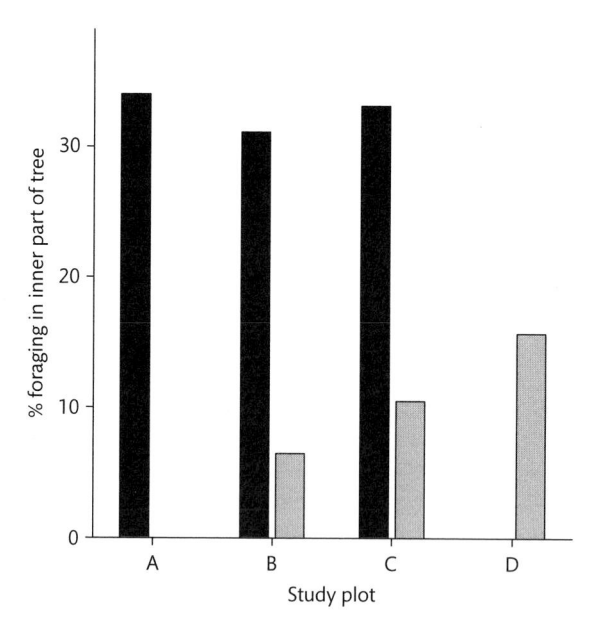

Fig. 10.4 Difference in foraging location of coal tits in years in which crested and willow tits are present (grey bars) and years in which crested and willow tit numbers have been experimentally reduced (black bars). Coal tits spend a higher proportion of their time foraging in the inner part of the pine trees when the larger species have been removed. In Plots B and C observations were made in two different winters with different treatments. After Alatalo *et al.* (1985b, Table 3).

foraged more inside the tree and used the niche space made available by the absence of the larger tits (Fig. 10.4) (Alatalo *et al.* 1985b, see also Chapter 7).

Andy Gosler in Ireland and in England and Lars Gustafsson in Sweden also compared the morphology of coal tit populations in sites in which more or fewer competitors had been present for extended time periods. Andy Gosler compared coal tit morphological variation in Wytham woods, England (where it coexists with marsh, blue, and great tits) to that in Ireland where it has fewer competitors. He found that in Irish populations interindividual bill variation was greater than in England, especially because of an increase in sexual dimorphism in bill measurements (Fig 10.5) (Gosler and Carruthers 1994). Lars Gustafsson, on the other hand, did not observe an increase in sexual dimorphism in Gotland coal tit foraging niches nor in bill measurements. Population niche width nevertheless increased because of a large difference in individual niches especially occurring with age. Adults foraged

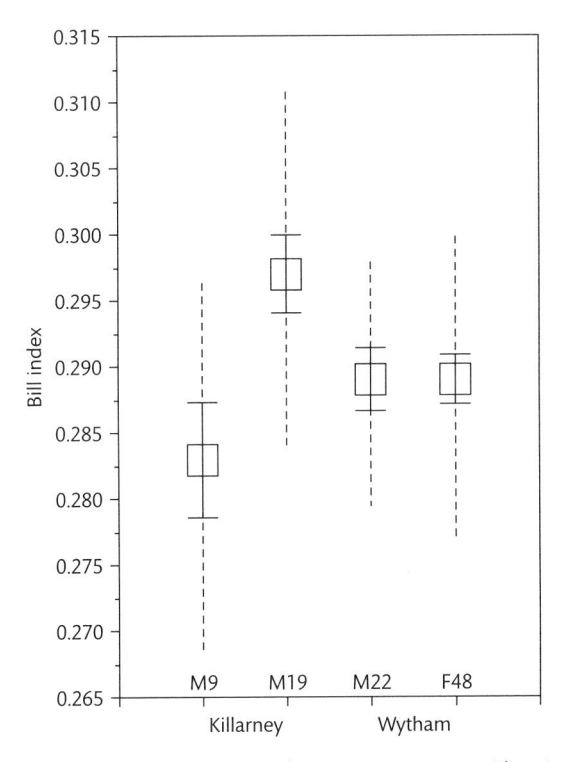

Fig. 10.5 Sexual difference in bill shape (index) of coal tits in England (Wytham) and Ireland (Killarney). In the Irish populations bill shape differs between males and females, while in England there is no such difference. These differences are reflected in the foraging niches of these populations as illustrated in Fig. 10. 2. From Gosler and Carruthers (1994), with permission from John Wiley and Sons.

mostly in the middle of the tree, while juveniles mostly foraged either on the outside, among needles, or on the inside, close to the trunk and on thick branches. As concerns the age effects he suggested that adults, being dominant, optimized the part of the tree in which they foraged. Optimization was for a combination of food availability and predator avoidance. Subordinate juveniles were forced to forage in sites that were more exposed to predators. Comparing foraging niches of individuals between years showed no change for adults, while juvenile foraging niches became more similar to those of adults. The release from interspecific competition resulted in an increase in population size and hence in an increase in intraspecific competition. This resulted in an increase in interindividual variation. Part of the increase in population niche width though was linked to age-independent differences in tarsus length, wing shape, and body mass (Fig 10.6). Birds with more rounded wings, longer tarsi, and higher mass foraged more on branches, while birds with longer wings, shorter tarsi, and lower mass foraged more on needles, independent of age. Population niche width clearly increased because of more morphological differences between individuals.

It is interesting to observe that the way population niche width widened differed between Ireland and Gotland. In Ireland it came about primarily through sex-linked differences, while in Gotland it came about through a combination of age effects and an increase in variation in morphological traits that influenced

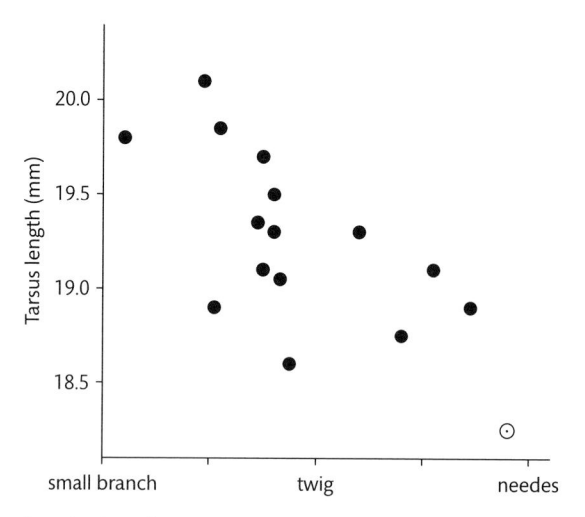

Fig. 10.6 Tarsus length plotted against average substrate used for individual adult coal tits on Gotland. The open circle represents the mean value of the mainland population. Note that on Gotland, in the absence of the larger willow tit, coal tits are larger and forage less on needles and more on the inner part of the tree. After Gustafsson (1988a), with permission from Elsevier.

foraging sites. Perhaps this is linked to the very high coal tit density in Gotland compared to the Swedish mainland, resulting in severe intraspecific competition, while the difference in density between Ireland and England is not so pronounced.

Peter Grant formulated six criteria that need to be met to ascertain that ecological character displacement occurs (Grant 1972; Schluter and McPhail 1992). We can test to what extent the above case meets these criteria.

(1) The pattern could not occur by chance.

As pointed out by Alatalo *et al.* (1986) all seven cases in which members of the tit foraging guild were compared between a site in sympatry and one in allopatry showed niche changes in the direction expected if interspecific competition was more intense in the more species-rich site. All three studies discussed above showed the same pattern (one was included in Alatalo's comparison), suggesting that these responses are non-random.

(2) Phenotypic differences should have a genetic basis.

Bill measurements, in general, are heritable in birds. The differences in bill sizes between those on the Swedish mainland and those on Gotland have a genetic component as shown through reciprocal exchanges (Alatalo and Gustafsson 1988).

(3) Enhanced differences should result from actual evolutionary shifts.

Given (2), and given the increased interindividual variation on Gotland compared to the Swedish mainland, this condition is satisfied.

(4) Morphological differences should reflect differences in resource use.

This was clearly the case (Fig. 10.2, 10.3, 10.5).

(5) Sites of sympatry and allopatry should not differ greatly in food, climate, or other environmental features affecting the phenotype.

This is certainly true for the Swedish and Danish study, perhaps less so in the Ireland/England comparison, because coal tits in the species-poor environment had expanded their habitat use.

(6) There must be independent evidence for competition.

This was demonstrated through the experimental removals of putative competitors in Sweden (Alatalo *et al.* 1985b).

10.2.2 Plethodontid salamanders

A recent study of plethodonthid salamanders addressed the six criteria listed above to evaluate to what extent differences between populations in sympatry and in allopatry could be the result of interspecific competition (Adams and Rohlf 2000). They found that various morphological traits were more dissimilar in sympatry than in allopatry. What was particularly interesting in their study was that they were able to link differences in the jaw closing mechanism to the relative length of the squamosal and the dentary and to differences in the forces involved in jaw closure and hence in food taken. Thus in sympatry the larger species that eats larger prey (*Plethodon hoffmani*) has evolved a faster but weaker jaw,

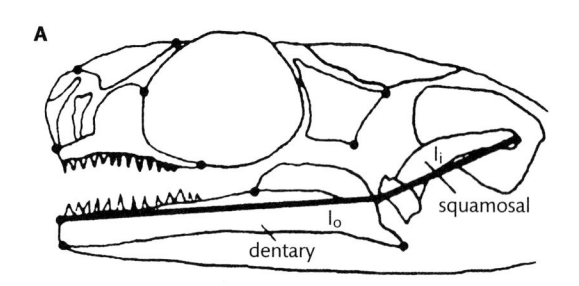

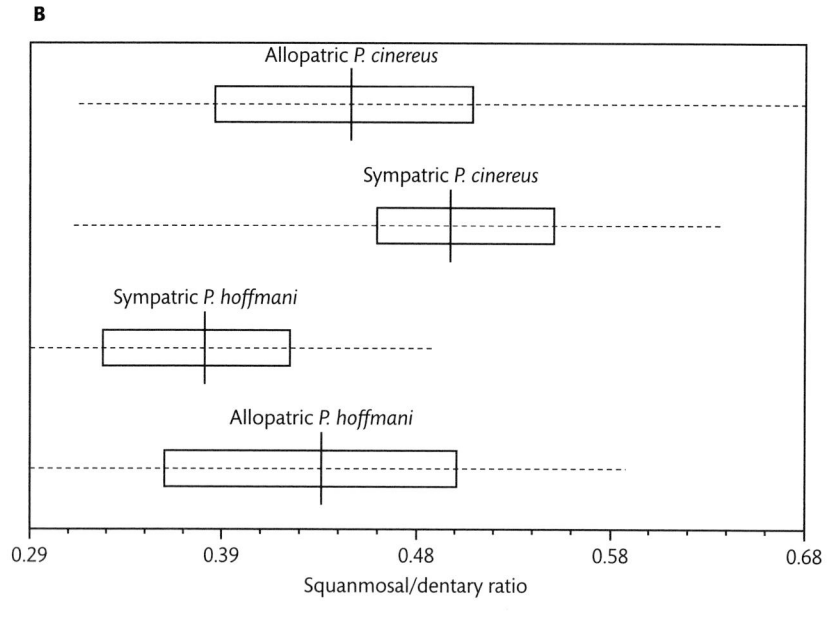

Fig. 10.7 (A) Biomechanical representation of jaw closing mechanisms in plethodontid salamanders; (B) plots of ratios of squamosal length to dentary length. The ratio between these two affects jaw closure rate and power. In sympatry the ratios of these bones in the two species are more different than in allopatry, resulting in larger differences in food taken in sympatry than in allopatry. From Adams and Rohlf (2000), with permission from the National Academy of Sciences of the United States of America.

while the smaller species that eats smaller prey (*P. cinereus*) has evolved a slower, more powerful jaw. The changes in jaw closure are thus associated with changes in prey consumption. Five of the six criteria for character displacement were supported in their work.

10.3 How rapidly can interspecific competition cause evolutionary changes in morphology? Observational data

The above observations show that, given 'sufficient time', populations that are exposed to or released from interspecific competition can evolve in an adaptive fashion. But what is 'sufficient time'?

A small number of observational studies of birds and other animals provide some answers.

Comparing two species of myzomelid honeyeaters (the scarlet-bibbed Myzomela and the ebony Myzomela) in the Bismarck Archipelago and New Guinea, Jared Diamond *et al.* (1989) concluded that a substantial amount of ecological character displacement occurred over a period of at most 250 years on Long Island after the species reinvaded the island following a volcanic explosion about three centuries ago. This resulted in a change in body size of 1.0–3.2 standard deviations (depending on the putative source of the ebony Myzomela) and of 0.3–0.9 standard deviations for the scarlet-bibbed Myzomela. Note also that in sympatry sexual dimorphism was reduced compared to the same species in allopatry, underlining again the important potential role of balance between intraspecific competition (favouring increased sexual dimorphism) and interspecific competition (favouring decreased sexual dimorphism).

Species introductions can also document the rate at which character displacement can occur, as illustrated by Simberloff *et al.* (2000). The introduction of the small Indian mongoose into islands around the world from an area where it lived in sympatry with congeners has led, on all islands to which it has been introduced, to an increase in male size and in sexual dimorphism in just 100–200 generations (Simberloff *et al.* 2000). The small Indian mongoose shows variation consistent with ecological release from competition with its congeners. In a review Dayan and Simberloff (2005) concluded that '*Character displacement research in the past two decades provides sound statistical support for the hypothesis in a wide variety of taxa, albeit with a phylogenetically skewed representation*'.

Some of the most rapid evolutionary changes in free-living populations have been documented in Darwin's finches. After an extreme drought in 1977, causing a decline of 85% in the population of the medium ground finch on Daphne

Island, strong selection favouring large-beaked individuals resulted in a substantial increase in finch beak size in this population. This rapid evolutionary change was caused by an increase of intraspecific competition, resulting from a sharp decline in food availability because of the drought (Boag and Grant 1981). In 1982 a small population of the large ground finch became established on Daphne Island. The two ground finch species coexisted happily for 22 years, specializing on seeds of different sizes, although having overlapping diets. Large-beaked medium ground finches especially were able to eat the large seeds preferred by the large ground finch, albeit less efficiently. The numbers of large ground finches increased from the founding population of 5 in 1982, to a maximum of 354 individuals in 2003. In 2003 and 2004 two successive very dry seasons caused numbers in both species to decline drastically because of a severe food shortage, but this time there was strong selection against large-beaked medium ground finches, because of interspecific competition with the large ground finch. Grant and Grant (2006) listed four lines of evidence to support the causal role of the large ground finch in the evolutionary change in beak morphology of the medium ground finch. We can thus conclude that interspecific competition for food between the medium and the large ground finch resulted in an evolutionary change that took between 1 and 22 years, depending on how one looks at the data (Fig. 10.8).

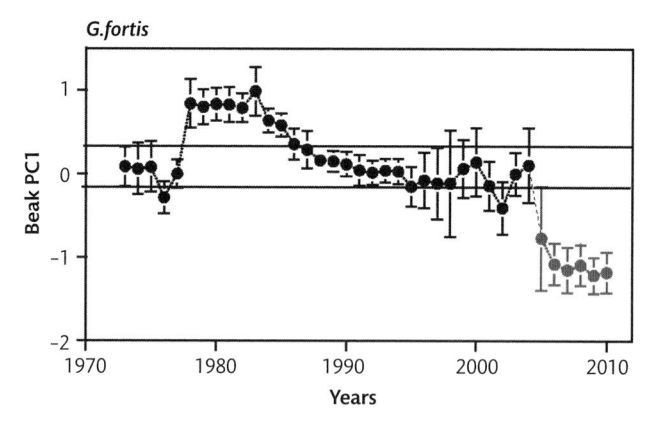

Fig. 10.8 Variation in beak size of the medium ground finch on Daphne Island, Galápagos, between 1973 and 2010. In the absence of large ground finches, during the early years of the study, droughts caused an increase in medium ground finch beak size, resulting from an increase in intraspecific competition. After large ground finches became established on Daphne, beaks of medium ground finches rapidly became smaller during a drought, providing evidence for a powerful effect of interspecific competition. From Grant and Grant (2010) with permission from the National Academy of Sciences of the United States of America.

10.4 How rapidly can interspecific competition cause evolutionary changes in morphology? Experimental data on selection pressures and evolutionary change

So far none of the examples listed has stemmed from controlled experiments. Although there are a large number of experiments documenting the existence of interspecific competition in birds (Chapter 8) and experiments that lead to character shifts in other animals (see for example Gray and Robinson 2002; Gray *et al.* 2005; Pfennig and Martin 2010; Pfennig *et al.* 2007) few permit estimation of the rate of change in controlled conditions.

10.4.1 Sticklebacks

In British Columbia, Canada, threespine sticklebacks are abundant in landlocked lakes. A benthic form (feeding mainly on benthic invertebrates in the littoral zone) and a limnetic form (feeding mainly on plankton) can co-occur in the same lake, although in some lakes only one, usually intermediate form, is found that exploits both food resources. In sympatry the two forms are character displaced (Schluter and McPhail 1992). The two forms have diverged rapidly since the melting of glaciers 15,000 years ago, and not only differ greatly in morphology (spine length, number of armour plates and gill raker number) but are also genetically sufficiently different to be considered two different species (McPhail 1984, 1989; Peichel *et al.* 2001) Nevertheless hybridization is frequent (McPhail 1992). This system of closely related species makes it possible to experimentally study the role of interspecific competition on morphology when food is a limiting resource.

Dolph Schluter measured selection by comparing responses in small artificial ponds when the two species were held alone or together for three months (Schluter 1994). An intermediate solitary species was held either alone or together with a limnetic species from a lake in which both a limnetic and a benthic species lived in sympatry. He found that the presence of the limnetic species altered natural selection on the target species. While in the control pond all intermediate individuals grew at a similar rate over a 90-day period, independently of phenotype, this was not the case in the experimental pond. The more individuals of the intermediate species differed from the limnetic phenotype, the faster they grew. Schluter's experiment demonstrates elegantly that interspecific competition for food can cause selection pressures sufficient to eventually result in speciation. He estimated that, given the selection pressures measured in his experiment, it would take only 500 generations to produce the observed differences between the benthic and limnetic species (Schluter 1994).

10.4.2 Interspecific competition and blue tit body size

I performed a long-term field experiment in which I manipulated the density of great and blue tit populations independently over five-year periods (see Chapter 9). One of the objectives of this experiment was to determine if populations subjected to different intensities of intra- and interspecific competition would evolve to become different compared to control populations. In Chapter 9 I reported that when blue tits were exposed to high densities of both blue and great tit, body mass of both fledglings and adults were lower than in a site in which densities of blue tit (Appendix 3, Table 3) or of great tit (Appendix 3, Table 4) were lower. Both intra- and interspecific competition thus affected,

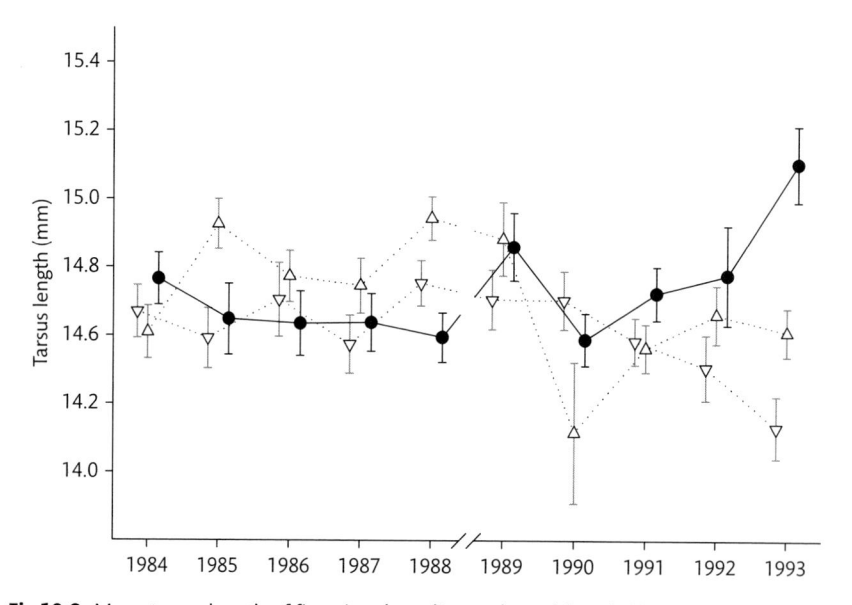

Fig 10.9 Mean tarsus length of first-time breeding male and female blue tits in plots in which density was manipulated. In plots B and C (control plots, open symbols) the nest-box configurations remained unchanged throughout the study (1984–93): both large- and small-holed nest-boxes were present resulting in high densities of both great and blue tit in Plot B; in Plot C small-holed boxes with a few large-holed boxes were present throughout, resulting in a high blue tit and a low great tit density. In Plot T (experimental plot, filled symbols) nest-box configuration was changed from small-holed boxes only in the period 1984–88 (low great tit density, high blue tit density) to a combination of large- and small-holed nest-boxes in the period 1989–93 (high blue tit density, high great tit density). The difference between the two periods is in great tit density, hence in the intensity of interspecific competition. The increase in interspecific competition not only resulted in a decrease in nestling and adult body mass (as reported in Chapter 9) but also in a significant *increase* in blue tit tarsus length. From Dhondt and Adriaensen, unpublished.

among other things, fledgling body mass. I was therefore very surprised to find that together with this reduced body mass the fledglings had longer tarsi and were thus structurally larger. Given that blue tit tarsus length is heritable (as shown from cross-fostering experiments (Dhondt 1982)) this suggested a genetic change in tarsus length resulting from effects of interspecific competition. To verify this, I also looked at measurements of first-time breeders (Fig. 10.9) and found that they too were larger in Plot T (where the intensity of interspecific competition had been increased in 1989) compared to plots B and C (in which the intensity of competition had remained unchanged throughout the study). The genetic basis of these differences is shown in Fig. 10.10 in which I plotted mid-parent against mid-offspring tarsus length. While in 1988 (the final year of the configuration in plot T with low interspecific competition), the measurements of both parents were very similar in the three study populations in Antwerp, five years later tarsus measurements of parents and their offspring in plot T had strongly increased compared to both other plots (Dhondt and Adriaensen, unpublished).

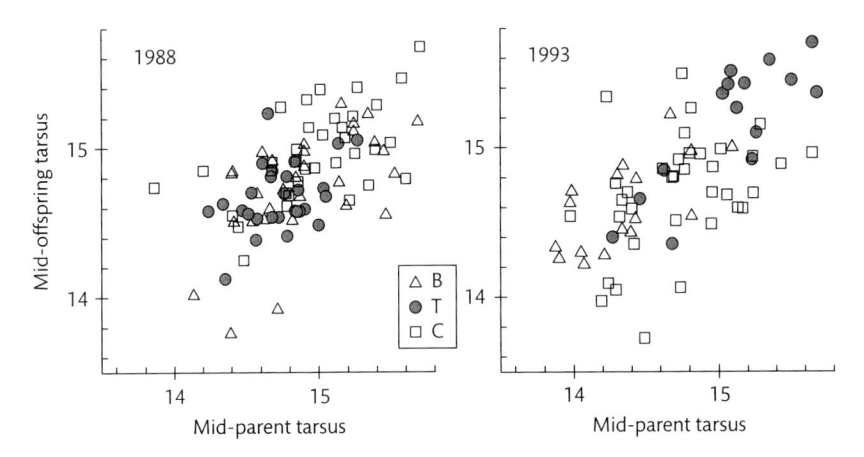

Fig. 10.10 Blue tit mid-parent tarsus length plotted against mid-offspring tarsus length in three Antwerp populations in 1988 (the last year of the five-year pre-treatment period) and 1993 (the last year of the five-year experimental period). Note how in 1988 adult and offspring tarsus length were very similar in the three populations, while in 1993 in plot T (in which the intensity of interspecific competition had been increased for five years) most adults, and therefore their offspring, were clearly larger than those in the two populations in which nest-box configurations had not been changed. Mid-parent/mid-offspring heritability estimates of tarsus length are: in 1988 0.62 ± SE 0.191 in Plot C, 0.37 ± SE 0.142 in Plot B and 0.37 ± SE 0.167 in Plot T; in 1993: 0.41 ± SE 0.158 in Plot C, 0.66 ± SE 0.144 in Plot B and 0.91 ± SE 0.179 in plot T). All P-values <0.04). From Dhondt and Adriaensen, unpublished.

In this open population there was thus a measurable micro-evolutionary change within five years. What was surprising was that blue tits, that are smaller and subordinate to great tit, became larger, thus closer in size to the great tit. The way this seems to have come about is that (1) local recruits gradually increased in size, (2) but immigrant recruits also increased in size (Dhondt, unpublished). The possible mechanism could be that when great tit density was increased this also resulted in an increase of intraspecific competition among blue tits during winter, especially for roosting sites, so that larger individuals, whatever their origin, were better able to recruit into the breeding population.

10.5 Community composition and interspecific competition

Peter and Rosemary Grant pointed out that in the 1960s it was assumed that only palaeontologists could study evolution (Grant and Grant 2008). Today, especially thanks to their groundbreaking work, we know that evolutionary changes can happen rapidly, at least in isolated island populations. Using examples from various taxa I believe I have been able to illustrate that interspecific competition can not only be documented, but that it can also exert strong selection on morphological traits (so as to decrease the intensity of interspecific competition), or can even be shown to result in an evolutionary change (so as to reduce the continued existence of interspecific competition). Field experiments to document this are a real challenge, especially in open populations, since one not only needs to be able to document the existence of interspecific competition, but must also be able to manipulate the intensity of competition for extended periods in replicated populations. Had the Grants not continued their very detailed study of all individuals in populations of different species for more than 30 years, they would never have been able to provide conclusive evidence for the very rapid evolutionary changes first caused by intra- and later caused by interspecific competition, illustrating that 'natural experiments' can also help.

A question that remains, however, is how general effects of interspecific competition can be in open populations. Long-term experiments to attempt to observe such changes in open populations are rare, for many reasons. We can, however, revisit the hotly debated idea from the 1970s and 1980s, that community structure is influenced by interspecific competition. The question asked was whether bird communities (later animal and plant communities in general) were assembled randomly (as supported by Simberloff, see (Simberloff and Boecklen 1981)), or were structured (as supported by Diamond). This latter author claimed that interspecific competition was largely responsible for the non-random assembly of bird communities, and formulated 'assembly rules'. Gotelli and McCabe (2002) summarized the debate and, using 96 data sets, evaluated to what extent Diamond's assembly rules are supported by the data. They concluded:

'Our findings settle a long-standing controversy over community patterns: in most natural communities of plants and nonparasitic animals, there is less species co-occurrence than expected by chance, in accordance with the predictions of Diamond's (1975) assembly rules model.... Our analyses demonstrate that species co-occurrence, measured for a variety of taxa at many different spatial scales, is usually less than expected by chance.'

I feel no need to repeat this discussion, as John Wiens, for example, has written an excellent book about this (Wiens 1989). Note, however, that although Diamond's assembly rules are supported by this meta analysis this does not prove that interspecific competition is the sole factor causing this (see for example Losos *et al.* (2004); Calsbeek and Cox 2010; Wiens *et al.* 1986). This is clearly emphasized by Gotelli and McCabe (2002) when they write:

'Although our results confirm the basic predictions of Diamond's (1975) model, they should not be construed as a definitive test of the model, because some important alternative hypotheses can also produce nonrandomness in the directions we observed. First, some species may exhibit "habitat checkerboards" and segregate because of affinities for nonoverlapping habitats, not because of competitive interactions.... Second, some species may exhibit "historical checkerboards" and co-occur infrequently because of allopatric speciation and other events that reflect biogeographic and evolutionary history.... Moreover, historical, habitat, and ecological checkerboards may not represent mutually exclusive hypotheses if evolutionary change reflects species interactions (the "ghost of competition past").'

Nevertheless, this study, combined with results such as those presented above, supports the idea that interspecific competition does often influence community composition, and is therefore a generally important interaction.

Another scenario by which competing species can coexist is provided by the 'storage effect'. This theory proposes that the reason why similar species with overlapping generations can coexist in a community is that in an environment that changes in time and space no single species can be the best in all conditions. This is explored in great detail by Peter Chesson (Chesson and Warner 1981; Warner and Chesson 1985; Chesson 1994), and supported in the real world by studies of, amongst others, water fleas (Caceres 1997, 1998) and various plant communities (Adler *et al.* 2006; Angert *et al.* 2009).

10.6 Interspecific competition and life-history traits

Phenotypic selection has been documented in hundreds of populations in nature (Kingsolver and Pfennig 2007; Kingsolver *et al.* 2001). In most cases the agent of selection was predation through natural predators or through human exploitation often causing rapid evolutionary changes in life history traits (Allendorf and Hard 2009; Carlson *et al.* 2007; Edeline *et al.* 2007; Edeline *et al.* 2008; Edeline *et al.* 2009; Reznick *et al.* 1990).

Disease epidemics can also result in life-history changes, as illustrated in the Tasmanian devil (Jones *et al.* 2008). In this large marsupial, endemic to Tasmania, the devil facial tumor disease (DFTD) was first observed in 1996. DFTD is an infectious cancer that is consistently fatal, and has caused population declines up to 89%. It has spread to most of Tasmania, thereby causing a massive reduction in population size. DFTD has caused an abrupt transition from iteroparity toward single breeding. This change in life history is associated with almost complete mortality of individuals from this infectious cancer past their first year of adult life. Devils have shown their capacity to respond to this disease-induced increased adult mortality with a sixteen-fold increase in the proportion of individuals exhibiting precocious sexual maturity (Jones *et al.* 2008). Furthermore, the rapid population decline of Tasmanian devils caused by DFTD has also resulted in a significant change in female dispersal (Lachish *et al.* 2011).

Given that intra- and interspecific competition occur frequently one would have expected that competition would frequently have been identified as the selection agent. This seems not to be the case, except in situations where introduced species have invaded new areas. In such cases especially interspecific competition often does not lead to coexistence but to extinction (Cheke and Hume 2008). Such examples, though, do not really document an evolutionary change in a life-history trait, but rather an inability of the subordinate species to adjust sufficiently rapidly when suddenly exposed to intense interspecific competition. Furthermore, native and invasive species can often coexist without evidence of clear interspecific competition because successful invaders preferentially use habitats disturbed by man, which are often avoided by native species (Blackburn *et al.* 2009).

Derek Roff and Stephen Stearns have summarized and discussed heritabilities of life-history traits and selection pressures that cause them to evolve (Roff 2002; Stearns 1992). They list many examples of non-zero heritabilities of life-history traits such as clutch size, survival, and dispersal, and discuss selection pressures operating on them, but it is interesting to note that neither provided documentation that interspecific competition would impact life-history traits.

Competition, whether intra- or interspecific, has an effect on per-capita growth rate. Per-capita growth rate itself, however, is the algebraic sum of fecundity and mortality rates, and of immigration and emigration. It is on these individual rates that selection operates, not on per-capita growth rate itself. When interspecific competition is added to intraspecific competition the equilibrium value of the population is reduced (Chapter 2). Therefore at least one of the four rates affecting per-capita growth rate has changed. One can thus imagine that the optimal investment in reproduction has decreased or that in survival has increased, potentially leading to evolutionary changes in the population's life-history traits caused by changes in trade-offs. I illustrated this in one of my Antwerp blue tit populations: as predation pressure increased optimal clutch sized decreased (Dhondt 2001). Selection pressures through changes in predation pressure can result in rapid evolutionary changes as illustrated, for example, by work on guppies (Gordon *et al.* 2009; Reznick and Ghalambor 2005).

10.6.1 Selection pressures on clutch size

One of David Lack's important contributions to life-history theory was to document that clutch size was under constant selection pressure. The optimal clutch size, that is the clutch size that produced the largest number of offspring surviving to breed, varied between years or habitats depending on food conditions. The way he actually phrased this was not quite how we think today. Thus he wrote: '…*we have advanced the view that, through natural selection, the normal clutch-size corresponds with the average number of young that the parents can raise*' (Lack and Lack 1951). Summarizing his thinking, he wrote in his 1966 book: '*The idea that the clutch-size of each species of bird has been evolved through natural selection is critical to the later argument of this book, so it may be illustrated at the start by the species in which the point can be seen most clearly, the Swift*' (Lack 1966), p. 3. What is interesting about Lack's theory is that it generated mostly studies to determine to what extent clutch size was optimal and the extent to which clutch size was heritable rather than what selection pressures moulded clutch size.

A particularly nice experiment concerning clutch size optimization was carried out by Göran Högstedt, who proposed that in a bird population there is not one optimal clutch size but many and that the optimal strategy for females is to be highly flexible in the number of eggs they lay so as to adaptively vary clutch size with available resources (Högstedt 1980c). He manipulated brood size of magpies either up or down when the young were 0–3 days old, but kept track of the original clutch laid, which varied between five and eight. The result, illustrated in Fig. 10.11 makes two points. First the clutches in which the original clutch size was maintained produced the largest number of fledglings, indicating that females optimized clutch sizes individually. Second, females that laid the largest clutch raised the largest number of young at any manipulation. About 85% of variance in clutch size was attributed to territory quality, while about 15% was due to differences between females, possibly due to hereditary differences.

The reason I elaborate on this is that selection pressures on clutch size must vary with food conditions when raising young. Given that intraspecific, but also interspecific, competition often impacts nesting success and reproductive rate (see Chapter 6 and Chapter 9) we can assume that variation in density, which is related to variation in competition intensity, could impact selection pressure and hence potentially result in evolutionary changes. I am fully aware of the fact that many factors, not just density, exert selection pressures, but the question I need to address here is if there is evidence for variation in selection pressure with density.

Let us return to the species whose populations have been studied in most detail; the tits. Clobert *et al.* (1988) documented that great tit survival is not only related to conspecific density and beech mast but is also influenced by blue tit density. An evolutionary change in survival would require a non-zero heritability of survival, which so far has not been found in birds (Papaix *et al.* 2010).

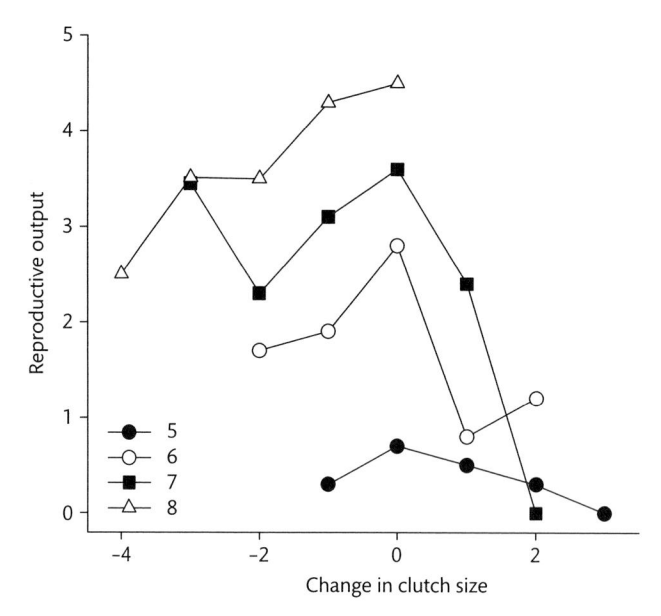

Fig. 10.11 Effects of change in clutch size in Swedish magpies. Unmodified broods (change of 0) were most successful, regardless of the initial clutch size; nests in which the initial clutch was largest were most successful when brood sizes were changed experimentally to be the same size. Thus a brood of 8 fledged 4.5 young when the initial clutch size had been 8 (a change of 0), 2.4 young when the initial clutch size had been 7 (+1), 1.2 young when the initial clutch had been 6 (+2), and no young when the initial clutch had been 5 (+3). After Högstedt (1980b).

Dispersal may have a heritable component, but is extremely difficult to estimate precisely and repeatedly, especially because of the unknown proportion of recruits that move out of the study zone (Winkler *et al.* 2005). This leaves us with clutch size.

Clutch size, in many bird species, including the great tit (Perrins and Jones 1974; Dhondt and Charmantier, unpublished) is heritable. If we want to determine the existence of an evolutionary change (or the potential for a change) resulting from a change in the intensity of interspecific competition, our best bet is to look at clutch size.

Dany Garant and colleagues quantified selection on clutch size, laying date, and egg mass of the Oxford great tits with data from over 5,000 nests using animal models (Garant *et al.* 2007). The part of their results which is relevant in this chapter is the observation that selection pressures on clutch size and egg mass varied with density, meaning that intraspecific competition influenced what selection pressure was observed. At low density, selection on clutch size and egg mass was directional, while at high density selection on clutch size was stabilizing (Fig. 10.12).

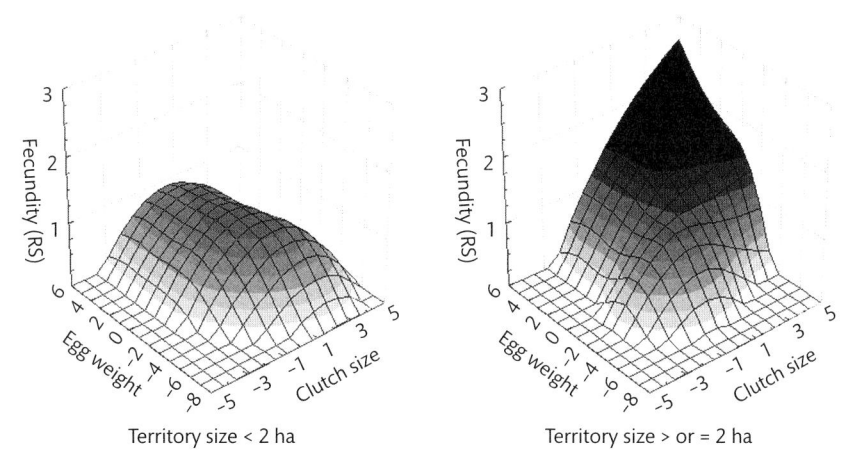

Fig. 10.12 Fecundity fitness surfaces for the combination of egg weight and clutch size depending on the territory size for each female. For illustrative purposes, territories were separated into being either less than or greater than or equal to 2 ha, but in all analyses territory size was treated as a continuous variable. Note that in small territories selection is stabilizing, while in large territories selection is directional. From Garant *et al.* (2007), with permission from John Wiley and Sons.

Given that intraspecific competition (along with other factors such as laying date, habitat quality, etc. (Dhondt *et al.* 1990b)) exerts selection pressures on clutch size, one could think that interspecific competition could have a similar effect. As I performed a long-term manipulation of the breeding density of great and blue tit density (see Chapter 9) it was possible to determine if an increase in interspecific competition by blue tits on great tits would cause a change in selection pressure on great tit clutch size. I found that in a high-quality plot selection on great tit clutch size was directional (favouring larger clutches) at low blue tit density, but became stabilizing at high blue tit density (Fig 10.13), an effect very similar to that obtained by Garant *et al.* (2007). Although selection pressure changed, there was no concomitant evolutionary response because of too much gene flow between study plots differing in quality and because the experiment covered only two to three generations (Dhondt *et al.* 1990b; Postma and van Noordwijk 2005).

10.7 Conclusions

We have now come a full circle. I believe I have presented sufficient evidence that, although interspecific competition can result in the exclusion of one of the com-

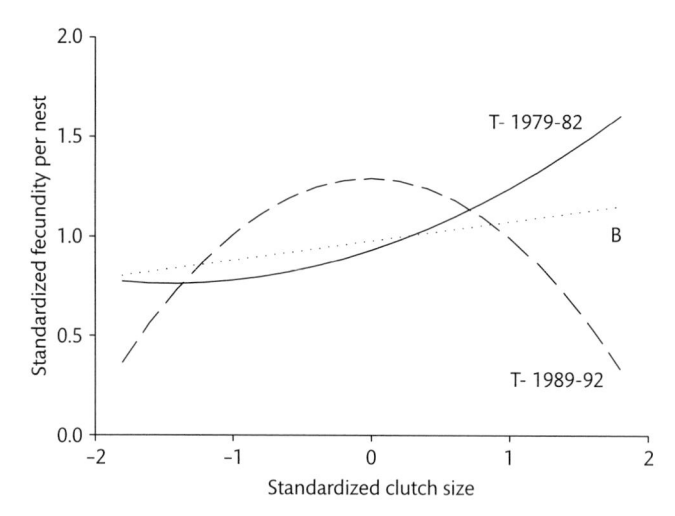

Fig. 10.13 Models illustrating the relationship between clutch size, normalized by year, and fecundity. In the control Plot B (dotted line) there was no significant period effect and hence a single model illustrates directional selection for large clutches. In the experimental plot there was similar selection for large clutches during the four-year period (1979–82) when blue tit density was low (continuous line). During the four-year period (1989–92) when blue tit density was experimentally increased (hatched line) selection changed to stabilizing selection (Dhondt, unpublished).

peting species and would therefore be transient, interspecific competition also occurs regularly and continuously between coexisting bird species. This competition can exert selection pressures on various morphological traits (especially those related to the foraging or habitat niche), but also on life-history traits such as clutch size. The former effect, especially, has generated a lot of research. Most examples, though, provided only indirect evidence for selection pressures from interspecific competition on morphological traits linked to foraging. Given that in the majority of species traits such as structural size or clutch size are heritable, changes in selection pressures should lead to evolutionary changes if gene flow is limited (but see Price et al. 1988). Note, however, that heritabilities of life-history traits are smaller than of morphological traits (Gustafsson 1986; Roff 2002; Stearns 1992). We expect, therefore, that effects of interspecific competition (if any) will be easier to detect in morphological traits than in traits linked to fitness, and among those traits will be easier to detect on fecundity than in traits related to viability.

The evidence, be it still limited, that interspecific competition can cause rapid evolutionary change, provides the mechanism needed to support the hypothesis that interspecific competition can impact community structure by allowing stable coexistence of species with similar requirements. Competing species can coexist if the importance of interspecific competition can be sufficiently reduced. It does not require, though, that interspecific competition be completely abolished (MacArthur 1972). Jared Diamonds's conclusions that interspecific competition impacts community structure is now more widely supported (Gotelli and McCabe 2002; Gotelli *et al.* 2010). Two recent analyses of large-scale data sets separated effects of species interactions from those caused by historical effects, dispersal barriers, and habitat selection. Nicholas Gotelli and colleagues studied factors influencing the large-scale spatial distribution of breeding birds in Denmark at different scales. They concluded that non-random spatial aggregation at relatively large scales would, in part, be caused by an interplay of conspecific attraction and heterospecific avoidance (Gotelli *et al.* 2010). Irby Lovette and Wesley Hochachka studied co-occurrence of parulid warbler species in North America and concluded that species co-occurrence at a given site increased with time from common ancestry, and that this was especially true when the species pair had become differentiated in fundamental ecological and behavioural traits (Lovette and Hochachka 2006).

Coexistence of species is thus the result of both currently ongoing interspecific competition, and of interspecific competition in the past.

11
Concluding thoughts

In this book, I started by listing the arguments against the existence of interspecific competition among animals in general and birds in particular (Section 1.2), before systematically examining the merits of these arguments. I have shown that the necessary conditions for the possible existence of interspecific competition are fulfilled in a very large number of bird populations: resources are often limiting (Chapters 3, 4, and 5); intraspecific competition can often be documented for a diversity of demographic variables (Chapter 6); and resource use between potential competitors often overlaps (Chapter 7). It is worth underlining that many of the results stem from carefully carried out field experiments and from detailed, long-term studies. Finally, in Chapters 8 and 9 I reviewed field experiments used to directly test for interspecific competition in birds described. Based on this direct evidence, I have shown that we can conclude that:

(1) There is sufficient experimental evidence that interspecific competition is currently ongoing among many different bird species in various habitats and regions. Interspecific competition between birds and other taxa can also influence bird numbers, growth, reproduction, nest site choice, foraging site, or habitat use. The experiments provide direct (not circumstantial) evidence for ongoing interspecific competition when differences between experimental and control sites are observed. This happened in the majority of experiments (Table 8.11; Chapter 9).

(2) The criticism that the taxa used to test for the existence of interspecific competition were systematically biased in favour of species in which one expected interspecific competition is not really supported, although clearly a number of experiments were carried out to follow up on earlier observations or experiments. As already mentioned in Chapter 1, John Wiens commented on this and pointed out that one would not normally invest in all the work to carry out a field experiment unless there was at least some reason to believe that interspecific competition would occur (Wiens 1989). I fully agree with his conclusion that counting the frequency with which studies report the existence of interspecific competition is '*of doubtful value*'. That is the reason I was careful not to over-interpret the counts in Chapter 8. What is more

interesting is to carry out experiments that will elucidate the resource for which competition is occurring, the mechanism involved, the demographic parameter impacted, and variability in time and space in the intensity of competition.

(3) There is, however, a different bias that appears when reviewing experimental studies of interspecific competition: a preponderance of studies used cavity-nesting birds. Researchers have studied both secondary cavity nesters in nest-boxes (great and blue tit, pied and collared flycatcher, nuthatch, treecreeper, some woodpeckers, and some tropical cavity nesters), but also birds in natural cavities, a more natural situation. These latter involved both primary cavity nesters (that excavate their own cavities) and secondary cavity nesters (that use holes that already exist). One needs to ask whether the results obtained by studying cavity-nesting species can be generalized. In a large proportion of the experiments that did not involve cavity-nesting birds, evidence for interspecific competition was found (Sections 8.5, 8.6, 8.8, 8.9), as was true also for most, but not all, studies involving cavity nesting bird species. I suggest therefore that the frequent existence of interspecific competition among birds can be generalized to non-cavity-nesting species. Nevertheless the resources over which competition occurs may be somewhat different. Removal experiments (Section 8.5) and experiments using playback (Section 8.8) test the existence of competition for space, primarily mediated through aggressive behaviour; experiments involving insects (Sections 8.6.1 and 8.6.2) and fish (Section 8.6.3) primarily looked at competition for food. Experiments involving cavity nesters also explored the limitation of these two resources, but more often studied nest site or roosting site limitation. I believe we can, therefore, conclude that interspecific competition for food and space occurs generally, regardless of the nest type used, while the importance of nest site limitation in cup-nesting species will require more careful studies, such as the exemplary study of Martin and Martin (2001).

(4) Kathy Martin's application of the nest web perspective to various natural systems (see Sections 5.4 and 8.3) is very informative and promising, because the breadth of interactions is explicitly recognized to include interactions between primary and secondary cavity nesters, and it is attempted to comprehensively study variations in numbers of all vertebrate species using cavities. Combining the nest web approach with detailed studies on the fate of marked individuals, while logistically challenging, would generate fascinating results.

(5) The density of cavity nesters can influence that of open-nesting species negatively (Sections 8.3, 8.4), positively (Section 8.9), or not at all (Sections 8.3, 8.4).

(6) A large proportion of field experiments performed in the last 10–20 years were of high quality: experiments usually had many replicates, many controls, and were carried out for extended periods, some longer than 10 years. The criticism that most field experiments were of poor quality is no longer valid.

(7) At least some of the field experiments tested for the existence of both intra- and interspecific competition and were therefore 'comprehensive'. Thus, it is only partially valid to criticize experiments testing the existence of interspecific competition for failure to be comprehensive. Some experiments, especially when placed in the broader context of a long-term research programme, indeed provide a comprehensive examination of interspecific competition in nature. It remains true though that only few experiments have attempted to study effects of interspecific competition both on population size and on demographic parameters, while identifying which resources were limiting at what time in the annual cycle, and which mechanisms were involved in the competitive interactions. Even fewer attempted to use interspecific competition to evaluate the extent to which relatively rapid evolutionary changes might occur when the intensity of interspecific competition was changed experimentally. It will be a challenge to identify the best study systems in which to carry out such comprehensive studies.

(8) Conspecific and heterospecific attraction may be more important in structuring communities than previously thought (Section 8.9). Birds are able to balance precisely the benefits and the costs of cohabitation with other species that use similar resources, as illustrated by the experiments carried out by Forsman *et al.* (2008).

(9) The conclusion that interspecific competition is currently ongoing in many systems, and the observation that it can induce relatively rapid micro-evolutionary changes (Sections 10.3, 10.4, 10.6) greatly strengthens the validity of many of the studies that have concluded that interspecific competition is important in shaping communities solely through using indirect evidence (allopatric distribution of closely related species; differences in habitat and niche use of closely related species that live in sympatry; size ratios in coexisting species; etc.).

Overall I believe we can conclude that in birds interspecific competition occurs regularly, or at least intermittently, in extant populations. During the annual cycle, resource availability and resource use vary, as does population size. It is therefore obvious that the intensity of competition, or even its existence, will vary throughout the year (Oksanen 1987): competition for nest sites will only occur during the breeding season, competition for roost sites mostly during winter, and competition for food during periods of shortage relative to population size.

Given the evidence presented in this book, it is clear that interspecific competition is a generally important force, and is involved in structuring avian communities. As such, interspecific competition needs to be taken into account when modelling changes in community composition in the face of global change.

Appendix 1

Common and scientific names of bird species mentioned in the text

Common name	Scientific name
American kestrel	*Falco parverius*
American redstart	*Setophaga ruticilla*
aquatic warbler	*Acrocephalus paludicola*
ash-throated flycatcher	*Myiarchus cinerascens*
Audouin's gull	*Larus audouini*
band-rumped storm petrel	*Oceanodroma castro*
barn swallow	*Hirundo rustica*
bearded vulture	*Gypaetus barbatus*
bell miner	*Manorina melanophrys*
Bermuda petrel	*Pterodroma cahow*
black-bellied seed cracker	*Pyrenestes ostrinus*
blackcap	*Sylvia atricapilla*
black-capped chickadee	*Parus (Poecile) atricapillus*
black-capped vireo	*Vireo atricapilla*
black-necked swan	*Cygnus melanocoryphus*
black-throated blue warbler	*Dendroica caerulescens*
blue tit	*Parus (Cyanistes) caeruleus*
Bonelli's eagle	*Aquila fasciata*
booted eagle	*Aquila pennata*
brambling	*Fringilla montifringilla*
bridled titmouse	*Parus (Baeolophus) wollweberi*
brown-headed nuthatch	*Sitta pusilla*
cactus groundfinch	*Geospiza scandens*
cahow	*Pterodroma cahow*
Carolina chickadee	*Parus (Poecile) carolinensis*
chaffinch	*Fringilla coelebs*
chestnut-backed chickadee	*Parus (Poecile) rufescens*
chestnut-capped brush-finch	*Buarremon brunneinuchus*
chestnut-collared longspur	*Calcarius ornatus*
chiffchaff	*Phylloscopus collybita*
coal tit	*Parus (Periparus) ater*

collared flycatcher	*Ficedula albicollis*
common guillemot	*Uria aalge*
cormorant	*Phalacrocorax carbo*
Cory's shearwater	*Calonectris diomedea*
crested tit	*Parus (Lophophanes) cristatus*
downy woodpecker	*Picoides pubescens*
dunnock	*Prunella modularis*
dusky indigobird	*Vidua funerea*
Eastern bluebird	*Sialia sialis*
Eastern screech owl	*Otus asio*
ebony Myzomela	*Myzomela pammelaena*
Eurasian blackbird	*Turdus merula*
Eurasian nuthatch	*Sitta europaea*
Eurasian treecreeper	*Certhia familiaris*
European robin	*Erithacus rubecula*
European starling	*Sturnus vulgaris*
ferruginous pygmy-owl	*Glaucidium brasilianum*
field sparrow	*Spizella pusilla*
garden warbler	*Sylvia borin*
Gila woodpecker	*Melanerpes uropygialis*
goldcrest	*Regulus regulus*
golden-cheeked warbler	*Dendroica chrysoparia*
golden-crowned sparrow	*Zonotrichia atricapilla*
goldeneye	*Bucephala clangula*
goshawk	*Accipiter gentilis*
great crested grebe	*Podiceps cristatus*
great grey skrike	*Lanius excubitor*
great reed warbler	*Acrocephalus arundinaceus*
great spotted woodpecker	*Dendrocopos major*
great tit	*Parus (Parus) major*
grey partridge	*Perdrix perdrix*
Griffon vulture	*Gyps fulvus*
hooded crow	*Corvus cornix*
house finch	*Carpodacus mexicanus*
house sparrow	*Passer domesticus*
house wren	*Troglodytes aedon*
Hume's ground tit	*Pseudopodoces humilis*
jackdaw	*Corvus monedula*
Jambandu indigobird	*Vidua raricola*
jay	*Garrulus glandarius*
junco	*Junco hiemalis*

kittiwake	*Rissa tridactyla*
large ground finch	*Geospiza magnirostris*
least flycatcher	*Empidonax minimus*
lesser sheathbill	*Chionis minor*
little shearwater	*Puffinus assimilis*
long-tailed hermit	*Phaethornis superciliousus*
Lucy's warbler	*Oreothlypis luciae*
Madeiran storm petrel	*Oceanodroma castro*
magpie	*Pica pica*
mallard	*Anas platyrhynchos*
marsh tit	*Parus (Poecile) palustris*
marsh warbler	*Acrocephalus palustris*
medium ground finch	*Geospiza fortis*
moorhen	*Gallinula chloropus*
mountain bluebird	*Sialia currucoides*
mountain chickadee	*Parus (Poecile) gambeli*
moustached warbler	*Acrocephalus melanopogon*
New Zealand robin	*Petroica australis*
northern flicker	*Colaptes auratus*
nuthatch	*Sitta europaea*
orange-crowned warbler	*Vermivora celata*
parasitic jaeger	*Stercocarius parasiticus*
pheasant	*Phasianus colchicus*
pied flycatcher	*Ficedula hypoleuca*
pine warbler	*Dendroica pinus*
Planalto woodcreeper	*Dendrocolaptes platyrostris*
pygmy nuthatch	*Sitta pygmaea*
pygmy owl	*Glaucidium passerinum*
red-backed skrike	*Lanius collurio*
red-bellied woodpecker	*Melanerpes carolinus*
red-breasted nuthatch	*Sitta canadensis*
red-cockaded woodpecker	*Picoides borealis*
red-eyed vireo	*Vireo olivaceus*
red-headed quelea	*Quelea erythrops*
red-headed woodpecker	*Melanerpes erythrocephalus*
redwing	*Turdus iliacus*
reed warbler	*Acrocephalus scirpaceus*
ring-necked parakeet	*Psittacula krameri*
ruby-throated hummingbird	*Archilochus colubris*
rufous-collared sparrow	*Zonotrichia capensis*
saddleback	*Philesturnus carunculatus*

San Clemente loggerhead shrike	*Lanius ludovicianus mearnsi*
scarlet-bibbed Myzomela	*Myzomela sclateri*
scarlet macaw	*Ara macao*
sedge warbler	*Acrocephalus schoenobaenus*
shag	*Phalacrocorax aristotelis*
song sparrow	*Melospiza melodia*
song thrush	*Turdus philomelos*
south polar skua	*Stercorarius maccormicki*
southern black tit	*Parus (Melaniparus) niger*
Spanish imperial eagle	*Aquila adalberti*
sparrowhawk	*Accipiter nisus*
stitchbird	*Notiomystis cincta*
stonechat	*Saxicola torquata axillaris*
stripe-breasted tit	*Parus (Melaniparus) fasciiventer*
stripe-headed brush-finch	*Buarremon torquatus*
swift	*Apus apus*
Tengmalm's owl	*Aegolius funereus*
tree swallow	*Tachycineta bicolor*
tufted duck	*Aythya fuligula*
tufted titmouse	*Parus (Baeolophus) bicolor*
variable indigobird	*Vidua funerea*
varied tit	*Parus (Cyanistes) varius*
Virginia warbler	*Vermivora virginiae*
white-breasted nuthatch	*Sitta carolinensis*
white-tailed tropicbird	*Phaethon lepturus*
willow tit	*Parus (Poecile) montanus*
willow warbler	*Phylloscopus trochilus*
wood duck	*Aix sponsa*
woodpigeon	*Columba palumbus*

For species belonging to the family Paridae (tits, chickadees, titmice) I provide both the old genus name (*Parus*) and the new genus name (*Poecile, Baeolophus,* etc.)

Appendix 2

Common and scientific names of other species mentioned in the text

Other animal species

Common name	Scientific name
brook stickleback	*Culaea inconstans*
flying fox (Madagascar)	*Pteropus rufus*
golden Northern bumble bee	*Bombus fervidus*
guppy	*Poecilia reticulata*
half-black bumble bee	*Bombus vagans*
Mexican spadefoot toad	*Spea multiplicata*
ninespine stickleback	*Pungitius pungitius*
Northern flying squirrel	*Glaucomys sabrinus*
Plains spadefoot toad	*Spea bombifrons*
red squirrel	*Tamasciurus hudsonicus*
roach	*Rutilus rutilus*
small Indian mongoose	*Herpestes javanicus*
Southern flying squirrel	*Glaucomys volans*
stingless bee	*Trigona spp*
Tasmanian devil	*Sarcophilus harrisii*
Texas rat snake	*Elaphe obsoleta*
Verreaux's sifaka	*Propithecus verreauxi*

Plant species

Common name	Scientific name
beech tree	*Fagus sylvatica*
broccoli	*Brassica oleracea*
jewelweed	*Impatiens biflora*
kapok tree	*Ceiba pentandra*
longleaf pine	*Pinus palustris*
ponderosa pine	*Pinus ponderosa*
Saguaro cactus	*Carnegiea gigantea*

Appendix 3
Detailed results of analyses summarized in Chapter 9. All pertain to the Ghent and Antwerp study sites in Belgium

Table 1. Effects of con- and heterospecific density on reproduction of great and blue tit. *Intraspecific effect*: correlation coefficient between species density and same-species demographic variable; *additional interspecific effect*: partial correlation coefficient between density of species A and demographic variable of species B keeping density of species B constant; *combined effect*: Multiple correlation coefficient between densities of both species and a demographic variable. One-tailed probabilities + P<0.05; * P<0.025; ** P<0.005 (Data from Table 1 in Dhondt 1977) Clutch size: mean number of eggs in complete first clutches; nest success: proportion eggs giving rise to fledgling in successful first broods; reproductive rate: mean number of fledgling per female, per season, including first, repeat, and second broods.

Variable	Intraspecific effect	Additional interspecific effect	Combined effect
Great tit clutch size	−0.54*	−0.33	0.61*
Great tit nest success	−0.55*	−0.55*	0.71**
Great tit reproductive rate	−0.69**	−0.64**	0.86**
Blue tit clutch size	−0.54*	−0.38	0.62*
Blue tit nest success	−0.65**	−0.43+	0.73**
Blue tit reproductive rate	−0.46+	−0.34+	0.55

Table 2. Study plots and their nest-box configuration. The numbers in brackets represent the breeding seasons for which data were available in each plot. If nest-box configuration was changed in a plot between years, the plots have the same letter, followed by a number. Thus in Plot T the nest-box configuration differed between three periods, and T is listed as T1, T2, and T3

	Antwerp	Ghent
Large-holed	T1 (1979–83), L (1980–90)	SO1(1965–77;1979–86); HP (1964–2000), ZE (1960–86)***; MA1 (1960–78); GT1 (1972–76; 1983–84)
Small-holed *	C (1979–93), T2(1984–88)**	GT2 (1977–82); SO2 (1978)
Large- and small-holed	B(1979–2001), T3(1989–93)	MA2 (1979–86)

* in some cases, a few large-holed boxes; ** in 1984: a small number of large-holed boxes; *** three years data missing.

Table 3. Five-year means (± SE) of various reproductive traits in blue tit populations experimentally subjected to different intensities of *intraspecific competition*. The experimental plots are MA and T (changed from large-holed boxes in the first period to large-and small-holed boxes in the second period); the control plots are SO (large-holed boxes throughout) and B (large- and small-holed boxes throughout). P-values are two-tailed for main effects, and one-tailed for the plot*period interaction term. At Ghent n = 203 for laying date and clutch-sze; n = 173 for the other variables. At Antwerp n = 309 for male body mass; n = 359 for female body mass; n = 448 for the other variables. Significance of nesting success calculated on arcsine transformed values (after Dhondt and Adriaensen 1999) *: effect as expected if intraspecific competition influences demographic variable. A significant plot*period interaction proves the existence of intraspecific competition. Nestling body mass was measured 15 days after hatching; adult body mass was measured when parents were trapped feeding nestling aged 7-10 days. The difference between male and female response in body mass was shown by a significant three-way sex*plot*period interaction (P=0.003) (Dhondt and Adriaensen 1999).

	Parameter estimates from PROC MIXED					p-values of PROC MIXED		
Ghent	MA 73–77	MA 79–83	SO 73–77	SO 79–83	plot	period	interaction
Blue tit density	low	high	low	low			
Laydate	107.0± 2.1	107.6±2.0	113.4±2.1	110.7±2.1	0.0001	Ns	0.02 *
Clutch size	10.35±0.39	10.58±0.30	10.55±0.40	10.72±0.36	ns	ns	Ns
Nest success	0.73±0.05	0.73±0.04	0.77±0.05	0.89±0.05	0.0001	ns	0.03 *
Productivity	7.54±0.75	7.81±0.69	8.16±0.76	9.40±0.74	0.003	ns	Ns

Antwerp	T 79–83	T 89–93	B 79–83	B 89–93	plot	period	interaction
Blue tit density	Low	High	High	High			
Laydate	108.0±1.7	102.4±1.7	112.1±1.7	104.1±1.6	0.0001	0.02	0.02 *
Clutch size	12.01±0.22	11.13±0.20	11.43±0.19	10.88±0.18	0.01	0.01	Ns
Nest success	0.86±0.02	0.88±0.02	0.85±0.02	0.82±0.02	0.009	ns	Ns
productivity	10.34±0.33	9.81±0.31	9.82±0.29	8.98±0.28	0.004	0.09	Ns
Body mass pulli	11.17±0.12	10.88±0.12	11.17±0.12	11.18±0.12	0.01	ns	0.005 *
Mass ad male	10.76±0.08	10.85±0.08	10.94±0.07	10.87±0.07	ns	ns	Ns
Mass ad female	10.82±0.08	10.50±0.08	10.69±0.07	10.73±0.07	ns	ns	0.001 *

Table 4. Five-year means (± SE) of various reproductive traits in blue tit populations experimentally subjected to different intensities of *interspecific competition*. P-values are two-tailed for main effects, and one-tailed for the plot*period interaction term. The experimental plot is T where during the first period only small-holed boxes were available, while both box types were available during the second period. In Plot B both box types were available throughout. At Antwerp n = 397 for female body mass; n = 348 for male body mass; n = 500 for the other variables. Significance of nesting success was calculated on arcsine transformed values. P-values are two-tailed for plot and for period, but one-tailed for the plot*period interaction term, because we expect that reproduction in plot T will be less successful in the period with a high great tit density (after Dhondt and Adriaensen 1999)

Antwerp	T 84-88	T 89-93	B 84-88	B 89-93	Plot	Period	Interaction
Great tit density	low	High	High	High			
Laydate	113.9±1.8	102.4±1.8	115.3±1.8	104.1±1.8	0.002	0.02	Ns
Clutch size	11.46±0.16	11.12±0.17	10.75±0.16	10.88±0.15	0.003	Ns	0.07
Nest success	0.88±0.02	0.89±0.02	0.86±0.02	0.82±0.02	0.001	Ns	Ns
Productivity	10.12±0.26	9.79±0.28	9.24±0.26	8.97±0.2	0.001	Ns	Ns
Body mass pulli	11.16±0.12	10.88±0.12	11.19±0.12	11.18±0.12	0.003	ns	0.006 *
Mass ad male	11.07±0.07	10.85±0.08	11.02±0.07	10.87±0.07	ns	0.07	Ns
Mass ad female	10.93±0.06	10.50±0.08	10.95±0.07	10.73±0.07	0.01	0.003	0.02 *

Table 5. Correlation between per-capita growth rate and log(N) and results of randomization test indicating the probability that this correlation is caused by chance, using Pollard et al. (1985) randomization test. If P < 0.05 the correlation is not caused by chance. Plots with a [1] are from around Antwerp; those with a [2] are from around Ghent.

Plot	Duration (years)	Nest-box configuration	Correlation coefficient R	P of R caused by chance
Blue tit				
B[1]	15	26 + 32	-0.76	0.013
L[1]	9	32	-0.83	0.057
C[1]	15	26	-0.81	0.015
T[1]	15	32,26,32+26	-0.52	0.019
SO[2]	20	32 (1 yr 26)	-0.55	0.002
GT[2]	13	32,26	-0.67	0.031
ZE[2]	26	32	-0.46	<0.0001
MA[2]	28	32,32+26	-0.42	0.001
HP[2]	35	32	-0.45	<0.0001
Great tit				
B[1]	15	26 + 32	-0.88	0.012
L[1]	9	32	-0.79	0.044
SO[2]	11	32	-0.57	0.044
ZE[2]	26	32	-0.73	<0.0001
MA[2]	28	32,32+26	-0.76	0.001
HP[2]	35	32	-0.58	<0.0001

Table 6. Relationship between great and blue tit per-capita growth rate and conspecific and heterospecific density. Data for a few years are missing in HP and ZE. Plots with a [1] are from around Antwerp; those with a [2] are from around Ghent.

Study site	Period	Intraspecific effect	Interspecific effect	Interaction	F-value	df
Blue tit						
ZE [2]	1960–84	p=0.01	ns	ns	F=7.81	1,22
B [1]	1979–2001	p=0.0005	p=0.002	ns	F=17.67 F=12.76	1,19
MA [21]	1960–86	P=0.02	ns	ns	F=5.73	1,15
HP [2]	1964–1998	P=0.009	P=0.01	P=0.017	F=7.89 F=7.53 F=6.54	1,27
Great tit						
ZE [2]	1960–84	P<0.0001	ns	ns	F=22.96	1,22
B [1]	1979–2001	P=0.0002	ns	ns	F=20.22	1,20
MA [2]	1960–85	P<0.0001	ns	ns	F=37.49	1,25
HP [2]	1964–1998	P=0.048	P=0.059	P=0.09	F=4.32 F=3.89 F=3.07	1,27

Table 7. Great tit breeding density in different experimental configurations (L= large entrance hole; S: small entrance hole; L+S: high densities of both nest box types available). None of the interaction terms in the two-way ANOVA comparing the densities between Plots are significant: Plot MA with Plot SO: plot*period: F= 2.13, df 1,16; P=0.16; Plot T with Plot B: plot*period: F= 0.22, df 1,16; P=0.65.

Area	Periods	Plot	Box types	Mean density	period	Box types	Mean density
Ghent	1973–77	SO	L	1.93± SE 0.23	1979–83	L	1.40± SE 0.41
		MA	L	1.34 ± SE		L+S	1.56 ± SE 0.16
Antwerp	1979–83	T	L	2.34 ± SE 0.257	1989–93	L+S	2.80 ± SE 0.26
		B	L+S	3.15 ± SE 0.433		L+S	3.28 ± SE 0.39

Table 8. Effect of interspecific competition on great tit demographic variables in the Ghent plots (parameter estimates from PROC MIXED). Blue tit density was manipulated by providing only large-holed boxes (density: low), or both small-and large-holed nest-boxes (density: high) during 5-year periods. If interspecific competition occurs we expect a reduced reproduction of great tits during the period/plot in which blue tit numbers were increased, compared to the period in which blue tit numbers were lower; we expected no change in the control plot in which nest box configuration remained the same in both 5-year periods. A significant plot*treatment interaction term indicates interspecific competition. I therefore used one-tailed P-values for the plot*period interaction, but two-tailed P-values for possible plot or period effects (Dhondt, unpublished).

Plot and period	MA 1973-77	SO 1973-77	MA 1979-83	SO 1979-83		p-values of proc mixed		
Blue tit density	low	low	high	low		Plot	Period	Plot*Period
Julian laydate	107.8±1.08	111.9±1.23	107.3±0.70	113.9±0.78		<0.0001	Ns	ns
Clutch size	9.10±0.21	8.25±0.20	8.41±0.19	8.28±0.17		0.01	Ns	**0.04**
Nest success*	0.66±0.037	0.64±0.035	0.65±0.031	0.76±0.037		0.02	Ns	**0.017**
Reproductive rate	5.69±1.38	5.33±1.41	5.96±1.20	7.11±0.96		ns	Ns	ns

*arcsine for calculations.

Table 9. Effect of interspecific competition on great tit demographic variables in the Antwerp plots (parameter estimates from PROC MIXED). Blue tit density was manipulated by providing only large-holed boxes (density: low), or both small- and large-holed nest-boxes (density: high) during 5-year periods. If interspecific competition occurs we expect a reduced reproduction of great tits during the period/plot in which blue tit numbers were increased, compared to the period in which blue tit numbers were lower; we expected no change in the control plot in which nest box configuration remained the same in both 5-year periods. A significant plot*treatment interaction term indicates interspecific competition. I therefore used one-tailed P-values for the plot*period interaction, but two-tailed P-values for possible plot or period effects (Dhondt, unpublished).

Plot and period	T 1979–83	T 1989–93	B 1979–83	B 1989–93	p-values of proc mixed		
Blue tit density	low	high	high	high	Plot	Period	Plot*Period
Julian laydate	111.1±0.56	106.8±0.76	113.4±0.52	107.9±0.59	<0.0001	Ns	ns
Clutch size	9.41±0.17	8.39±0.16	9.18±0.14	8.60±0.14	Ns	0.001	ns (0.076)
Nest success*	0.91±0.012	0.83±0.019	0.88±0.013	0.83±0.017	Ns	Ns	ns
Reproductive rate	8.51±0.38	7.17±0.38	8.09±0.37	7.22±0.37	Ns	Ns	ns
Body mass pulli	17.72±0.08	16.91±0.13	17.71±0.06	17.12±0.11	Ns	Ns	ns
Body condition (pulli)	1.004± 0.004	0.960± 0.0058	1.004± 0.0032	0.992± 0.0056	0.0003	Ns	**0.0002**
Body condition (adult)**	1.0154± 0.0040	0.991± 0.0035	1.019± 0.0036	1.015± 0.0034	0.0002	0.0001	**0.003**

*(arcsine for calculations); ** Adult body condition: no sex effect.

Table 10. Percentage local born great tits in the breeding population at low and high blue tit density (as resulting from nest box manipulations; see Table 9). In brackets the total number of different individuals trapped in each 4-year period. A GENMOD analysis showed no treatment*plot interaction (females: $\chi^2= 0.52$, 1 df, P=0.47; males $\chi^2= 1.09$, 1 df, P=0.29), very significant plot effects (females: $\chi^2= 27.36$, 3 df, P<0.0001; males $\chi^2= 51.54$, 3 df, P<0.0001), and no treatment effect for females ($\chi^2= 0.03$, 1 df, P=0.87) but a strong treatment effect for males ($\chi^2= 7.02$, 1 df, P<0.0001). More great tit males recruit locally in plots with high blue tit density (Dhondt, ms).

Antwerp	T 1980-83	T 1990-93	B B1980-83	B 1990-93
Blue tit density	low	high	high	high
♂	9.8% (82)	18.9% (90)	31.6% (114)	31.3% (112)
♀	7.1% (84)	9.7% (93)	16.1% (112)	17.4% (121)

Ghent	MA 1975-78	MA 1980-83	HP 1975-78	HP 1980-83
Blue tit density	low	high	low	low
GT ♂	5.3% (19)	27.6% (29)	35.0 (60)	46.7% (90)
GT ♀	8.3% (24)	3.2% (31)	27.6% (76)	33.7% (83)

Table 11. Percentage surviving adult great tit dispersing between plots in the Antwerp study at low and high blue tit density. Dispersal is reported for birds breeding in Plots B and T (600 m apart), and recovered breeding in these plots and in plot L, which was adjacent to plot T (after Dhondt, manuscript); both: large and small-holed nest-boxes; large: large-holed boxes only. BT high: blue tit density high; BT low: blue tit density low.

Plot	Treatment	Breeding seasons	Gender	Total adults	Number surviving	% dispersed of surviving
B	both (BT high)	1979–82	♀	142	64	3.1
B	both (BT high)	1989–92	♀	146	62	3.2
B	both (BT high)	1979–82	♂	136	63	0.0
B	both (BT high)	1989–92	♂	126	42	0.0
T	large (BT low)	1979–82	♀	112	64	14.1
T	both (BT high)	1989–92	♀	122	61	0.0
T	large (BT low)	1979–82	♂	108	56	7.1
T	both (BT high)	1989–92	♂	104	41	0.0

Table 12. Percentage local born blue tit in the breeding population at low and high blue (Blue) and great tit (Great) density. In brackets the total number of different individuals trapped in each period. The comparison of the percentages local born male blue tit in plots T and B, between periods A and C, shows a significant effect of intraspecific competition on dispersal (three-way interaction in three-way log linear analysis G^2 = 15.92, 4 df, P=0.003). More blue tits recruit locally when blue tit density is increased, an effect of either the presence of small-holed nest boxes and/or of conspecific attraction. The comparison of the percentages local born male blue tit in plots T and B, between periods B and C, shows a significant effect of interspecific competition on dispersal (three-way interaction in three-way log linear analysis (G^2 = 17.0, 4 df, P=0.002). Fewer blue tits recruit locally when great tit density is increased, an effect of interspecific competition. Percentages local born female blue tit in the breeding population do not differ in either comparison.

Treatment-period	% local born blue tit ♂		Density in plot T		% local born blue tit ♀	
	Plot T	Plot B	Blue	Great	Plot T	Plot B
A 1980–83	1.8% (57)	21.7% (83)	Low	High	2.7% (74)	2.9% (104)
B 1985–88	5.8% (103)	4.5% (110)	High	Low	2% (99)	3.1% (96)
C 1990–93	14.7% (68)	19.6% (92)	High	High	5.7% (87)	6.7% (105)

The very low proportion local recruits in period B, and this in both plots, is the result of several extremely cold winters that caused increased mortality in blue tits.

References

Adams, D. C. and Rohlf, F. J. (2000). Ecological character displacement in *Plethodon*: Biomechanical differences found from a geometric morphometric study. *Proceedings of the National Academy of Sciences of the United States of America* **97**, 4106–11.

Adler, P. B., HilleRisLambers, J., Kyriakidis, P. C., Guan, Q. F., and Levine, J. M. (2006). Climate variability has a stabilizing effect on the coexistence of prairie grasses. *Proceedings of the National Academy of Sciences of the United States of America* **103**, 12793–8.

Adriaensen, F., Dhondt, A. A., Van Dongen, S., Lens, L., and Matthysen, E. (1998). Stabilizing selection on blue tit fledgling mass in the presence of sparrowhawks. *Proceedings of the Royal Society of London Series B-Biological Sciences* **265**, 1011–16.

Aho, T., Kuitunen, M., Suhonen, J., Jantti, A., and Hakkari, T. (1997). Behavioural responses of Eurasian treecreepers, *Certhia familiaris,* to competition with ants. *Animal Behaviour* **54**, 1283–90.

Aho, T., Kuitunen, M., Suhonen, J., Jantti, A., and Hakkari, T. (1999). Reproductive success of Eurasian Treecreepers, *Certhia familiaris*, lower in territories with wood ants. *Ecology* **80**, 998–1007.

Aitken, K. E. H. and Martin, K. (2008). Resource selection plasticity and community responses to experimental reduction of a critical resource. *Ecology* **89**, 971–80.

Aitken, K. E. H. and Martin, K. (2011). Experimental test of nest cavity limitation in mature mixed forest of Central British Columbia, Canada. *Journal of Wildlife Management*, in press.

Alatalo, R. V. (1982). Evidence for Interspecific Competition among European Tits *Parus spp* - a Review. *Annales Zoologici Fennici* **19**, 309–17.

Alatalo, R. V. and Lundberg, A. (1983). Laboratory experiments on habitat separation and foraging efficiency in Marsh and Willow Tits. *Ornis Scandinavica* **14**, 115–22.

Alatalo, R. V. and Moreno, J. (1987). Body size, interspecific interactions, and use of foraging sites in tits (Paridae). *Ecology* **68**, 1773–7.

Alatalo, R. V. and Gustafsson, L. (1988). Genetic component of morphological differentiation in coal tits under competitive release. *Evolution* **42**, 200–3.

Alatalo, R. V., Gustafsson, L., and Lundberg, A. (1986). Interspecific competition and niche changes in tits (*Parus spp*): evaluation of nonexperimental data. *American Naturalist* **127**, 819–34.

Alatalo, R. V., Gustafsson, L., Lundberg, A., and Ulfstrand, S. (1985a). Habitat shift of the Willow Tit *Parus montanus* in the absence of the Marsh Tit *Parus palustris*. *Ornis Scandinavica* **16**, 121–8.

Alatalo, R. V., Gustafsson, L., Linden, M., and Lundberg, A. (1985b). Interspecific competition and niche shifts in tits and the goldcrest: an experiment. *Journal of Animal Ecology* **54**, 977–84.

Alatalo, R. V., Eriksson, D., Gustafsson, L., and Larsson, K. (1987). Exploitation competition influences the use of foraging sites by tits: experimental evidence. *Ecology* **68**, 284–90.

Alerstam, T., Nilsson, S. G., and Ulfstrand, S. (1974). Niche differentiation during winter in woodland birds in southern Sweden and the island of Gotland. *Oikos* **25**, 321–30.

Allendorf, F. W. and Hard, J. J. (2009). Human-induced evolution caused by unnatural selection through harvest of wild animals. *Proceedings of the National Academy of Sciences of the United States of America* **106**, 9987–94.

Andrewartha, H. G. and Birch, L. C. (1954). *The Distribution and Abundance of Animals.* The University of Chicago Press, Chicago.

Angert, A. L., Huxman, T. E., Chesson, P., and Venable, D. L. (2009). Functional tradeoffs determine species coexistence via the storage effect. *Proceedings of the National Academy of Sciences of the United States of America* **106**, 11641–5.

Anonymous (1944). Easter Meeting. *Journal of Animal Ecology* **13**, 176–7.

Aparicio, J. M. (1994). The effect of variation in the laying interval on proximate determination of clutch size in the European Kestrel. *Journal of Avian Biology* **25**, 275–80.

Arcese, P. and Smith, J. N. M. (1988). Effects of population density and supplemental food on reproduction in song sparrows. *Journal of Animal Ecology* **57**, 119–36.

Arcese, P., Smith, J. N. M., Hochachka, W. M., Rogers, C. M., and Ludwig, D. (1992). Stability, regulation, and the determination of abundance in an insular song sparrow population. *Ecology* **73**, 805–22.

Ardia, D. (2002). Energetic consequences of sex-related habitat segregation in wintering American kestrels (*Falco sparverius*). *Canadian Journal of Zoology - Revue Canadienne de Zoologie* **80**, 516–23.

Ardia, D. R. and Bildstein, K. L. (1997). Sex-related differences in habitat selection in wintering American kestrels, *Falco sparverius*. *Animal Behaviour* **53**, 1305–11.

Armstrong, D. P., Castro, I., and Griffiths, R. (2007). Using adaptive management to determine requirements of re-introduced populations: the case of the New Zealand hihi. *Journal of Applied Ecology* **44**, 953–62.

Armstrong, D. P., Davidson, R. S., Dimond, W. J., et al. (2002). Population dynamics of reintroduced forest birds on New Zealand islands. *Journal of Biogeography* **29**, 609–21.

Askenmo, C., von Brömssen, A., Ekman, J., and Jansson, C. (1977). Impact of some wintering birds on spider abundance in spruce. *Oikos* **28**, 90–4.

Ayala, F. J. (1969). Experimental invalidation of principle of competitive exclusion. *Nature* **224**, 1076–9.

Ayala, F. J. (1970). Invalidation of principle of competitive exclusion defended. *Nature* **227**, 89–90.

Ayala, F. J., Gilpin, M. E., and Ehrenfeld, J. G. (1973). Competition between species: theoretical models and experimental tests. *Theoretical Population Biology* **4**, 331–56.

Baeyens, G. (1979). Description of the social behavior of the Magpie (*Pica pica*). *Ardea* **67**, 28–41.

Banks, W. E., d'Errico, F., Peterson, A. T., Kageyama, M., Sima, A., and Sanchez-Goni, M. F. (2008). Neanderthal extinction by competitive exclusion. *PloS ONE* **3(12)**, e3972.

Beauchamp, G. (1998). The effect of group size on mean food intake rate in birds. *Biological Reviews* **73**, 449–72.

Bennett, A. T. D., Lloyd, M. H., and Cuthill, I. C. (1996). Ant-derived formic acid can be toxic for birds. *Chemoecology* **7**, 189–90.

Bennett, P. A. and Owens, I. P. F. (2002). *Evolutionary Ecology of Birds.* Oxford University Press, Oxford.

Bennett, W. A. (1990). Scale of investigation and the detection of competition: an example from the house sparrow and house finch introductions in North-America. *American Naturalist* **135**, 725–47.

Berndt, R. (1941). Über die Einwirkung der strengen Winter 1928/29 und 1939/40 und den Einfluss der Winterfütterung auf den Brutbestand der Meisen. *Die Gefiederte Welt* **70**, 59–65, 80–81, 91–92, 101–03, 117–18.

Berndt, R. and Frantzen, M. (1964). Vom Einfluss des strengen Winters 1962/63 auf dem Brutbestand der Hohlenbruter bei Braunschweig. *Ornithologische Mitteilungen* **16**, 126–30.

Best, L. B. (1978). Field Sparrow reproductive success and nesting ecology. *Auk* **95**, 9–22.

Bijnens, L. and Dhondt, A. A. (1984). Vocalizations in a Belgian blue tit *Parus c. caeruleus* population. *Gerfaut - Giervalk* **74**, 243–69.

Blackburn, T. M., Lockwood, J. L., and Cassey, P. (2009). *Avian Invasions*. Oxford University Press, Oxford.

Blake, J. G. and Loiselle, B. A. (2008). Estimates of apparent survival rates for forest birds in eastern Ecuador. *Biotropica* **40**, 485–93.

Blanc, L. A. and Walters, J. R. (2008). Cavity excavation and enlargement as mechanisms for indirect interactions in an avian community. *Ecology* **89**, 506–14.

Blondel, J., Thomas, D. W., Charmantier, A., Perret, P., Bourgault, P., and Lambrechts, M. M. (2006). A thirty-year study of phenotypic and genetic variation of blue tits in Mediterranean habitat mosaics. *Bioscience* **56**, 661–73.

Boag, P. T. and Grant, P. R. (1981). Intense natural selection in a population of Darwin finches (Geospizinae) in the Galapagos. *Science* **214**, 82–5.

Bock, C. E. and Fleck, D. C. (1995). Avian response to nest box addition in 2 forests of the Colorado Front Range. *Journal of Field Ornithology* **66**, 352–62.

Bock, C. E., Cruz, A., Grant, M. C., Aid, C. S., and Strong, T. R. (1992). Field experimental evidence for diffuse competition among Southwestern riparian birds. *American Naturalist* **140**, 815–28.

Bolger, D. T., Patten, M. A., and Bostock, D. C. (2005). Avian reproductive failure in response to an extreme climatic event. *Oecologia* **142**, 398–406.

Bolnick, D. I., Ingram, T., Stutz, W. E., Snowberg, L. K., Lau, O. L., and Paull, J. S. (2010). Ecological release from interspecific competition leads to decoupled changes in population and individual niche width. *Proceedings of the Royal Society B-Biological Sciences* **277**, 1789–97.

Bolnick, D. I., Svanback, R., Fordyce, J. A., *et al.* (2003). The ecology of individuals: Incidence and implications of individual specialization. *American Naturalist* **161**, 1–28.

Bolton, M., Medeiros, R., Hothersall, B., and Campos, A. (2004). The use of artificial breeding chambers as a conservation measure for cavity-nesting procellariiform seabirds: a case study of the Madeiran storm petrel (*Oceanodroma castro*). *Biological Conservation* **116**, 73–80.

Bose, M. and Sarrazin, F. (2007). Competitive behaviour and feeding rate in a reintroduced population of Griffon Vultures *Gyps fulvus*. *Ibis* **149**, 490–501.

Both, C. (1998). Density dependence of clutch size: habitat heterogeneity or individual adjustment? *Journal of Animal Ecology* **67**, 659–66.

Both, C. (2000). Density dependence of avian clutch size in resident and migrant species: is there a constraint on the predictability of competitor density? *Journal of Avian Biology* **31**, 412–17.

Both, C. and Visser, M. E. (2000). Breeding territory size affects fitness: an experimental study on competition at the individual level. *Journal of Animal Ecology* **69**, 1021–30.

Boyle, W. A., Ganong, C. N., Clark, D. B., and Hast, M. A. (2008). Density, distribution, and attributes of tree cavities in an old-growth tropical rain forest. *Biotropica* **40**, 241–5.

Brawn, J. D. and Balda, R. P. (1988). Population biology of cavity nesters in Northern Arizona: do nest sites limit breeding densities. *Condor* **90**, 61–71.

Brawn, J. D., Boecklen, W. J., and Balda, R. P. (1987). Investigations of density interactions among breeding birds in ponderosa pine forests: correlative and experimental evidence. *Oecologia* **72**, 348–57.

Brenowitz, G. L. (1978). Gila woodpecker agonistic behavior. *Auk* **95**, 49–58.

Bried, J. and Jouventin, P. (1998). Why do lesser Sheathbills *Chionis minor* switch territory? *Journal of Avian Biology* **29**, 257–65.

Brightsmith, D. J. (2005). Competition, predation and nest niche shifts among tropical cavity nesters: ecological evidence. *Journal of Avian Biology* **36**, 74–83.

Brittingham, M. C. and Temple, S. A. (1988). Impacts of supplemental feeding on survival rates of black-capped chickadees. *Ecology* **69**, 581–9.

Broggi, J. and Brotons, L. (2001). Coal Tit fat-storing patterns during the non-breeding season: the role of residence status. *Journal of Avian Biology* **32**, 333–7.

Bronstein, J. L. (1994). Conditional outcomes in mutualistic interactions. *Trends in Ecology and Evolution* **9**, 214–17.

Brook, B. W. and Bradshaw, C. J. A. (2006). Strength of evidence for density dependence in abundance time series of 1198 species. *Ecology* **87**, 1445–51.

Brown, J. H. and Munger, J. C. (1985). Experimental manipulation of a desert rodent community: food addition and species removal. *Ecology* **66**, 1545–63.

Brown, J. L. (1969). Territorial behavior and population regulation in birds. *Wilson Bulletin* **81**, 293–329.

Brown, W. L. and Wilson, E. O. (1956). Character displacement. *Systematic Zoology* **5**, 49–64.

Brush, T. (1983). Cavity use by secondary cavity-nesting birds and response to manipulations. *Condor* **85**, 461–6.

Byholm, P. and Kekkonen, M. (2008). Food regulates reproduction differently in different habitats: experimental evidence in the goshawk. *Ecology* **89**, 1696–702.

Caceres, C. E. (1997). Temporal variation, dormancy, and coexistence: A field test of the storage effect. *Proceedings of the National Academy of Sciences of the United States of America* **94**, 9171–5.

Caceres, C. E. (1998). Seasonal dynamics and interspecific competition in Oneida Lake Daphnia. *Oecologia* **115**, 233–44.

Cadena, C. D. and Loiselle, B. A. (2007). Limits to elevational distributions in two species of emberizine finches: disentangling the role of interspecific competition, autoecology, and geographic variation in the environment. *Ecography* **30**, 491–504.

Calsbeek, R. and Cox, R. M. (2010). Experimentally assessing the relative importance of predation and competition as agents of selection. *Nature* **465**, 613–16.

Camprodon, J., Salvanya, J., and Soler-Zurita, J. (2008). The abundance and suitability of tree cavities and their impact on hole-nesting bird populations in beech forests of NE Iberian Peninsula. *Acta Ornithologica* **43**, 17–31.

Cardiff, S. W. and Dittmann, D. L. (2002). Ash-throated flycatcher (*Myiarchus cinerascens*). In A. Poole, ed. *The Birds of North America Online* Vol. 664. Cornell Lab of Ornithology, Ithaca.

Carlson, A., Sandstrom, U., and Olsson, K. (1998). Availability and use of natural tree holes by cavity nesting birds in a Swedish deciduous forest. *Ardea* **86**, 109–19.

Carlson, S. M., Edeline, E., Vollestad, L. A., *et al.* (2007). Four decades of opposing natural and human-induced artificial selection acting on Windermere pike (*Esox lucius*). *Ecology Letters* **10**, 512–21.

Carpenter, F. L. (1979). Competition between hummingbirds and insects for nectar. *American Zoologist* **19**, 1105–14.

Catzeflis, F. (1979). Etude qualitative et quantitative de l'avifaune de la pessière jurassienne du Chalet à Roch, Vaud. *Nos Oiseaux* **35**, 75–84.

Cawthorne, R. A. and Marchant, J. H. (1980). The effects of the 1978–79 winter on British bird populations. *Bird Study* **27**, 163–72.

Cederholm, G. and Ekman, J. (1976). Removal experiment on Crested Tit *Parus cristatus* and Willow Tit *P. montanus* in the breeding season. *Ornis Scandinavica* **7**, 207–13.

Chalfoun, A. D. and Martin, T. E. (2009). Habitat structure mediates predation risk for sedentary prey: experimental tests of alternative hypotheses. *Journal of Animal Ecology* **78**, 497–503.

Chamberlain, D. E., Gosler, A. G., and Glue, D. E. (2007). Effects of the winter beechmast crop on bird occurrence in British gardens. *Bird Study* **54**, 120–6.

Cheke, A. and Hume, J. (2008). *Lost Land of the Dodo: An Ecological History of Mauritius, Réunion and Rodrigues*. T & A.D. Poyser, London.

Chesson, P. (1994). Multispecies competition in variable environments. *Theoretical Population Biology* **45**, 227–76.

Chesson, P. L. and Warner, R. R. (1981). Environmental variability promotes coexistence in lottery competitive systems. *American Naturalist* **117**, 923–43.

Chitty, D. (1967). Reviews - What Regulates Bird Populations. *Ecology* **48**, 698–701.

Christman, B. J. (2001). Influences on the social organization of two Southwestern titmice. Ph.D., Cornell University, Ithaca, NY.

Christman, B. J. and Dhondt, A. A. (1997). Nest predation in black-capped chickadees: How safe are cavity nests? *Auk* **114**, 769–73.

Christman, B. J. and Gaulin, S. J. C. (1998). Unambiguous evidence of helping at the nest in Bridled Titmice. *Wilson Bulletin* **110**, 567–9.

Cimprich, D. A. and Grubb, T. C. (1994). Consequences for Carolina chickadees of foraging with tufted titmice in winter. *Ecology* **75**, 1615–25.

Clabaut, C., Herrel, A., Sanger, T. J., Smith, T. B., and Abzhanov, A. (2009). Development of beak polymorphism in the African seedcracker, *Pyrenestes ostrinus*. *Evolution and Development* **11**, 636–46.

Clobert, J., Perrins, C. M., McCleery, R. H., and Gosler, A. G. (1988). Survival rate in the great tit *Parus major* in relation to sex, age, and immigration status. *Journal of Animal Ecology* **57**, 287–306.

Cockle, K. L. and Bodrati, A. A. (2009). Nesting of the Planalto Woodcreeper (*Dendrocolaptes platyrostris*). *Wilson Journal of Ornithology* **121**, 789–95.

Cockle, K. L., Martin, K., and Drever, M. C. (2010). Supply of tree-holes limits nest density of cavity-nesting birds in primary and logged subtropical Atlantic forest. *Biological Conservation* **143**, 2851–7.

Connell, J. H. (1980). Diversity and the coevolution of competitors, or the Ghost of Competition Past. *Oikos* **35**, 131–8.

Connell, J. H. (1983). On the prevalence and relative importance of interspecific competition: evidence from field experiments. *American Naturalist* **122**, 661–96.

Connor, E. F. and Simberloff, D. (1979). The assembly of species communities: chance or competition. *Ecology* **60**, 1132–40.

Cooper, C. B., Hochachka, W. M., and Dhondt, A. A. (2007). Contrasting natural experiments confirm competition between house finches and house sparrows. *Ecology* **88**, 864–70.

Cornelius, C., Cockle, K., Politi, N., *et al.* (2008). Cavity-nesting birds in Neotropical forests: Cavities as a potentially limiting resource. *Ornitologia Neotropical* **19**, 253–68.

Cramp, S. (1963). Movements of tits in Europe in 1959 and after. *British Birds* **56**, 237–63.

Cramp, S., Pettet, A. and Sharrock, J. T. R. (1960). The irruption of tits in autumn 1957. *British Birds* **53**, 49–77, 99–117, 176–92.

Cresswell, W. (2003). Testing the mass-dependent predation hypothesis: in European blackbirds poor foragers have higher overwinter body reserves. *Animal Behaviour* **65**, 1035–44.

Cucco, M., Guasco, B., Malacarne, G., and Ottonelli, R. (2006). Effects of beta-carotene supplementation on chick growth, immune status and behaviour in the grey partridge, *Perdix perdix. Behavioural Processes* **73**, 325–32.

Danchin, E., Boulinier, T., and Massot, M. (1998). Conspecific reproductive success and breeding habitat selection: implications for the study of coloniality. *Ecology* **79**, 2415–28.

Danchin, E., Giraldeau, L. A., Valone, T. J., and Wagner, R. H. (2004). Public information: From nosy neighbors to cultural evolution. *Science* **305**, 487–91.

Davis, J. (1973). Habitat preferences and competition of wintering juncos and golden-crowned sparrows. *Ecology* **54**, 174–80.

Davis, S. E., Nager, R. G., and Furness, R. W. (2005). Food availability affects adult survival as well as breeding success of parasitic jaegers. *Ecology* **86**, 1047–56.

Dayan, T. and Simberloff, D. (2005). Ecological and community-wide character displacement: the next generation. *Ecology Letters* **8**, 875–94.

De Laet, J. F. and Dhondt, A. A. (1989). Weight loss of the female during the 1st brood as a factor influencing 2nd brood initiation in Great Tits *Parus major* and Blue Tits *Parus caeruleus. Ibis* **131**, 281–9.

De Laet, J. F., Dhondt, A. A. and De Boever, J. G. (1985). Circannual plasma androgen levels in free-living male great tits (*Parus major major* L.). *General and Comparative Endocrinology* **59**, 277–86.

De Neve, L., Soler, J. J., Soler, M., and Perez-Contreras, T. (2004). Nest size predicts the effect of food supplementation to magpie nestlings on their immunocompetence: an experimental test of nest size indicating parental ability. *Behavioral Ecology* **15**, 1031–6.

Desrochers, A., Hannon, S. J., and Nordin, K. E. (1988). Winter survival and territory acquisition in a northern population of Black-capped Chickadees. *Auk* **105**, 727–36.

Dhondt, A. A. (1966). A method to establish boundaries of bird territories. *Gerfaut - Giervalk* **56**, 404–8.

Dhondt, A. A. (1970). Sex ratio of nestling great tits. *Bird Study* **17**, 282–6.

Dhondt, A. A. (1971). The regulation of numbers in Belgian populations of Great Tits. In J. Den Boer and G. R. Gradwell, eds. *Proceedings of the Advanced Study Institute on 'Dynamics of Numbers in Populations' Oosterbeek 1970*, pp. 532–47. PUDOC, Wageningen.

Dhondt, A. A. (1977). Interspecific competition between Great and Blue Tit. *Nature* **268**, 521–3.

Dhondt, A. A. (1979). Summer dispersal and survival of juvenile Great Tits in southern Sweden. *Oecologia* **42**, 139–57.

Dhondt, A. A. (1982). Heritability of Blue Tit tarsus length from normal and cross-fostered broods. *Evolution* **36**, 418–19.

Dhondt, A. A. (1985a). The effect of interspecific competition on numbers in bird populations. *Acta XVIII Congressus Internationalis Ornithologicus*, Moscow, 792–6.

Dhondt, A. A. (1985b). Wildlife population dynamics and regulation: the implications of non linear models. *XVIIth Congress of the International Union of Game Biologists*, Ministry of Agriculture, Brussels, 163–72.

Dhondt, A. A. (1987a). Polygynous Blue Tits and monogamous Great Tits - Does the polygyny-threshold model hold. *American Naturalist* **129**, 213–20.

Dhondt, A. A. (1987b). Resultaten van 25 jaar mezenonderzoek in Vlaanderen. Deel 1 - De studie van aantalsschommelingen bij mezen. *Oriolus* **53**, 101–09.

Dhondt, A. A. (1988). Carrying-capacity - a confusing concept. *Acta Oecologica-Oecologia Generalis* **9**, 337–46.

Dhondt, A. A. (1989a). Blue tit. In I. Newton, ed. *Lifetime Reproduction in Birds*, pp. 15–33. Academic Press, London.

Dhondt, A. A. (1989b). Ecological and evolutionary effects of interspecific competition in tits. *Wilson Bulletin* **101**, 198–216.

Dhondt, A. A. (2001). Trade-offs between reproduction and survival in Tits. *Ardea* **89**, 155–66.

Dhondt, A. A. (2007). What drives differences between North American and Eurasian tit studies? In K. A. Otter, ed. *Ecology and Behavior of Chickadees and Titmice: An Integrated Approach.*, pp. 299–310. Oxford University Press, Oxford.

Dhondt, A. A. (2010a). Broodedness, not latitude, affects the response of reproductive timing of birds to food supplementation. *Journal of Ornithology* **151**, 955–7.

Dhondt, A. A. (2010b). Effects of competition on great and blue tit reproduction: intensity and importance in relation to habitat quality. *Journal of Animal Ecology* **79**, 257–65.

Dhondt, A. A. Effect of heterospecific density on dispersal in a resident species: an experimental study. Submitted.

Dhondt, A. A. and Hublé, J. (1968). Fledging date and sex in relation to dispersal in young Great Tits. *Bird Study* **15**, 127–234.

Dhondt, A. A. and Eyckerman, R. (1980). Competition between the great tit and the blue tit outside the breeding season in field experiments. *Ecology* **61**, 1291–6.

Dhondt, A. A. and Olaerts, G. (1981). Variations in survival and dispersal with ringing date as shown by recoveries of Belgian Great Tits *Parus major*. *Ibis* **123**, 96–8.

Dhondt, A. A. and Schillemans, J. (1983). Reproductive Success of the Great Tit in Relation to its Territorial Status. *Animal Behaviour* **31**, 902–12.

Dhondt, A. A. and Adriaensen, F. (1999). Experiments on competition between Great and Blue Tit: effects on Blue Tit reproductive success and population processes. *Ostrich* **70**, 39–48.

Dhondt, A. A. and Hochachka, W. M. (2001). Variations in calcium use by birds during the breeding season. *Condor* **103**, 592–8.

Dhondt, A. A., Eyckerman, R., and Hublé, J. (1979). Will great tits become little tits? *Biological Journal of the Linnean Society* **11**, 289–94.

Dhondt, A. A., Schillemans, J., and Delaet, J. (1982). Blue Tit territories in populations at different density levels. *Ardea* **70**, 185–8.

Dhondt, A. A., Kempenaers, B., and De Laet, J. (1991). Protected winter roosting sites as a limiting resource for Blue Tits. *Acta XX Congressus Internationalis Ornithologici,* New Zealand Ornithological Congress Trust Board, Christchurch, New Zealand, 1436–43.

Dhondt, A. A., Kempenaers, B., and Adriaensen, F. (1992). Density-dependent clutch size caused by habitat heterogeneity. *Journal of Animal Ecology* **61**, 643–8.

Dhondt, A. A., Kast, T. L., and Allen, P. E. (2002). Geographical differences in seasonal clutch size variation in multi-brooded bird species. *Ibis* **144**, 646–51.

Dhondt, A. A., Blondel, J., and Perret, P. (2010). Why do Corsican Blue Tits *Cyanistes caeruleus ogliastrae* not use nest boxes for roosting?. *Journal of Ornithology* **151**, 95–101.

Dhondt, A. A., Matthysen, E., Adriaensen, F., and Lambrechts, M. M. (1990a). Population dynamics and regulation of a high density Blue Tit population. In J. Blondel, A. Gosler, J.-D. Lebreton, and R. McCleery, eds. *Population Biology of Passerine Birds*, pp. 39–53. Springer Verlag, Berlin, Heidelberg.

Dhondt, A. A., Adriaensen, F., Matthysen, E., and Kempenaers, B. (1990b). Nonadaptive clutch sizes in tits. *Nature* **348**, 723–5.

Diamond, J. M. (1974). Colonization of exploded volcanic islands by birds: supertramp strategy. *Science* **184**, 803–6.

Diamond, J. M. (1975). Assembly of species communities. In M. L. Cody and J. M. Diamond, eds. *Ecology and Evolution of Communities*, pp. 342–444. Harvard University Press, Cambridge, MA.

Diamond, J. M., Pimm, S. L., Gilpin, M. E., and Lecroy, M. (1989). Rapid evolution of character displacement in myzomelid honeyeaters. *American Naturalist* **134**, 675–708.

Dingemanse, N. J., Both, C., van Noordwijk, A. J., Rutten, A. L., and Drent, P. J. (2003). Natal dispersal and personalities in great tits (*Parus major*). *Proceedings of the Royal Society of London Series B-Biological Sciences* **270**, 741–7.

Dixon, K. L. (1954). Some ecological relations of chickadees and titmice in Central California. *Condor* **56**, 113–24.

Doherty, P. F. and Grubb, T. C., Jr (2002). Survivorship of permanent-resident birds in a fragmented forested landscape. *Ecology* **83**, 844–57.

Doherty, P. F. and Grubb, T. C. (2003). Relationship of nutritional condition of permanent-resident woodland birds with woodlot area, supplemental food, and snow cover. *Auk* **120**, 331–6.

Dolby, A. S. and Grubb, T. C. (1998). Benefits to satellite members in mixed-species foraging groups: an experimental analysis. *Animal Behaviour* **56**, 501–9.

Dolby, A. S. and Grubb, T. C. (2000). Social context affects risk taking by a satellite species in a mixed-species foraging group. *Behavioral Ecology* **11**, 110–14.

Doligez, B., Cadet, C., Danchin, E., and Boulinier, T. (2003). When to use public information for breeding habitat selection? The role of environmental predictability and density dependence. *Animal Behaviour* **66**, 973–88.

Doutrelant, C. and Lambrechts, M. M. (2001). Macrogeographic variation in song: a test of competition and habitat effects in blue tits. *Ethology* **107**, 533–44.

Doutrelant, C., Leitao, A., Otter, K., and Lambrechts, M. M. (2000a). Effect of blue tit song syntax on great tit territorial responsiveness: an experimental test of the character shift hypothesis. *Behavioral Ecology and Sociobiology* **48**, 119–24.

Doutrelant, C., Blondel, J., Perret, P., and Lambrechts, M. M. (2000b). Blue Tit song repertoire size, male quality and interspecific competition. *Journal of Avian Biology* **31**, 360–6.

Drent, P. J. (1984). Mortality and dispersal in summer and its consequences for the density of Great Tits *Parus major* at the onset of autumn. *Ardea* **72**, 127–62.

Drent, P. J. (1987). The importance of nestboxes for territory settlement, survival and density of the Great Tit. *Ardea* **75**, 59–71.

Dunn, E. (1977). Predation by weasels (*Mustela nivalis*) on breeding tits (*Parus spp*) in relation to density of tits and rodents. *Journal of Animal Ecology* **46**, 633–52.

East, M. L. and Perrins, C. M. (1988). The effect of nestboxes on breeding populations of birds in broadleaved temperate woodlands. *Ibis* **130**, 393–401.

Edeline, E., Ben Ari, T., Vollestad, L. A., *et al.* (2008). Antagonistic selection from predators and pathogens alters food-web structure. *Proceedings of the National Academy of Sciences of the United States of America* **105**, 19792–6.

Edeline, E., Le Rouzic, A., Winfield, I. J., *et al.* (2009). Harvest-induced disruptive selection increases variance in fitness-related traits. *Proceedings of the Royal Society B-Biological Sciences* **276**, 4163–71.

Edeline, E., Carlson, S. M., Stige, L. C., *et al.* (2007). Trait changes in a harvested population are driven by a dynamic tug-of-war between natural and harvest selection. *Proceedings of the National Academy of Sciences of the United States of America* **104**, 15799–804.

Egan, E. S. and Brittingham, M. C. (1994). Winter survival rates of a southern population of Black-capped Chickadees. *Wilson Bulletin* **106**, 514–21.

Einarsen, A. S. (1945a). Some factors affecting ring-necked pheasant population density. II. *Murrelet* **26**, 39–44.

Einarsen, A. S. (1945b). Some factors affecting ring-necked pheasant population density. *Murrelet* **26**, 2–7.

Ekman, J. (1979). Non-territorial Willow Tits *Parus montanus* in late summer and early autumn. *Ornis Scandinavica* **10**, 262–7.

Ekman, J. (1984a). Stability and persistence of an age-structured avian population in a seasonal environment. *Journal of Animal Ecology* **53**, 135–46.

Ekman, J. (1984b). Density-dependent seasonal mortality and population fluctuations of the temperate zone willow tit (*Parus montanus*). *Journal of Animal Ecology* **53**, 119–34.

Ekman, J. (1986). Tree use and predator vulnerability of wintering passerines. *Ornis Scandinavica* **17**, 261–7.

Ekman, J. (1989). Ecology of non-breeding social systems of *Parus*. *Wilson Bulletin* **101**, 263–88.

Ekman, J., Cederholm, G., and Askenmo, C. (1981). Spacing and survival in winter groups of willow tit *Parus montanus* and crested tit *P. cristatus*: a removal study. *Journal of Animal Ecology* **50**, 1–9.

Elmberg, J., Pöysä, H., Sjöberg, K., and Nummi, P. (1997). Interspecific interactions and coexistence in dabbling ducks: observations and an experiment. *Oecologia* **111**, 129–36.

Emlen, S. T. (1972). Experimental analysis of parameters of bird song eliciting species recognition. *Behaviour* **41**, 130–71.

Enemar, A. (1987). Nesting, clutch-size and breeding success of the dunnock *Prunella modularis* in mountain birch forest in Swedish Lapland. *Acta Regiae Societatis Scientiarum et Litterarum Gothoburgensis Zoologica* **14**, 29–35.

Enemar, A. and Sjöstrand, B. (1972). Effects of the introduction of pied flycatchers *Ficedula hypoleuca* on the composition of a passerine bird community. *Ornis Scandinavica* **3**, 79–87.

Eriksson, M. O. G. (1979). Competition between freshwater fish and goldeneyes *Bucephala clangula* (L.) for common prey. *Oecologia* **41**, 99–107.

Fenoglio, S., Cucco, M., and Malacarne, G. (2002). The effect of a carotenoid-rich diet on immunocompetence and behavioural performances in moorhen chicks. *Ethology, Ecology and Evolution* **14**, 149–56.

Fernandez-Juricic, E., Beauchamp, G., and Bastain, B. (2007). Group-size and distance-to-neighbour effects on feeding and vigilance in brown-headed cowbirds. *Animal Behaviour* **73**, 771–8.

Fernandez, C., Azkona, P., and Donazar, J. A. (1998). Density-dependent effects on productivity in the Griffon vulture *Gyps fulvus*: the role of interference and habitat heterogeneity. *Ibis* **140**, 64–9.

Ferrer, M. and Donazar, J. A. (1996). Density-dependent fecundity by habitat heterogeneity in an increasing population of Spanish imperial eagles. *Ecology* **77**, 69–74.

Ferrer, M., Newton, I., and Casado, E. (2006). How to test different density-dependent fecundity hypotheses in an increasing or stable population. *Journal of Animal Ecology* **75**, 111–17.

Fletcher, R. J. (2007). Species interactions and population density mediate the use of social cues for habitat selection. *Journal of Animal Ecology* **76**, 598–606.

Fonstad, T. (1984). Reduced territorial overlap between the willow warbler *Phylloscopus trochilus* and the brambling *Fringilla montifringilla* in heath birch forest: competition or different habitat preferences? *Oikos* **42**, 314–22.

Forsman, J. T. and Mönkkönen, M. (2003). The role of climate in limiting European resident bird populations. *Journal of Biogeography* **30**, 55–70.

Forsman, J. T., Seppanen, J. T., and Mönkkönen, M. (2002). Positive fitness consequences of interspecific interaction with a potential competitor. *Proceedings of the Royal Society of London Series B-Biological Sciences* **269**, 1619–23.

Forsman, J. T., Thomson, R. L., and Seppanen, J. T. (2007). Mechanisms and fitness effects of interspecific information use between migrant and resident birds. *Behavioral Ecology* **18**, 888–94.

Forsman, J. T., Hjernquist, M. B., and Gustafsson, L. (2009). Experimental evidence for the use of density based interspecific social information in forest birds. *Ecography* **32**, 539–45.

Forsman, J. T., Mönkkönen, M., Helle, P., and Inkeroinen, J. (1998). Heterospecific attraction and food resources in migrants' breeding patch selection in northern boreal forest. *Oecologia* **115**, 278–86.

Forsman, J. T., Hjernquist, M. B., Taipale, J., and Gustafsson, L. (2008). Competitor density cues for habitat quality facilitating habitat selection and investment decisions. *Behavioral Ecology* **19**, 539–45.

Fort, K. T. and Otter, K. A. (2004a). Territorial breakdown of black-capped chickadees, *Poecile atricapillus*, in disturbed habitats? *Animal Behaviour* **68**, 407–15.

Fort, K. T. and Otter, K. A. (2004b). Effects of habitat disturbance on reproduction in black-capped chickadees (*Poecile atricapillus*) in northern British Columbia. *Auk* **121**, 1070–80.

Fowler, C. W. (1981). Density dependence as related to life history strategy. *Ecology* **62**, 602–10.

Fretwell, S. D. and Lucas, H. L. (1969). On territorial behavior and other factors influencing habitat distribution in birds. 1. Theoretical development. *Acta Biotheoretica* **19**, 16–36.

Garant, D., Kruuk, L. E. B., McCleery, R. H., and Sheldon, B. C. (2007). The effects of environmental heterogeneity on multivariate selection on reproductive traits in female great tits. *Evolution* **61**, 1546–59.

Garcia, E. F. J. (1983). An experimental test of competition for space between blackcaps *Sylvia atricapilla* and garden warblers *Sylvia borin* in the breeding-season. *Journal of Animal Ecology* **52**, 795–805.

Garcia, P. F. J., Merkle, M. S., and Barclay, R. M. R. (1993). Energy allocation to reproduction and maintenance in mountain bluebirds (*Sialia currucoides*): a food supplementation experiment. *Canadian Journal of Zoology-Revue Canadienne de Zoologie* **71**, 2352–7.

Gasparini, J., Roulin, A., Gill, V. A., Hatch, S. A., and Boulinier, T. (2006). In kittiwakes food availability partially explains the seasonal decline in humoral immunocompetence. *Functional Ecology* **20**, 457–63.

Gause, G. F. (1934). *The Struggle for Existence* The Williams & Wilkins company, Baltimore.

Gause, G. F. (1970). Criticism of invalidation of principle of competitive exclusion. *Nature* **227**, 89.

Geer, T. (1981). Factors affecting the delivery of prey to nestling sparrowhawks (*Accipiter nisus*). *Journal of Zoology* **195**, 71–80.

Geer, T. A. (1982). The selection of tits *Parus spp* by sparrowhawks *Accipiter nisus*. *Ibis* **124**, 159–67.

Gerhardt, R. P. (2004). Cavity nesting in raptors of Tikal National Park and Vicinity, Peten, Guatemala. *Ornitologia Neotropical* **15**, 477–83.

Gibb, J. A. (1954). Feeding ecology of tits, with notes on Treecreeper and Goldcrest. *Ibis* **96**, 513–43.

Gibb, J. A. (1958). Predation by tits and squirrels on the eucosmid *Ernarmonia conicolana* (Heyl). *Journal of Animal Ecology* **27**, 375–96.

Gibb, J. A. (1960). Populations of Tits and Goldcrests and their food supply in the pine plantations. *Ibis* **102**, 163–208.

Gibbs, J. P., Hunter, M. L., and Melvin, S. M. (1993). Snag availability and communities of cavity-nesting birds in tropical versus temperate forests. *Biotropica* **25**, 236–41.

Giles, N. (1994). Tufted duck (*Aythya fuligula*) habitat use and brood survival increases after fish removal from gravel-pit lakes. *Hydrobiologia* **280**, 387–92.

Gill, F. B., Mack, A. L., and Ray, R. T. (1982). Competition between hermit hummingbirds Phaethorninae and insects for nectar in a Costa Rican rain forest. *Ibis* **124**, 44–9.

Gilpin, M. E. and Justice, K. E. (1972). Reinterpretation of invalidation of principle of competitive exclusion. *Nature* **236**, 273–301.

Gilpin, M. E., Case, T. J., and Ayala, F. J. (1976). θ-Selection. *Mathematical Biosciences* **32**, 131–9.

Glas, P. (1960). Factors governing density in the Chaffinch (*Fringilla coelebs*) in different types of wood. *Archives Néerlandaises de Zoologie* **13**, 466–72.

Goodbody, I. M. (1952). The post-fledging dispersal of juvenile titmice. *British Birds* **45**, 279–85.

Gordon, S. P., Reznick, D. N., Kinnison, M. T., et al. (2009). Adaptive changes in life history and survival following a new guppy introduction. *American Naturalist* **174**, 34–45.

Gorissen, L., Gorissen, M., and Eens, M. (2006). Heterospecific song matching in two closely related songbirds (*Parus major* and *P. caeruleus*): great tits match blue tits but not vice versa. *Behavioral Ecology and Sociobiology* **60**, 260–9.

Gosler, A. G. (1996). Environmental and social determinants of winter fat storage in the great tit *Parus major*. *Journal of Animal Ecology* **65**, 1–17.

Gosler, A. G. (2002). Strategy and constraint in the winter fattening response to temperature in the great tit *Parus major*. *Journal of Animal Ecology* **71**, 771–9.

Gosler, A. G. and Carruthers, T. D. (1994). Bill size and niche breadth in the Irish coal tit *Parus ater hibernicus*. *Journal of Avian Biology* **25**, 171–7.

Gosler, A. G., Greenwood, J. J. D., and Perrins, C. (1995). Predation risk and the cost of being fat. *Nature* **377**, 621–3.

Gotelli, N. J. and McCabe, D. J. (2002). Species co-occurrence: A meta-analysis of J. M. Diamond's assembly rules model. *Ecology* **83**, 2091–6.

Gotelli, N. J., Graves, G. R., and Rahbek, C. (2010). Macroecological signals of species interactions in the Danish avifauna. *Proceedings of the National Academy of Sciences of the United States of America* **107**, 5030–5.

Granbom, M. and Smith, H. G. (2006). Food limitation during breeding in a heterogeneous landscape. *Auk* **123**, 97–107.

Grant, B. R. and Grant, P. R. (2010). Songs of Darwin's finches diverge when a new species enters the community. *Proceedings of the National Academy of Sciences of the United States of America* **107**, 20156–63.

Grant, P. R. (1972). Convergent and divergent character displacement. *Biological Journal of the Linnean Society* **4**, 39–68.

Grant, P. R. and Grant, B. R. (2006). Evolution of character displacement in Darwin's finches. *Science* **313**, 224–6.

Grant, P. R. and Grant, B. R. (2008). *How and Why Species Multiply.* Princeton University Press, Princeton and Oxford.

Graveland, J. and Drent, R. H. (1997). Calcium availability limits breeding success of passerines on poor soils. *Journal of Animal Ecology* **66**, 279–88.

Gray, S. M. and Robinson, B. W. (2002). Experimental evidence that competition between stickleback species favours adaptive character divergence. *Ecology Letters* **5**, 264–72.

Gray, S. M., Robinson, B. W., and Parsons, K. J. (2005). Testing alternative explanations of character shifts against ecological character displacement in brook sticklebacks (*Culaea inconstans*) that coexist with ninespine sticklebacks (*Pungitius pungitius*). *Oecologia* **146**, 25–35.

Grether, G. F., Losin, N., Anderson, C. N., and Okamoto, K. (2009). The role of interspecific interference competition in character displacement and the evolution of competitor recognition. *Biological Reviews* **84**, 617–35.

Grieco, F. (2003). Greater food availability reduces tarsus asymmetry in nestling Blue Tits. *Condor* **105**, 599–603.

Groom, J. D. and Grubb, T. C. (2006). Patch colonization dynamics in Carolina Chickadees (*Poecile carolinensis*) in a fragmented landscape: a manipulative study. *Auk* **123**, 1149–60.

Grubb, T. C. (1989). Ptilochronology: feather growth bars as indicators of nutritional status. *Auk* **106**, 314–20.

Grubb, T. C. and Cimprich, D. A. (1990). Supplementary food improves the nutritional condition of wintering woodland birds: evidence from ptilochronology. *Ornis Scandinavica* **21**, 277–81.

Grubb, T. C. and Pravasudov, V. V. (1994). Tufted Titmouse (*Baeolophus bicolor*),. In A. Poole, ed. *The Birds of North America Online*. Cornell Lab of Ornithology, Ithaca.

Gunnarsson, B. (2007). Predation on spiders: ecological mechanisms and evolutionary consequences. *Journal of Arachnology* **35**, 509–29.

Gustafsson, L. (1986). Lifetime reproductive success and heritability: empirical support for Fisher's fundamental theorem. *American Naturalist* **128**, 761–4.

Gustafsson, L. (1987). Interspecific competition lowers fitness in collared flycatchers *Ficedula albicollis*: an experimental demonstration. *Ecology* **68**, 291–6.

Gustafsson, L. (1988a). Foraging behavior of individual coal tits, *Parus ater*, in relation to their age, sex and morphology. *Animal Behaviour* **36**, 696–704.

Gustafsson, L. (1988b). Interspecific and intraspecific competition for nest holes in a population of the collared flycatcher *Ficedula albicollis*. *Ibis* **130**, 11–16.

Haas, K., Kohler, U., Diehl, S., *et al.* (2007). Influence of fish on habitat choice of water birds: a whole system experiment. *Ecology* **88**, 2915–25.

Haemig, P. D. (1992). Competition between ants and birds in a Swedish Forest. *Oikos* **65**, 479–83.

Haemig, P. D. (1994). Effects of ants on the foraging of birds in spruce trees. *Oecologia* **97**, 35–40.

Haemig, P. D. (1996). Interference from ants alters foraging ecology of great tits. *Behavioral Ecology and Sociobiology* **38**, 25–9.

Haemig, P. D. (1999). Predation risk alters interactions among species: competition and facilitation between ants and nesting birds in a boreal forest. *Ecology Letters* **2**, 178–84.

Haftorn, S. (1954). Contribution to the food biology of tits especially about storing of surplus food. Part I. The crested tit (*Parus c. cristatus* L.). *Kongelige Norske Videnskabers Selskabs Skrifter* **1953**, 1–123.

Haftorn, S. (2000). Rank-dependent winter fattening in the willow tit *Parus montanus*. *Ornis Fennica* **77**, 49–56.

Hahn, B. A. and Silverman, E. D. (2007). Managing breeding forest songbirds with conspecific song playbacks. *Animal Conservation* **10**, 436–41.

Hairston, N. G. (1980). The experimental test of an analysis of field distributions: competition in terrestrial salamanders. *Ecology* **61**, 817–26.

Hairston, N. G. (1989). *Ecological Experiments: Purpose, Design, and Execution.* Cambridge University Press, Cambridge.

Hampton, S. E. (2005). Increased niche differentiation between two *Conochilus* species over 33 years of climate change and food web alteration. *Limnology and Oceanography* **50**, 421–6.

Hanson, M. A. and Butler, M. G. (1994). Responses to food-web manipulation in a shallow waterfowl lake. *Hydrobiologia* **280**, 457–66.

Harrison, T. J. E., Smith, J. A., Martin, G. R., et al. (2010). Does food supplementation really enhance productivity of breeding birds?. *Oecologia* **164**, 311–20.

Hartley, P. H. T. (1953). An ecological study of the feeding habits of the english titmice. *Journal of Animal Ecology* **22**, 261–88.

Heath, S. R., Kershner, E. L., Cooper, D. M., et al. (2008). Rodent control and food supplementation increase productivity of endangered San Clemente Loggerhead Shrikes (*Lanius ludovicianus mearnsi*). *Biological Conservation* **141**, 2506–15.

Heeb, P., Werner, I., Mateman, A. C., et al. (1999). Ectoparasite infestation and sex-biased local recruitment of hosts. *Nature* **400**, 63–5.

Heegaard, E. and Vandvik, V. (2004). Climate change affects the outcome of competitive interactions: an application of principal response curves. *Oecologia* **139**, 459–66.

Hinsley, S. A., Carpenter, J. E., Broughton, R. K., et al. (2007). Habitat selection by Marsh Tits *Poecile palustris* in the UK. *Ibis* **149**, 224–33.

Hochachka, W. M. and Dhondt, A. A. (2000). Density-dependent decline of host abundance resulting from a new infectious disease. *Proceedings of the National Academy of Sciences of the United States of America* **97**, 5303–6.

Hogstad, O. (1975a). Structure of small passerine communities in subalpine birch forest in Fennoscandia. In F. E. Wielgolaski, ed. *Fennoscandian tundra ecosystems, part* 2. , pp. 94–104. Springer, Berlin.

Hogstad, O. (1975b). Quantitative relations between hole-nesting and open-nesting species within a passerine breeding community. *Norwegian Journal of Zoology* **23**, 261–7.

Hogstad, O. (1978). Differentiation of foraging niche among tits, *Parus spp*, in Norway during winter. *Ibis* **120**, 139–46.

Hogstad, O. (1988). Advantages of social foraging of Willow Tits *Parus montanus*. *Ibis* **130**, 275–83.

Hogstad, O. (1989a). Subordination in Mixed-Age Bird Flocks - a Removal Study. *Ibis* **131**, 128–34.

Hogstad, O. (1989b). The role of juvenile willow tits, *Parus montanus*, in the regulation of winter flock size - an experimental study. *Animal Behaviour* **38**, 920–5.

Hogstad, O. (1990). Winter floaters in willow tits *Parus montanus*: a matter of choice or making the best of a bad situation? In J. Blondel, A. Gosler, J.-D. Lebreton, and R. McCleery, eds. *Population Biology of Passerine Birds*, pp. 415–21. Springer-Verlag, Berlin Heidelberg.

Hogstad, O. (1999). Territory acquisition during winter by juvenile Willow Tits *Parus montanus*. *Ibis* **141**, 615–20.

Hogstad, O. (2003). Strained energy budget of winter floaters in the Willow Tit as indicated by ptilochronology. *Ibis* **145**, E19-E23.

Högstedt, G. (1980a). Prediction and test of the effects of interspecific competition. *Nature* **283**, 64–6.

Högstedt, G. (1980b). Resource partitioning in magpie *Pica pica* and jackdaw *Corvus monedula* during the breeding season. *Ornis Scandinavica* **11**, 110–15.

Högstedt, G. (1980c). Evolution of clutch size in birds: adaptive variation in relation to territory quality. *Science* **210**, 1148–50.

Högstedt, G. (1981). Effect of additional food on reproductive success in the magpie (*Pica pica*). *Journal of Animal Ecology* **50**, 219–29.

Holleback, M. (1974). Behavioral interactions and dispersal of family in Black-capped Chickadees. *Wilson Bulletin* **86**, 466–8.

Holmes, R. T. (2007). Understanding population change in migratory songbirds: long-term and experimental studies of Neotropical migrants in breeding and wintering areas. *Ibis* **149**, 2–13.

Holmes, R. T., Schultz, J. C. and Nothnagle, P. (1979). Bird predation on forest insects: exclosure experiment. *Science* **206**, 462–3.

Holt, R. D. (1977). Predation, apparent competition, and structure of prey communities. *Theoretical Population Biology* **12**, 197–229.

Hooks, C. R. R., Pandey, R. R., and Johnson, M. W. (2003). Impact of avian and arthropod predation on lepidopteran caterpillar densities and plant productivity in an ephemeral agroecosystem. *Ecological Entomology* **28**, 522–32.

Howe, F. P., Knight, R. L., McEwen, L. C., and George, T. L. (1996). Direct and indirect effects of insecticide applications on growth and survival of nestling passerines. *Ecological Applications* **6**, 1314–24.

Hromada, M., Antczak, M., Valone, T. J., and Tryjanowski, P. (2008). Settling decisions and heterospecific social information use in shrikes. *PLoS ONE* **3**, e3930.

Ingold, D. J. (1994a). Influence of nest site competition between European Starlings and Woodpeckers. *Wilson Bulletin* **106**, 227–41.

Ingold, D. J. (1994b). Nest-site characteristics of red-bellied and red-headed woodpeckers and northern flickers in east-central Ohio. *Ohio Journal of Science* **94**, 2–7.

Jaksic, F. M., Feinsinger, P., and Jimenez, J. E. (1993). A long-term study on the dynamics of guild structure among predatory vertebrates at a semiarid neotropical site. *Oikos* **67**, 87–96.

James, H. F., Ericson, P. G. P., Slikas, B., Lei, F. M., Gill, F. B., and Olson, S. L. (2003). *Pseudopodoces humilis*, a misclassified terrestrial tit (Paridae) of the Tibetan Plateau: evolutionary consequences of shifting adaptive zones. *Ibis* **145**, 185–202.

Jansson, C., Ekman, J., and von Brömssen, A. (1981). Winter mortality and food supply in tits *Parus spp*. *Oikos* **37**, 313–22.

Jantti, A., Aho, T., Hakkarainen, H., Kuitunen, M., and Suhonen, J. (2001). Prey depletion by the foraging of the Eurasian treecreeper, *Certhia familiaris*, on tree-trunk arthropods. *Oecologia* **128**, 488–91.

Jantti, A., Suorsa, P., Hakkarainen, H., Sorvari, J., Huhta, E., and Kuitunen, M. (2007). Within territory abundance of red wood ants *Formica rufa* is associated with the body condition of nestlings in the Eurasian treecreeper *Certhia familiaris*. *Journal of Avian Biology* **38**, 619–24.

Jones, M. E., Cockburn, A., Hamede, R., *et al*. (2008). Life-history change in disease-ravaged Tasmanian devil populations. *Proceedings of the National Academy of Sciences of the United States of America* **105**, 10023–7.

Jullien, M. and Clobert, J. (2000). The survival value of flocking in neotropical birds: reality or fiction? *Ecology* **81**, 3416–30.

Källander, H. (1981). The effects of provision of food in winter on a population of the great tit *Parus major* and the blue tit *Parus caeruleus*. *Ornis Scandinavica* **12**, 244–8.

Kaneko, S. (2005). Seasonal population changes of five parasitoids attacking the scale insect *Nipponaclerda biwakoensis* on the common reed, with special reference to predation by wintering birds. *Entomological Science* **8**, 323–9.

Kappes, J. J. and Davis, J. M. (2008). Evidence of positive indirect effects within a community of cavity-nesting vertebrates. *Condor* **110**, 441–9.

Keddy, P. A. (1989). *Competition*. Chapman and Hall, London, New York

Keddy, P. A. (2001). *Competition* 2nd Ed. Kluwer, Dordrecht.

Kempenaers, B. and Dhondt, A. A. (1991). Competition between Blue and Great Tit for Roosting Sites in Winter - an Aviary Experiment. *Ornis Scandinavica* **22**, 73–5.

Kempenaers, B. and Dhondt, A. A. (1992). Experimental test of an hypothesis explaining density dependent clutch-size in tits *Parus spp*. *Ibis* **134**, 192–4.

Kempenaers, B., Adriaensen, F., van Noordwijk, A. J., and Dhondt, A. A. (1996). Genetic similarity, inbreeding and hatching failure in blue tits: are unhatched eggs infertile? *Proceedings of the Royal Society of London Series B-Biological Sciences* **263**, 179–85.

Kingsolver, J. G. and Pfennig, D. W. (2007). Patterns and power of phenotypic selection in nature. *Bioscience* **57**, 561–72.

Kingsolver, J. G., Hoekstra, H. E., Hoekstra, J. M., *et al*. (2001). The strength of phenotypic selection in natural populations. *American Naturalist* **157**, 245–61.

Klomp, H. (1967). Boekbespreking: Lack D. 1966. Population Studies of Birds. *Ardea* **55**, 156–7.

Klomp, H. (1980). Fluctuations and stability in Great Tit populations. *Ardea* **68**, 205–24.

Kluijver, H. N. (1950). Daily routines of the Great Tit, *Parus m. major* L. *Ardea* **38**, 99–135.

Kluijver, H. N. (1951). The population ecology of the great tit, *Parus m. major* L. *Ardea* **39**, 1–135.

Kluijver, H. N. and Tinbergen, L. (1953). Territory and the regulation of density in titmice. *Archives Néerlandaises de Zoologie* **10**, 265–89.

Kluyver, H. N. (1952). Notes on body weight and time of breeding in the Great Tit, *Parus m. major* L. *Ardea* **40**, 123–41.

Kluyver, H. N. (1957). Roosting habits, sexual dominance and survival in the great tit. *Cold Spring Harbor Symposia on Quantitative Biology* **22**, 281–5.

Kluyver, H. N. (1966). Regulation of a bird population. *Ostrich. Suppl.* **No. 6**, 389–96.

Kluyver, H. N. (1971). Regulation of numbers in populations of great tits (*Parus m. major*). In P. J. den Boer and G. R. Gradwell, eds. *Proceedings of the Advanced Study Institute on 'Dynamics of Numbers in Populations' Oosterbeek* 1970. PUDOC, Wageningen.

Kluyver, H. N., Van Balen, J. H., and Cavé, A. J. (1977). The occurrence of time-saving mechanisms in the breeding biology of the Great Tit. In B. Stonehouse and C. M. Perrins, eds. *Evolutionary Ecology*, pp. 154–69. Macmillan, London.

Knapton, R. W. and Krebs, J. R. (1974). Settlement patterns, territory size, and breeding density in song sparrow (*Melospiza melodia*). *Canadian Journal of Zoology-Revue Canadienne de Zoologie* **52**, 1413–20.

Koenig, W. D. and Dickinson, J. L. (2004). *Ecology and Evolution of Cooperative Breeding in Birds*. Cambridge University Press, Cambridge.

Koh, L. P. (2008). Birds defend oil palms from herbivorous insects. *Ecological Applications* **18**, 821–5.

Kokko, H. and Lopez-Sepulcre, A. (2007). The ecogenetic link between demography and evolution: can we bridge the gap between theory and data? *Ecology Letters* **10**, 773–82.

Korner-Nievergelt, F., Baader, E., Fischer, L., Schaffner, W., Korner-Nievergelt, P., and Kestenholz, M. (2008). Do birds during irruption years differ from birds during 'normal' years? *Vogelwarte* **46**, 207–16.

Kothbauer-Hellmann, R. (1990). Molesting in juvenile tits (*Parus spp*): on the formation of dominance order. *Journal für Ornithologie* **131**, 421–7.

Krams, I. (1998). Rank-dependent fattening strategies of Willow Tit *Parus montanus* and Crested Tit *P. cristatus* mixed flock members. *Ornis Fennica* **75**, 19–26.

Krams, I. (2000). Length of feeding day and body weight of great tits in a single- and a two-predator environment. *Behavioral Ecology and Sociobiology* **48**, 147–53.

Krams, I. (2001). Seeing without being seen: a removal experiment with mixed flocks of Willow and Crested Tits *Parus montanus* and *cristatus*. *Ibis* **143**, 476–81.

Krams, I. (2002). Mass-dependent take-off ability in wintering great tits (*Parus major*): comparison of top-ranked adult males and subordinate juvenile females. *Behavioral Ecology and Sociobiology* **51**, 345–9.

Krams, I. A. (1996). Predation risk and shifts of foraging sites in mixed willow and crested tit flocks. *Journal of Avian Biology* **27**, 153–6.

Krebs, J. R. (1970). Regulation of numbers in great tit (Aves: Passeriformes). *Journal of Zoology* **162**, 317–33.

Krebs, J. R. (1971). Territory and breeding density in great tit, *Parus major* L. *Ecology* **52**, 2–22.

Krebs, J. R. (1977). Song and territory in the great tits *Parus major*. In B. P. Stonehouse, C. M. Perrins, eds. *Evolutionary Ecology*, pp. 43-62. MacMillan, London.

Kress, S. W. (1983). The use of decoys, sound recordings, and gull control for re-establishing a tern colony in Maine. *Colonial Waterbirds* **6**, 185–96.

Kress, S. W. and Nettleship, D. N. (1988). Re-establishment of Atlantic puffins (*Fratercula arctica*) at a former breeding site in the Gulf of Maine. *Journal of Field Ornithology* **59**, 161–70.

Kricher, J. C. (1983). Correlation between house finch increase and house sparrow decline. *American Birds* **37**, 358–60.

Krüger, O. and Lindström, J. (2001). Habitat heterogeneity affects population growth in goshawk *Accipiter gentilis*. *Journal of Animal Ecology* **70**, 173–81.

Kubota, H. and Nakamura, M. (2000). Effects of supplemental food on intra- and inter-specific behaviour of the Varied Tit *Parus varius*. *Ibis* **142**, 312–19.

Kullberg, C. and Ekman, J. (2000). Does predation maintain tit community diversity? *Oikos* **89**, 41–5.

Kullberg, C., Fransson, T., and Jakobsson, S. (1996). Impaired predator evasion in fat blackcaps (*Sylvia atricapilla*). *Proceedings of the Royal Society of London Series B-Biological Sciences* **263**, 1671–5.

Lachish, S., Miller, K. J., Storfer, A., Goldizen, A. W., and Jones, M. E. (2011). Evidence that disease-induced population decline changes genetic structure and alters dispersal patterns in the Tasmanian devil. *Heredity* **106**, 172–82.

Lack, D. (1944). Ecological aspects of species-formation in passerine birds. *Ibis* **86**, 260–86.

Lack, D. (1945). The ecology of closely related species with special reference to cormorant (*Phalacrocorax carbo*) and shag (*P. aristotelis*). *Journal of Animal Ecology* **14**, 12–16.

Lack, D. (1947). *Darwin's Finches*. The University Press, Cambridge.

Lack, D. (1954). *Population Regulation of Birds*. Clarendon Press, Oxford.

Lack, D. (1966). *Population Studies of Birds*. Clarendon Press, Oxford.

Lack, D. (1971). *Ecological Isolation in Birds*. Harvard University Press, Cambridge.

Lack, D. and Lack, E. (1951). The breeding biology of the Swift *Apus apus*. *Ibis* **93**, 544–6.

Lahti, K., Orell, M., Rytkonen, S., and Koivula, K. (1998). Time and food dependence in Willow Tit winter survival. *Ecology* **79**, 2904–16.

Lambrechts, M. and Dhondt, A. A. (1988). Male quality and territory quality in the great tit *Parus major*. *Animal Behaviour* **36**, 596–601.

Laverty, T. M. and Plowright, R. C. (1985). Competition between hummingbirds and bumble bees for nectar in flowers of *Impatiens biflora*. *Oecologia* **66**, 25–32.

Laves, K. S. and Loeb, S. C. (1999). Effects of southern flying squirrels *Glaucomys volans* on red-cockaded woodpecker *Picoides borealis* reproductive success. *Animal Conservation* **2**, 295–303.

Law, R. and Watkinson, A. R. (1989). Competition. In J. M. Cherrett, ed. *Ecological Concepts*, pp. 243–84. Blackwell Scientific Publications, Oxford.

Lebreton, J.-D. (2009). Assessing density-dependence: where are we left? In D. L. Thomson, E. G. Cooch, and M. J. Conroy, eds. *Demographic Processes in Marked Populations*, pp. 19–42. Springer Verlag, Berlin, Heidelberg.

Leisler, B. (1988). Interspecific interactions among European marsh-nesting passerines. In H. Ouellet, ed. *Acta XIX Congressus Internationalis Ornithologci*, Vol. 2, pp. 2635–44. Ottawa University Press, Ottawa.

Lens, L. (1996). Wind stress affects foraging site competition between Crested Tits and Willow Tits. *Journal of Avian Biology* **27**, 41–6.

Lens, L. and Dhondt, A. A. (1992). Variation in coherence of crested tit winter flocks: an example of multivariate optimization. *Acta Oecologica-International Journal of Ecology* **13**, 553–67.

Lens, L. and Wauters, L. A. (1996). Effects of population growth on Crested Tit *Parus cristatus* post-fledging settlement. *Ibis* **138**, 545–51.

Lilliendahl, K. and Solmundsson, J. (2006). Feeding ecology of sympatric European shags *Phalacrocorax aristotelis* and great cormorants *P. carbo* in Iceland. *Marine Biology* **149**, 979–90.

Lima, S. L. (1986). Predation risk and unpredictable feeding conditions: determinants of body-mass in birds. *Ecology* **67**, 377–85.

Lima, S. L. (1993). Ecological and evolutionary perspectives on escape from predatory attack: a survey of North-American birds. *Wilson Bulletin* **105**, 1–47.

Loeb, S. C. and Hooper, R. G. (1997). An experimental test of interspecific competition for red-cockaded woodpecker cavities. *Journal of Wildlife Management* **61**, 1268–80.

Lohmus, A. and Remm, J. (2005). Nest quality limits the number of hole-nesting passerines in their natural cavity-rich habitat. *Acta Oecologica-International Journal of Ecology* **27**, 125–8.

Löhrl, H. (1957). Populationsökologische Untersuchungen beim Halsbandschnapper (*Ficedula albicollis*). *Bonner Zoologische Beiträge* **8**, 130–77.

Löhrl, H. (1974). *Die Tannenmeise*. A. Ziemsen Verlag, Wittemberg Lutherstadt.

Löhrl, H. (1976). Die Sumpfmeise (*Parus palustris*) als Brutvogel des Fichtenwaldes im Vergleich zu Tannen-, Blau- und Kohlmeise (*P. ater, P. caeruleus* und *P. major*). *Vogelwelt* **97**, 217–23.

Löhrl, H. (1977). Nistökologische und ethologische Anpassungserscheinungen bei Höhlenbrütern. *Die Vogelwarte* **29**, 92–101.

Losos, J. B., Schoener, T. W., and Spiller, D. A. (2004). Predator-induced behaviour shifts and natural selection in field-experimental lizard populations. *Nature* **432**, 505–8.

Lotka, A. J. (1925). *Elements of Physical Biology*.

Lotka, A. J. (1932). The growth of mixed populations: two species competing for a common food supply. *Journal of the Washington Academy of Sciences* **22**, 461–9.

Lovette, I. J. and Hochachka, W. M. (2006). Simultaneous effects of phylogenetic niche conservatism and competition on avian community structure. *Ecology* **87**, S14–S28.

Loyn, R. H. (1985). Ecology, distribution and density of birds in Victorian forests. In A. Keast, H. F. Recher, H. Ford and D. Saunders, eds. *Birds of Eucalypt Forests and Woodlands: Ecology, Conservation, Management.*, pp. 33–46. Surrey Beatty & Sons, Chipping Norton, NSW.

Loyn, R. H., Runnalls, R. G., Forward, G. Y., and Tyers, J. (1983). Territorial bell miners and other birds affecting populations of insect prey. *Science* **221**, 1411–13.

Ludescher, F.-B. (1973). Sumpfmeise (*Parus p. palustris* L.) und Weidenmeise (*Parus montanus salicarius* Br.) als sympatrische Zwillingsarten. *Journal für Ornithologie* **114** 5–56.

Macarthur, R. H. (1958). Population ecology of some warblers of northeastern coniferous forests. *Ecology* **39**, 599–619.

MacArthur, R. H. (1972). *Geographical Ecology: Patterns in the Distribution of Species*. Harper and Row, New York.

Macleod, R., Gosler, A. G., and Cresswell, W. (2005). Diurnal mass gain strategies and perceived predation risk in the great tit *Parus major*. *Journal of Animal Ecology* **74**, 956–64.

Mahon, C. L. and Martin, K. (2006). Nest survival of chickadees in managed forests: habitat, predator, and year effects. *Journal of Wildlife Management* **70**, 1257–65.

Mairy, F. (1969). Abnormal developments in the range and nesting behavior of a population of Chaffinches *Fringilla coelebs* on the plane of the High Fagnes Belgian canton of Malmedy. *Gerfaut - Giervalk* **59**, 48–69.

Marler, P. (1960). Bird songs and mate selection. In W. E. Lanyon and W. N. Tavolga, eds. *Animal Sounds and Communication*, pp. 248–367. American Institute of Biological Sciences, Washington, D.C.

Marra, P. P. (2000). The role of behavioral dominance in structuring patterns of habitat occupancy in a migrant bird during the nonbreeding season. *Behavioral Ecology* **11**, 299–308.

Marra, P. P. and Holmes, R. T. (2001). Consequences of dominance-mediated habitat segregation in American Redstarts during the nonbreeding season. *Auk* **118**, 92–104.

Marshall, M. R., Cooper, R. J., DeCecco, J. A., Strazanac, J., and Butler, L. (2002). Effects of experimentally reduced prey abundance on the breeding ecology of the red-eyed vireo. *Ecological Applications* **12**, 261–80.

Martin, K. and Eadie, J. M. (1999). Nest webs: A community-wide approach to the management and conservation of cavity-nesting forest birds. *Forest Ecology and Management* **115**, 243–57.

Martin, K., Aitken, K. E. H., and Wiebe, K. L. (2004). Nest sites and nest webs for cavity-nesting communities in interior British Columbia, Canada: nest characteristics and niche partitioning. *Condor* **106**, 5–19.

Martin, P. A., Johnson, D. L., Forsyth, D. J., and Hill, B. D. (1998). Indirect effects of the pyrethroid insecticide deltamethrin on reproductive success of chestnut-collared longspurs. *Ecotoxicology* **7**, 89–97.

Martin, P. R. and Martin, T. E. (2001). Ecological and fitness consequences of species coexistence: A removal experiment with wood warblers. *Ecology* **82**, 189–206.

Martin, P. R., Fotheringham, J. R., Ratcliffe, L., and Robertson, R. J. (1996). Response of American redstarts (suborder Passeri) and least flycatchers (suborder Tyranni) to heterospecific playback: The role of song in aggressive interactions and interference competition. *Behavioral Ecology and Sociobiology* **39**, 227–35.

Martin, T. E. (1986). Competition in breeding birds. On the importance of considering processes at the level of the individual. In *Current Ornithology*, Vol. 4, pp. 181–10. Plenum Press, New York.

Martin, T. E. (1988). On the advantage of being different: nest predation and the coexistence of bird species. *Proceedings of the National Academy of Sciences of the United States of America* **85**, 2196–9.

Martin, T. E. (1992). Breeding productivity considerations: what are the appropriate habitat features for management? In J. M. Hagan and D. W. Johnston, eds. *Ecology and Conservation of Neotropical Migrants*, pp. 455–73. Smithsonian Institution Press, Washington D.C.

Martin, T. E. (1993). Nest predation and nest sites - new perspectives on old patterns. *Bioscience* **43**, 523–32.

Martinez, J. A., Calvo, J. F., Martinez, J. E., Zuberogoitia, I., Zabala, J., and Redpath, S. M. (2008). Breeding performance, age effects and territory occupancy in a Bonelli's Eagle *Hieraaetus fasciatus* population. *Ibis* **150**, 223–33.

Matthysen, E. (1990). Nonbreeding social organisation in *Parus*. In D. M. Power, ed. *Current Ornithology*, Vol. 7, pp. 209–49. Plenum Press, New York.

Matthysen, E. (1998). *The Nuthatches*. T & A D Poyser, London.

Matthysen, E. (2005). Density-dependent dispersal in birds and mammals. *Ecography* **28**, 403–16.

Matthysen, E., Adriaensen, F., and Dhondt, A. A. (2001). Local recruitment of great and blue tits (*Parus major, P. caeruleus*) in relation to study plot size and degree of isolation. *Ecography* **24**, 33–42.

Mattsson, B. J. and Niemi, G. J. (2008). Causes and consequences of distribution patterns in a migratory songbird across its geographic range. *Canadian Journal of Zoology-Revue Canadienne de Zoologie* **86**, 314–28.

Mazia, C. N., Chaneton, E. J., Kitzberger, T., and Garibaldi, L. A. (2009). Variable strength of top-down effects in *Nothofagus* forests: bird predation and insect herbivory during an ENSO event. *Austral Ecology* **34**, 359–67.

McCleery, R. H. and Clobert, J. (1990). Differences in recruitment of young by immigrant and resident great tits in Wytham Wood. In J. Blondel, A. Gosler, J.-D. Lebreton, and R. McCleery, eds. *Population Biology of Passerine Birds*, pp. 423–40. Springer Verlag, Berlin Heidelberg.

McCormack, J. E., Jablonski, P. G., and Brown, J. L. (2007). Producer-scrounger roles and joining based on dominance in a free-living group of Mexican jays (*Aphelocoma ultramarina*). *Behaviour* **144**, 967–82.

McKenzie, A. J., Petty, S. J., Toms, M. P., and Furness, R. W. (2007). Importance of Sitka Spruce *Picea sitchensis* seed and garden bird-feeders for Siskins *Carduelis spinus* and Coal Tits *Periparus ater*. *Bird Study* **54**, 236–47.

McPhail, J. D. (1984). Ecology and evolution of sympatric sticklebacks (*Gasterosteus*): morphological and genetic evidence for a species pair in Enos Lake, British-Columbia. *Canadian Journal of Zoology-Revue Canadienne de Zoologie* **62**, 1402–8.

McPhail, J. D. (1989). Status of the Enos Lake stickleback species pair, *Gasterosteus spp*. *Canadian Field Naturalist* **103**, 216–19.

McPhail, J. D. (1992). Ecology and evolution of sympatric sticklebacks (*Gasterosteus*): evidence for a species pair in Paxton Lake, Texada Island, British Columbia. *Canadian Journal of Zoology-Revue Canadienne de Zoologie* **70**, 361–9.

Merilä, J. and Wiggins, D. A. (1995). Interspecific competition for nest holes causes adult mortality in the Collared Flycatcher. *Condor* **97**, 445–50.

Miller, E. H. (1982). Character and variance shift in acoustic signals of birds. In D. E. Kroodsma and E. H. Miller, eds. *Acoustic Communication in Birds.*, Vol. 1, pp. 253–95. Academic Press, New York.

Minot, E. O. (1981). Effects of interspecific competition for food in breeding blue and great tits. *Journal of Animal Ecology* **50**, 375–85.

Minot, E. O. and Perrins, C. M. (1986). Interspecific interference competition - nest sites for blue and great tits. *Journal of Animal Ecology* **55**, 331–50.

Mols, C. M. M. and Visser, M. E. (2002). Great tits can reduce caterpillar damage in apple orchards. *Journal of Applied Ecology* **39**, 888–99.

Mönkkönen, L. and Forsman, J. (2002). Heterospecific attraction among forest birds: a review. *Ornithological Science* **1**, 41–51.

Mönkkönen, M., Helle, P., and Soppela, K. (1990). Numerical and behavioral responses of migrant passerines to experimental manipulation of resident tits (*Parus spp*): heterospecific attraction in northern breeding bird communities. *Oecologia* **85**, 218–25.

Mönkkönen, M., Forsman, J. T., and Helle, P. (1996). Mixed-species foraging aggregations and heterospecific attraction in boreal bird communities. *Oikos* **77**, 127–36.

Mönkkönen, M., Forsman, J. T., and Bokma, F. (2006). Energy availability, abundance, energy-use and species richness in forest bird communities: a test of the species-energy theory. *Global Ecology and Biogeography* **15**, 290–302.

Mönkkönen, M., Helle, P., Niemi, G. J., and Montgomery, K. (1997). Heterospecific attraction affects community structure and migrant abundances in northern breeding bird communities. *Canadian Journal of Zoology-Revue Canadienne de Zoologie* **75**, 2077–83.

Morse, D. H. (1967). Foraging relationships of brown-headed nuthatches and pine warblers. *Ecology* **48**, 94–103.

Morse, D. H. (1970). Ecological aspects of some mixed-species foraging flocks of birds. *Ecological Monographs* **40**, 119–68.

Morse, D. H. (1974). Niche breadth as a function of social dominance. *American Naturalist* **108**, 818–30.

Munger, J. C. and Brown, J. H. (1981). Competition in desert rodents: an experiment with semipermeable exclosures. *Science* **211**, 510–12.

Murray, B. G. (1976). Critique of interspecific territoriality and character convergence. *Condor* **78**, 518–25.

Murray, B. G. (1981). The origins of adaptive interspecific territorialism. *Biological Reviews of the Cambridge Philosophical Society* **56**, 1–22.

Murray, B. G. (1988). Interspecific territoriality in *Acrocephalus*: a critical review. *Ornis Scandinavica* **19**, 309–13.

Nager, R. G., Ruegger, C., and Van Noordwijk, A. J. (1997). Nutrient or energy limitation on egg formation: a feeding experiment in great tits. *Journal of Animal Ecology* **66**, 495–507.

Nagy, L. R. and Holmes, R. T. (2005). Food limits annual fecundity of a migratory songbird: an experimental study. *Ecology* **86**, 675–81.

Naugler, C. T. and Ratcliffe, L. (1994). Character release in bird song: a test of the acoustic competition hypothesis using American tree sparrows *Spizella arborea*. *Journal of Avian Biology* **25**, 142–8.

Newton, I. (1988). A key factor analysis of a sparrowhawk population. *Oecologia* **76**, 588–96.

Newton, I. (1991). Habitat variation and population regulation in Sparrowhawks. *Ibis* **133**, 76–88.

Newton, I. (1998). *Population Limitation in Birds*. Academic Press, San Diego, London.

Newton, I. (2006). Advances in the study of irruptive migration. *Ardea* **94**, 433–60.

Nicholson, A. J. (1933). The balance of animal populations. *Journal of Animal Ecology* **2**, 132–78.

Nicolaus, M., Both, C., Ubels, R., Edelaar, P., and Tinbergen, J. M. (2009). No experimental evidence for local competition in the nestling phase as a driving force for density-dependent avian clutch size. *Journal of Animal Ecology* **78**, 828–38.

Nielsen, C. R., Parker, P. G., and Gates, R. J. (2006). Intraspecific nest parasitism of cavity-nesting wood ducks: costs and benefits to hosts and parasites. *Animal Behaviour* **72**, 917–26.

Nilsson, J. A. (1991). Clutch size determination in the marsh tit (*Parus palustris*). *Ecology* **72**, 1757–62.

Nilsson, J. A. and Smith, H. G. (1988). Effects of dispersal date on winter flock establishment and social dominance in marsh tits *Parus palustris*. *Journal of Animal Ecology* **57**, 917–28.

Nilsson, J. A. and Cardmark, A. (2001). Sibling competition affects individual growth strategies in marsh tit, *Parus palustris*, nestlings. *Animal Behaviour* **61**, 357–65.

Nilsson, S. G. (1987). Limitation and regulation of population density in the nuthatch *Sitta europaea* (Aves) breeding in natural cavities. *Journal of Animal Ecology* **56**, 921–37.

Nocedal, J. and Ficken, M. S. (1998). Helpers in the bridled titmouse. *Southwestern Naturalist* **43**, 279–82.

Norberg, U. M. (1979). Morphology of the wings, legs and tail of 3 coniferous forest tits, the goldcrest, and the treecreeper in relation to locomotor pattern and feeding station selection. *Philosophical Transactions of the Royal Society of London Series B-Biological Sciences* **287**, 131–65.

Oksanen, L. (1987). Interspecific competition and the structure of bird guilds in boreal Europe: the importance of doing fieldwork in the right season. *Trends in Ecology and Evolution* **2**, 376–9.

Orell, M. and Ojanen, M. (1983). Effect of habitat, date of laying and density on clutch size of the great tit *Parus major* in northern Finland. *Holarctic Ecology* **6**, 413–23.

Orians, G. H. and Willson, M. F. (1964). Interspecific territories of birds. *Ecology* **45**, 736–45.

Oro, D., Margalida, A., Carrete, M., Heredia, R., and Donazar, J. A. (2008). Testing the goodness of supplementary feeding to enhance population viability in an endangered vulture. *PloS ONE* **3** (12) e4084.

Ortubay, S., Cussac, V., Battini, M., *et al.* (2006). Is the decline of birds and amphibians in a steppe lake of northern Patagonia a consequence of limnological changes following fish introduction? *Aquatic Conservation-Marine and Freshwater Ecosystems* **16**, 93–105.

Otter, K. and Ratcliffe, L. (1996). Female initiated divorce in a monogamous songbird: Abandoning mates for males of higher quality. *Proceedings of the Royal Society of London Series B-Biological Sciences* **263**, 351–5.

Papaix, J., Cubaynes, S., Buoro, M., Charmantier, A., Perret, P., and Gimenez, O. (2010). Combining capture-recapture data and pedigree information to assess heritability of demographic parameters in the wild. *Journal of Evolutionary Biology* **23**, 2176–84.

Parejo, D., Danchin, E., Silva, N., White, J. F., Dreiss, A. N., and Aviles, J. M. (2008). Do great tits rely on inadvertent social information from blue tits? A habitat selection experiment. *Behavioral Ecology and Sociobiology* **62**, 1569–79.

Park, Y.-S., Lee, W.-S., and Rhim, S.-J. (2004). Differences in breeding success of tits in artificial nest boxes between hog fat supplied and non-supplied coniferous forests. *Journal of Korean Forestry Society* **93**, 383–7.

Pattanavibool, A. and Edge, W. D. (1996). Single-tree selection silviculture affects cavity resources in mixed deciduous forests in Thailand. *Journal of Wildlife Management* **60**, 67–73.

Patten, M. A. (2007). Geographic variation in calcium and clutch size. *Journal of Avian Biology* **38**, 637–43.

Payne, R. B. and Groschupf, K. D. (1984). Sexual selection and interspecific competition: a field experiment on territorial behavior of nonparental finches (*Vidua spp*). *Auk* **101**, 140–5.

Pearl, R. and Reed, L. J. (1920). On the rate of growth of the population of the United States since 1790 and its mathematical representation. *Proceedings of the National Academy of Sciences of the United States of America* **6**, 275–88.

Pehrsson, O. (1984). Relationships of food to spatial and temporal breeding strategies of mallards in Sweden. *Journal of Wildlife Management* **48**, 322–39.

Peichel, C. L., Nereng, K. S., Ohgi, K. A., *et al*. (2001). The genetic architecture of divergence between threespine stickleback species. *Nature* **414**, 901–5.

Peiman, K. S. and Robinson, B. W. (2010). Ecology and evolution of resource-related heterospecific aggression. *Quarterly Review of Biology* **85**, 133–58.

Perdeck, A. C., Visser, M. E., and Van Balen, J. H. (2000). Great Tit *Parus major* survival, and the beech-crop cycle. *Ardea* **88**, 99–108.

Perrins, C. M. (1965). Population fluctuations and clutch-size in the great tit, *Parus major* L. *Journal of Animal Ecology* **34**, 601–47.

Perrins, C. M. (1966). The effect of beech crops on Great Tit populations and movements. *British Birds* **59**, 419–32.

Perrins, C. M. (1970). Timing of birds breeding seasons. *Ibis* **112**, 242–55.

Perrins, C. M. (1979). *British Tits*. Collins, London.

Perrins, C. M. and Jones, P. J. (1974). Inheritance of clutch size in Great Tit (*Parus major* L.). *Condor* **76**, 225–8.

Perrins, C. M. and McCleery, R. H. (1994). Competition and egg-weight in the Great Tit *Parus major*. *Ibis* **136**, 454–6.

Pfeifer, S. (1955). Ergebnisse zweier Versuche zur Steigerung der Siedlungsdichte der Vögel auf forstlicher Kleinfläche und benachbarter Grossfläche. *Waldhygiene* **1**, 76–8.

Pfeifer, S. and Keil, W. (1960). Weitere Ergebnisse des Versuches zur Steigerung der Siedlungsdichte hohlen-und freibrutender Vogelarten eines Eichen-Hainbuchen Waldes bei Frankfurt am Main. *Vogelwelt* **81**, 141–5.

Pfennig, D. W. and Martin, R. A. (2010). Evolution of character displacement in spadefoot toads: different proximate mechanisms in different species. *Evolution* **64**, 2331–41.

Pfennig, D. W., Rice, A. M., and Martin, R. A. (2007). Field and experimental evidence for competition's role in phenotypic divergence. *Evolution* **61**, 257–71.

Phillips, V. E. (1992). Variation in winter wildfowl numbers on gravel pit lakes at Great Linford, Buckinghamshire, 1974–79 and 1984–91, with particular reference to the effects of fish removal. *Bird Study* **39**, 177–85.

Pianka, E. R. (1970). R-selection and K-selection. *American Naturalist* **104**, 592–7.

Pimm, S. L. (1978). Experimental approach to the effects of predictability on community structure. *American Zoologist* **18**, 797–808.

Pimm, S. L., Rosenzweig, M. L. and Mitchell, W. (1985). Competition and food selection: field tests of a theory. *Ecology* **66**, 798–807.

Piovesan, G. and Adams, J. M. (2001). Masting behaviour in beech: linking reproduction and climatic variation. *Canadian Journal of Botany-Revue Canadienne de Botanique* **79**, 1039–47.

Politi, N., Hunter, M., and Rivera, L. (2009). Nest selection by cavity-nesting birds in subtropical montane forests of the Andes: implications for sustainable forest management. *Biotropica* **41**, 354–60.

Pollard, E., Lakhani, K. H., and Rothery, P. (1987). The detection of density-dependence from a series of annual censuses. *Ecology* **68**, 2046–55.

Postma, E. and van Noordwijk, A. J. (2005). Gene flow maintains a large genetic difference in clutch size at a small spatial scale. *Nature* **433**, 65–8.

Pravosudov, V. V. (2006). On seasonality in food-storing behaviour in parids: do we know the whole story? *Animal Behaviour* **71**, 1455–60.

Pravosudova, E. V., Grubb, T. C., Parker, P. G., and Doherty, P. F. (1999). Patch size and composition of social groups in wintering tufted titmice. *Auk* **116**, 1152–5.

Preston, K. L. and Rotenberry, J. T. (2006). Independent effects of food and predator-mediated processes on annual fecundity in a songbird. *Ecology* **87**, 160–8.

Price, T., Kirkpatrick, M., and Arnold, S. J. (1988). Directional selection and the evolution of breeding date in birds. *Science* **240**, 798–9.

Pulliam, H. R. and Caraco, T. (1984). Living in groups: is there an optimal group size? In J. R. Krebs and N. B. Davies, eds. *Behavioural Ecology: An Evolutionary Approach*, pp. 122–47. Blackwell, Oxford.

Rabol, J. (1987). Coexistence and competition between overwintering willow warblers *Phylloscopus trochilus* and local warblers at Lake Naivasha, Kenya. *Ornis Scandinavica* **18**, 101–21.

Raffel, T. R., Martin, L. B., and Rohr, J. R. (2008). Parasites as predators: unifying natural enemy ecology. *Trends in Ecology and Evolution* **23**, 610–18.

Ramos, J. A., Monteiro, L. R., Sola, E., and Moniz, Z. (1997). Characteristics and competition for nest cavities in burrowing procellariiformes. *Condor* **99**, 634–41.

Reed, T. M. (1982). Interspecific territoriality in the chaffinch and great tit on islands and the mainland of Scotland: playback and removal experiments. *Animal Behaviour* **30**, 171–81.

Remm, J., Lohmus, A., and Rosenvald, R. (2008). Density and diversity of hole-nesting passerines: dependence on the characteristics of cavities. *Acta Ornithologica* **43**, 83–91.

Remsen, J. V. and Graves, W. S. (1995). Distribution patterns of Buarremon brush-finches (Emberizinae) and interspecific competition in Andean birds. *Auk* **112**, 225–36.

Reynoldson, T. B. and Bellamy, P. E. (1971). The establishment of interspecific competition in field populations, with an example of competition in action between *Polycelis nigra* (Mull.) and *P. tenuis* (Ijima) (Turbellaria, Tricladida). In J. den Boer and G. R. Gradwell, eds. *Proceedings of the Advanced Study Institute on 'Dynamics of Numbers in Populations' Oosterbeek* 1970. pp. 282–97. PUDOC, Wageningen.

Reznick, D. A., Bryga, H., and Endler, J. A. (1990). Experimentally induced life-history evolution in a natural population. *Nature* **346**, 357–9.

Reznick, D. N. and Ghalambor, C. K. (2005). Selection in nature: experimental manipulations of natural populations. *Integrative and Comparative Biology* **45**, 456–62.

Ribaut, J.-P. (1964). The dynamics of a population of blackbirds, *Turdus merula* L. *Revue Suisse de Zoologie* **71**, 815–902.

Richards, F. J. (1959). A flexible growth function for empirical use. *Journal of Experimental Botany* **10**, 290–300.

Ritz, M. S. (2007). Sex-specific mass loss in chick-rearing South Polar Skuas *Stercorarius maccormicki* - stress induced or adaptive? *Ibis* **149**, 156–65.

Robb, G. N., McDonald, R. A., Chamberlain, D. E., and Bearhop, S. (2008a). Food for thought: supplementary feeding as a driver of ecological change in avian populations. *Frontiers in Ecology and the Environment* **6**, 476–84.

Robb, G. N., McDonald, R. A., Chamberlain, D. E., Reynolds, S. J., Harrison, T. J. E., and Bearhop, S. (2008b). Winter feeding of birds increases productivity in the subsequent breeding season. *Biology Letters* **4**, 220–23.

Robinson, S. K. and Terborgh, J. (1995). Interspecific aggression and habitat selection by Amazonian birds. *Journal of Animal Ecology* **64**, 1–11.

Rodenhouse, N. L. and Holmes, R. T. (1992). Results of experimental and natural food reductions for breeding black-throated blue warblers. *Ecology* **73**, 357–72.

Rodenhouse, N. L., Sherry, T. W., and Holmes, R. T. (1997). Site-dependent regulation of population size: A new synthesis. *Ecology* **78**, 2025–42.

Rodenhouse, N. L., Sillett, T. S., Doran, P. J., and Holmes, R. T. (2003). Multiple density-dependence mechanisms regulate a migratory bird population during the breeding season. *Proceedings of the Royal Society of London Series B-Biological Sciences* **270**, 2105–10.

Rodriguez, A., Jansson, G., and Andrén, H. (2007). Composition of an avian guild in spatially structured habitats supports a competition-colonization trade-off. *Proceedings of the Royal Society B-Biological Sciences* **274**, 1403–11.

Roff, D. A. (2002). *Life History Evolution*. Sinauer Associates, Inc., Sunderland, Massachusetts.

Rogers, C. M. (1995). Experimental evidence for temperature-dependent winter lipid storage in the dark-eyed junco (*Junco hyemalis oreganus*) and song sparrow (*Melospiza melodia morphna*). *Physiological Zoology* **68**, 277–89.

Rogers, C. M. and Smith, J. N. M. (1993). Life-history theory in the nonbreeding period: trade-offs in avian fat reserves. *Ecology* **74**, 419–26.

Rosa, S., Granadeiro, J. P., Vinagre, C., Franca, S., Cabral, H. N., and Palmeirim, J. M. (2008). Impact of predation on the polychaete *Hediste diversicolor* in estuarine intertidal flats. *Estuarine Coastal and Shelf Science* **78**, 655–64.

Saether, B. E. (1983). Mechanism of interspecific spacing out in a territorial system of the Chiffchaff *Phylloscopus collybita* and the Willow Warbler *Phlloscopus trochilus*. *Ornis Scandinavica* **14**, 154–60.

Saether, B. E., Engen, S., and Matthysen, E. (2002). Demographic characteristics and population dynamical patterns of solitary birds. *Science* **295**, 2070–3.

Saino, N., Calza, S., and Moller, A. P. (1997). Immunocompetence of nestling barn swallows in relation to brood size and parental effort. *Journal of Animal Ecology* **66**, 827–36.

Saitou, T. (1978). Ecological study of social organization in the Great Tit, *Parus major* L. I. Basic structure of the winter flocks. *Japanese Journal of Ecology* **28**, 199–214.

Saitou, T. (1979). Ecological study of social organization in the Great Tit, *Parus major* L. II. Formation of the basic flocks. *Journal of the Yamashina Institute for Ornithology* **11**, 137–48.

Saitou, T. (1982). Compound flocks as an aggregation of the flocks of constant composition in the Great Tit, *Parus major*. *Journal of the Yamashina Institute for Ornithology* **14**, 239–305.

Salewski, V. and Jones, P. (2006). Palearctic passerines in Afrotropical environments: a review. *Journal of Ornithology* **147**, 192–201.

Salewski, V., Bairlein, F., and Leisler, B. (2003). Niche partitioning of two Palearctic passerine migrants with Afrotropical residents in their West African winter quarters. *Behavioral Ecology* **14**, 493–502.

Samson, F. B. and Lewis, S. J. (1979). Experiments on population regulation in 2 North-american parids. *Wilson Bulletin* **91**, 222–33.

Sanchez, M. I., Green, A. J., and Alejandre, R. (2006). Shorebird predation affects density, biomass, and size distribution of benthic chironomids in salt pans: an exclosure experiment. *Journal of the North American Benthological Society* **25**, 9–18.

Sanchez, S., Cuervo, J. J., and Moreno, E. (2007). Suitable cavities as a scarce resource for both cavity and non-cavity nesting birds in managed temperate forests. A case study in the Iberian Peninsula. *Ardeola* **54**, 261–74.

Sasvari, L. and Orell, M. (1992). Breeding Success in a North and a Central-European Population of the Great Tit *Parus major*. *Ornis Scandinavica* **23**, 96–100.

Scheuerlein, A. and Gwinner, E. (2002). Is Food Availability a Circannual Zeitgeber in Tropical Birds? A Field Experiment on Stonechats in Tropical Africa. *Journal of Biological Rhythms* **17**, 171–80.

Schluter, D. (1994). Experimental evidence that competition promotes divergence in adaptive radiation. *Science* **266**, 798–801.

Schluter, D. and McPhail, J. D. (1992). Ecological character displacement and speciation in sticklebacks. *American Naturalist* **140**, 85–108.

Schmidt, K. H. and Wolff, S. (1985). Hat die Winterfütterung einen Einfluss auf Gewicht und Überlebensrate von Kohlmeisen (*Parus major*)? *Journal für Ornithologie* **126**, 175–80.

Schoech, S. J. and Hahn, T. P. (2008). Latitude affects degree of advancement in laying by birds in response to food supplementation: a meta-analysis. *Oecologia* **157**, 369–76.

Schoener, T. W. (1973). Population-growth regulated by intraspecific competition for energy or time: some simple representations. *Theoretical Population Biology* **4**, 56–84.

Schoener, T. W. (1974). Competition and form of habitat shift. *Theoretical Population Biology* **6**, 265–307.

Schoener, T. W. (1982). The controversy over interspecific competition. *American Scientist* **70**, 586–95.

Schoener, T. W. (1983). Field experiments on interspecific competition. *American Naturalist* **122**, 240–85.

Sebastian-Gonzalez, E., Sanchez-Zapata, J. A., Botella, F., and Ovaskainen, O. (2010). Testing the heterospecific attraction hypothesis with time-series data on species co-occurrence. *Proceedings of the Royal Society of London Series B-Biological Sciences* **277**, 2983–90.

Semel, B., Sherman, P. W., and Byers, S. M. (1988). Effects of brood parasitism and nest-box placement on Wood Duck breeding ecology. *Condor* **90**, 920–30.

Seppanen, J. T., Forsman, J. T., Mönkkönen, M., and Thomson, R. L. (2007). Social information use is a process across time, space, and ecology, reaching heterospecifics. *Ecology* **88**, 1622–33.

Shaw, P. (2003). Breeding activity and provisioning rates of Stripe-breasted Tits (*Parus fasciiventer*) at Bwindi impenetrable forest, Uganda. *Ostrich* **74**, 129–32.

Sherry, T. W. and Holmes, R. T. (1988). Habitat selection by breeding American Redstarts in response to a dominant competitor, the Least Flycatcher. *Auk* **105**, 350–64.

Sillett, T. S. and Holmes, R. T. (2002). Variation in survivorship of a migratory songbird throughout its annual cycle. *Journal of Animal Ecology* **71**, 296–308.

Sillett, T. S., Holmes, R. T., and Sherry, T. W. (2000). Impacts of a global climate cycle on population dynamics of a migratory songbird. *Science* **288**, 2040–2.

Sillett, T. S., Rodenhouse, N. L., and Holmes, R. T. (2004). Experimentally reducing neighbor density affects reproduction and behavior of a migratory songbird. *Ecology* **85**, 2467–77.

Simberloff, D. and Boecklen, W. (1981). Santa Rosalia reconsidered: size ratios and competition. *Evolution* **35**, 1206–28.

Simberloff, D., Dayan, T., Jones, C., and Ogura, G. (2000). Character displacement and release in the small Indian mongoose, *Herpestes javanicus*. *Ecology* **81**, 2086–99.

Simmons, R. E. (1993). Effects of supplementary food on density-reduced breeding in an African eagle: adaptive restraint or ecological constraint? *Ibis* **135**, 394–402.

Sinclair, A. R. E. (1989). Population regulation in animals. In J. M. Cherrett, ed. *Ecological Concepts*. Blackwell Scientific Publications, Oxford.

Siriwardena, G. M., Stevens, D. K., Anderson, G. Q. A., Vickery, J. A., Calbrade, N. A., and Dodd, S. (2007). The effect of supplementary winter seed food on breeding populations of farmland birds: evidence from two large-scale experiments. *Journal of Applied Ecology* **44**, 920–32.

Slagsvold, T. (1978). Competition between Great Tit *Parus major* and Pied Flycatcher *Ficedula hypoleuca*: an experiment. *Ornis Scandinavica* **9**, 46–50.

Smith, F. E. (1963). Population dynamics in *Daphnia magna* and a new model for population growth. *Ecology* **44**, 651–63.

Smith, J. N. M. (1974). Food searching behavior of 2 European thrushes .1. Description and analysis of search paths. *Behaviour* **48**, 276–302.

Smith, J. N. M., Taitt, M. J., and Zanette, L. (2002). Removing brown-headed cowbirds increases seasonal fecundity and population growth in song sparrows. *Ecology* **83**, 3037–47.

Smith, J. N. M., Montgomerie, R. D., Taitt, M. J., and Yomtov, Y. (1980). A winter feeding experiment on an island song sparrow population. *Oecologia* **47**, 164–70.

Smith, J. N. M., Keller, L. F., Marr, A. B., and Arcese, P. (2006). *Conservation and Biology of Small Populations: The Song Sparrow of Mandarte Island*. Oxford University Press, Oxford.

Smith, J. N. M., Marr, A. B., Arcese, P., and Keller, L. F. (2006). Fluctuations in numbers: population regulation and catastrophic mortality. In J. N. M. Smith, L. F. Keller, A. B. Marr, and P. Arcese eds. *Conservation and Biology of Small Populations: The Song Sparrows of Mandarte Island*, Chapter 4. Oxford University Press, Oxford

Smith, J. N. M., Grant, P. R., Grant, B. R., Abbott, I. J., and Abbott, L. K. (1978). Seasonal variation in feeding habits of Darwins ground finches. *Ecology* **59**, 1137–50.

Smith, S. M. (1967). Seasonal changes in the survival of the Black-capped Chickadee. *Condor* **69**, 344–59.

Smith, S. M. (1978). Underworld in a territorial sparrow - adaptive strategy for floaters. *American Naturalist* **112**, 571–82.

Smith, S. M. (1984). Flock switching in chickadees: why be a winter floater. *American Naturalist* **123**, 81–98.

Smith, S. M. (1987). Responses of floaters to removal experiments on wintering chickadees. *Behavioral Ecology and Sociobiology* **20**, 363–7.

Smith, S. M. (1989). Black-capped chickadee summer floaters. *Wilson Bulletin* **101**, 344–9.

Smith, T. B. (1990). Resource use by bill morphs of an African finch: evidence for intraspecific competition. *Ecology* **71**, 1246–57.

Smulders, T. V. (1998). A game theoretical model of the evolution of food hoarding: applications to the paridae. *American Naturalist* **151**, 356–66.

Snow, D. (1949). Jämförände studier över våra mesarters näringssökenda. *Vår Fågelvärld* **8**, 159–69.

Snow, D. W. (1955). Geographical variation of the Coal Tit, *Parus ater* L. *Ardea* **43**, 195–225.

Sørensen, M. F. L. (1997). Niche shifts of Coal Tits *Parus ater* in Denmark. *Journal of Avian Biology* **28**, 68–72.

Southern, H. N. (1970). Natural control of a population of tawny owls (*Strix aluco*). *Journal of Zoology* **162**, 197–285.

Sperry, J. H., Peak, R. G., Cimprich, D. A., and Weatherhead, P. J. (2008). Snake activity affects seasonal variation in nest predation risk for birds. *Journal of Avian Biology* **39**, 379–83.

Stamps, J. A. (1988). Conspecific attraction and aggregation in territorial species. *American Naturalist* **131**, 329–47.

Stearns, S. C. (1992). *The Evolution of Life Histories*. Oxford University Press, Oxford.

Steer, J. and Burns, K. C. (2008). Seasonal variation in male-female competition, cooperation and selfish hoarding in a monogamous songbird. *Behavioral Ecology and Sociobiology* **62**, 1175–83.

Stefanski, R. A. (1967). Utilization of breeding territory in Black-capped Chickadee. *Condor* **69**, 259–67.

Stenseth, N.-C., Durant, J. M., Fowler, M. S., *et al.* (ms). Climate-induced changes affect competitive relationships in a pair of bird species.

Strubbe, D. and Matthysen, E. (2009). Experimental evidence for nest-site competition between invasive ring-necked parakeets (*Psittacula krameri*) and native nuthatches (*Sitta europaea*). *Biological Conservation* **142**, 1588–94.

Studds, C. E. and Marra, P. P. (2005). Nonbreeding habitat occupancy and population processes: an upgrade experiment with a migratory bird. *Ecology* **86**, 2380–5.

Suhonen, J., Halonen, M., Mappes, T., and Korpimaki, E. (2007). Interspecific competition limits larders of pygmy owls *Glaucidium passerinum*. *Journal of Avian Biology* **38**, 630–4.

Svensson, E. (1995). Avian reproductive timing: when should parents be prudent. *Animal Behaviour* **49**, 1569–75.

Szekely, T., Szep, T., and Juhasz, T. (1989). Mixed species flocking of tits (*Parus spp*): a field experiment. *Oecologia* **78**, 490–5.

Tarboton, W. R. (1981). Cooperative breeding and group territoriality in the Black Tit. *Ostrich* **52**, 216–25.

Tavecchia, G., Pradel, R., Genovart, M., and Oro, D. (2007). Density-dependent parameters and demographic equilibrium in open populations. *Oikos* **116**, 1481–92.

Terborgh, J. and Weske, J. S. (1975). Role of competition in distribution of Andean birds. *Ecology* **56**, 562–76.

Thomson, R. L., Forsman, J. T., and Mönkkönen, M. (2003). Positive interactions between migrant and resident birds: testing the heterospecific attraction hypothesis. *Oecologia* **134**, 431–8.

Thomson, R. L., Forsman, J. T., Sarda-Palomera, F., and Mönkkönen, M. (2006). Fear factor: prey habitat selection and its consequences in a predation risk landscape. *Ecography* **29**, 507–14.

Tobias, J. (1997). Food availability as a determinant of pairing behaviour in the European robin. *Journal of Animal Ecology* **66**, 629–39.

Török, J. (1993). The predator-prey size hypothesis in 3 assemblages of forest birds. *Oecologia* **95**, 474–8.

Török, J. and Tóth, L. (1999). Asymmetric competition between two tit species: a reciprocal removal experiment. *Journal of Animal Ecology* **68**, 338–45.

Tubelis, D. P. (2007). Mixed-species flocks of birds in the Cerrado, South America: A review. *Ornitologia Neotropical* **18**, 75–97.

Turcotte, Y. and Desrochers, A. (2005). Landscape-dependent distribution of northern forest birds in winter. *Ecography* **28**, 129–40.

Ulfstrand, S. (1962). On the nonbreeding ecology and migratory movements of the Great Tit (*Parus major*) and the Blue Tit (*Parus caeruleus*) in Southern Sweden. *Vår Fågelvärld* **suppl. 3**, 1–145.

Underwood, A. J. (2009). Components of design in ecological field experiments. *Annales Zoologici Fennici* **46**, 93–111.

Underwood, T. (1986). The analysis of competition by field experiments. In J. Kikkawa and D. J. Anderson, eds. *Community Ecology: Pattern and Process*. Blackwell Scientific Publications, Oxford.

Vaclav, R., Hoi, H., and Blomqvist, D. (2003). Food supplementation affects extrapair paternity in house sparrows (*Passer domesticus*). *Behavioral Ecology* **14**, 730–5.

Van Bael, S. A. and Brawn, J. D. (2005). The direct and indirect effects of insectivory by birds in two contrasting Neotropical forests. *Oecologia* **145**, 658–68.

Van Bael, S. A., Brawn, J. D., and Robinson, S. K. (2003). Birds defend trees from herbivores in a Neotropical forest canopy. *Proceedings of the National Academy of Sciences of the United States of America* **100**, 8304–7.

Van Bael, S. A., Bichier, P., and Greenberg, R. (2007). Bird predation on insects reduces damage to the foliage of cocoa trees (*Theobroma cacao*) in western Panama. *Journal of Tropical Ecology* **23**, 715–19.

Van Balen, J. H. (1973). Comparative study of breeding ecology of Great Tit *Parus major* in different habitats. *Ardea* **61**, 1–93.

Van Balen, J. H. (1980). Population fluctuations of the Great Tit and feeding conditions in winter. *Ardea* **68**, 143–64.

Van Balen, J. H., Booy, C. J. H., Van Franeker, J. A., and Osieck, E. R. (1982). Studies on hole-nesting birds in natural nest sites. 1. Availability and occupation of natural nest sites. *Ardea* **70**, 1–24.

Van Gasteren, H., Monstert, K., Groot, H., and Van Ruiten, L. (1992). The irruption of the Coal Tit *Parus ater* in the autumn of 1989 in the Netherlands and Northwest Europe. *Limosa* **65**, 57–66.

Van Noordwijk, A. J. and Scharloo, W. (1981). Inbreeding in an island population of the great tit. *Evolution* **35**, 674–88.

Van Valen, L. (1965). Morphological variation and width of ecological niche. *American Naturalist* **99**, 377–90.

Verhulst, P.-F. (1838). Notice sur la loi que la population suit dans son accroissement. *Correspondance Mathématique et Physique* **10**, 113–21.

Verhulst, P.-F. (1845). Recherches mathématiques sur la loi d'accroissement de la population. *Mémoires de l'Academie Royale des Sciences, des Lettres et des Beaux-Arts de Belgique* **18**, 1–38.

Verhulst, P.-F. (1847). Deuxième Mémoire sur la loi d'accroissement de la population. *Mémoires de l'Academie Royale des Sciences, des Lettres et des Beaux-Arts de Belgique* **20**, 1–32.

Visser, M. E., Both, C., and Lambrechts, M. M. (2004). Global climate change leads to mistimed avian reproduction. *Birds and Climate Change* **35**, 89–110.

Visser, M. E., van Noordwijk, A. J., Tinbergen, J. M., and Lessells, C. M. (1998). Warmer springs lead to mistimed reproduction in great tits (*Parus major*). *Proceedings of the Royal Society of London Series B-Biological Sciences* **265**, 1867–70.

Visser, M. E., Adriaensen, F., van Balen, J. H., *et al.* (2003). Variable responses to large-scale climate change in European *Parus* populations. *Proceedings of the Royal Society of London Series B-Biological Sciences* **270**, 367–72.

Volterra, V. (1926). Fluctuations in the abundance of a species considered mathematically. *Nature* **118**, 558–60.

Walters, E. L. and James, F. C. (2010). Quantifying purported competition with individual- and population-level metrics. *Conservation Biology* **24**, 1569–77.

Warner, R. R. and Chesson, P. L. (1985). Coexistence mediated by recruitment fluctuations: a field guide to the storage effect. *American Naturalist* **125**, 769–87.

Waters, J. R., Noon, B. R., and Verner, J. (1990). Lack of nest site limitation in a cavity-nesting bird community. *Journal of Wildlife Management* **54**, 239–45.

Watson, A. and Jenkins, D. (1968). Experiments on population control by territorial behaviour in red grouse. *Journal of Animal Ecology* **37**, 595–614.

Weinzierl, H. (1957). Beitrag zur künstlichen Steigerung der Siedlungsdichte höhlenbrütender Singvögel im Auwald. *Waldhygiene* **2**, 105–12.

Wellenstein, G. (1968). Weitere Ergebnisse über die Auswirkungen einer planmässigen Ansiedlung von Waldameisen und höhlenbrütenden Vögel in Lehr - u. Versuchsrevier Schwetzingen. *Angewandte Ornithologie* **3**, 40–53.

Wesołowski, T. (2007). Lessons from long-term hole-nester studies in a primeval temperate forest. *Journal of Ornithology* **148**, S395-S405.

Wesołowski, T., Tomiałojć, L., and Stawarczyk, T. (1987). Why low numbers of *Parus major* in Białowieża Forest: removal experiments. *Acta Ornithologica* **23**, 303–16.

Wiens, J. A. (1977). Competition and variable environments. *American Scientist* **65**, 590–7.

Wiens, J. A. (1989). *The Ecology of Bird Communities.* Cambridge University Press, Cambridge.

Wiens, J. A. (1993). Fat times, lean times and competition among predators. *Trends in Ecology and Evolution* **8**, 348–9.

Wiens, J. A., Rotenberry, J. T., and Vanhorne, B. (1986). A lesson in the limitations of field experiments: shrubsteppe birds and habitat alteration. *Ecology* **67**, 365–76.

Wiktander, U., Olsson, O., and Nilsson, S. G. (2001). Annual and seasonal reproductive trends in the Lesser Spotted Woodpecker *Dendrocopos minor*. *Ibis* **143**, 72–82.

Wilkin, T. A., Garant, D., Gosler, A. G., and Sheldon, B. C. (2006). Density effects on life-history traits in a wild population of the great tit *Parus major*: analyses of long-term data with GIS techniques. *Journal of Animal Ecology* **75**, 604–15.

Williams, J. B. and Batzli, G. O. (1979a). Interference competition and niche shifts in the bark-foraging guild in central Illinois. *Wilson Bulletin* **91**, 400–11.

Williams, J. B. and Batzli, G. O. (1979b). Competition among bark-foraging birds in central Illinois: experimental evidence. *Condor* **81**, 122–32.

Willis, E. O. and Oniki, Y. (1978). Birds and army ants. *Annual Review of Ecology and Systematics* **9**, 243–63.

Wilson, A. G. and Arcese, P. (2008). Influential factors for natal dispersal in an avian island metapopulation. *Journal of Avian Biology* **39**, 341–7.

Wilson, W. H. (2001). The effects of supplemental feeding on wintering Black-capped Chickadees (*Poecile atricapilla*) in central Maine: population and individual responses. *Wilson Bulletin* **113**, 65–72.

Winfield, D. K. and Winfield, I. J. (1994). Possible competitive interactions between overwintering tufted duck (*Aythya fuligula* (L)) and fish populations of Lough Neagh, Northern Ireland: evidence from diet studies. *Hydrobiologia* **280**, 377–86.

Winfield, I. J., Winfield, D. K., and Tobin, C. M. (1992). Interactions between the Roach, *Rutilus rutilus*, and waterfowl populations of Lough Neagh, Northern Ireland. *Environmental Biology of Fishes* **33**, 207–14.

Wingate, D. B. (1977). Excluding competitors from Bermuda petrel nesting burrows. In S. A. Temple, ed. *Proceedings of Symposium on Management Techniques for Preserving Endangered Birds*, pp. 93–102. University of Wisconsin Press and Croom Helm, Madison

Winkler, D. W., Wrege, P. H., Allen, P. E., *et al.* (2005). The natal dispersal of tree swallows in a continuous mainland environment. *Journal of Animal Ecology* **74**, 1080–90.

Witter, M. S. and Cuthill, I. C. (1993). The ecological costs of avian fat storage. *Philosophical Transactions of the Royal Society of London Series B-Biological Sciences* **340**, 73–92.

Wootton, J. T. (1987). Interspecific competition between introduced house finch populations and 2 associated passerine species. *Oecologia* **71**, 325–31.

Wrege, P. H., Wikelski, M., Mandel, J. T., Rassweiler, T., and Couzin, I. D. (2005). Antbirds parasitize foraging army ants. *Ecology* **86**, 555–9.

Wright, R. M. and Phillips, V. E. (1990). Mallard duckling response to increased food supply. *Game Conservancy Annual Review* **21**, 105–8.

Wright, R. M. and Phillips, V. E. (1992). Changes in the aquatic vegetation of 2 gravel pit lakes after reducing the fish population density. *Aquatic Botany* **43**, 43–9.

Wyllie, I. and Newton, I. (1991). Demography of an increasing population of sparrowhawks. *Journal of Animal Ecology* **60**, 749–66.

Wynne-Edwards, V. C. (1962). *Animal Dispersion in Relation to Social Behaviour*. Oliver and Boyd, Edinburgh.

Zanette, L., Clinchy, M., and Smith, J. N. M. (2006). Combined food and predator effects on songbird nest survival and annual reproductive success: results from a bi-factorial experiment. *Oecologia* **147**, 632–40.

Zanette, L., Smith, J. N. M., van Oort, H., and Clinchy, M. (2003). Synergistic effects of food and predators on annual reproductive success in song sparrows. *Proceedings of the Royal Society of London Series B-Biological Sciences* **270**, 799–803.

Index

Acoustic Competition Hypothesis 35
acoustic interference 35–6
amensalism 13, 14
Anticipation Hypothesis 56
ants 137, 145–8
Antwerp tit studies
    detailed analyses 234–44
    evolutionary effects of IC 214–5, 218, 220–1
    long-term IC experiments 171–201
    population processes, effects of IC on 94, 100–1
army ants and birds 147–8
arthropod predation 63–5
assembly rules model 216–17
asymmetric competition 13

beak morphology
    ecological character release 206
    evolution 212
    in IC 111
    in niche overlap 114
beech masting
    and bird populations 43
    and food limitation 55, 102, 105
    and food supplementation 67
    and mass 53–4
    and predation 53–4
    and survival 219
    and tit populations 40–1, 42, 43
bees 150–1
bell miner (*Manorina melanophrys*) 140, 143
biotic interactions 1–2
bird names, common/scientific 229–32
blackbird, Eurasian (*Turdus merula*) 38, 52, 70
blackcap (*Sylvia atricapilla*) 140, 143–4
bluebird, eastern (*Sialia sialis*) 124
bluebird, mountain (*Sialia currucoides*)
    nest site competition 130, 131
    in nest web 78–9
    supplemental feeding effects 60
body fat reserves 50–5
brambling (*Fringilla montifringilla*) 140, 141–2
breeding density and food supplementation 46
brush-finch, chestnut-capped (*Buarremon brunneinuchus*) 113–4

brush-finch, stripe-headed (*Buarremon torquatus*) 113–4
BTO Garden BirdWatch 43
Buffer Hypothesis 25–8
burrow-nesting seabirds 156–7

cahow *see* petrel, Bermuda
calcium 57, 63
carotenes 63
cavity nesters
    cavity characteristics 74, 75, 80
    community-level effects 128–35
    diversity of habitats 72
    excavators *vs* non-excavators 72–3
    heterospecific attraction 161–2
    logging, effects of 75, 80
    nest box suitability 76
    nest site availability
        experimental manipulation 73–4, 126–7, 131
        limited 72–3
        in natural forests 74–7
        superabundance 73–7
    non-cavity nesters *see* cup-nesting species
    predation at nest site 72–3
    primary *vs* secondary 77, 78
    secondary 72, 73, 75–6
        nest availability 118, 126, 129, 130, 133
        in nest web communities 77, 78
    in termite mounds 75
    usurpation, protection from 73
chaffinch (*Fringilla coelebs*)
    and buffer hypothesis 27
    competition with cavity nesters 136
    competition with great tits 36–7
    heterospecific attraction 164
    removal experiments 140, 141
Character Shift Hypothesis 36
chickadee, black-capped (*Poecile atricapillus*)
    flock switcher (floater) 33, 101
    food and breeding 39
    food supplementation effects 45–8
    nest box suitability 76–7
    predation at nest site 72–3
    territoriality 25, 29, 31, 32–3

chickadee, Carolina (*Poecile carolinensis*)
  foraging niches 110–11
  mixed species foraging 45
  supplemental food, effects 44
  territoriality 31, 33
chickadee, chestnut-backed (*Poecile rufescens*) 73
chickadee, mountain (*Poecile gambeli*) 79, 130, 132–3, 135
chiffchaff (*Phylloscopus collybita*)
  competition with cavity nesters 136
  removal experiments 140, 141
climate change and IC 199–200
clutch size/quality
  changes with IC 218–21
  and density 85–6, 93–6, 98, 165, 184, 91, 193
  and fecundity 222
  and food availability 53–4, 59
  and food supplementation 55–8
Common Garden Census 44
common/scientific names
  birds 229–32
  other species 233
community composition 216–17
competition 1
  asymmetric 13
  categories 13
  definitions 13–15
    Newton's 14–15
    Wiens' 14
  diffuse 13
  exploitation (scramble) 13
  interference (contest) 13
  vs predation 2
Competition-Colonization Trade-Off (CCTO) 112
Constraints Hypothesis 55, 56
contest (interference) competition 13
cormorant (*Phalacrocorax carbo*) 8, 103
crow, hooded (*Corvus cornix*) 120
cup-nesting species 69–71

density dependence 9–11
  dispersal 194–6
  habitat heterogeneity in 92–4
  and intraspecific competition 83–4
    time series data 84
  in introduced populations 91–2
  processes affected by 84–90
    Audouin's gull 90
    song sparrow 85–6
    sparrowhawk 89–90
    titmice 86, 87, 94–101, 182–9, 190–2, 194–6
    warbler 86–9
devil facial tumor disease 218

Diamond's assembly rules model 216–17
diffuse competition 13
disease epidemics 217–18
dispersal
  heterospecific attraction 165
  IC in tits 188, 194–6
distribution 26, 27
dominance
  and body fat management 51–2
  and dispersal 99
  and foraging niches 106, 107–8, 109, 111
  and nest sites 70, 71, 126
  removal experiments 139–45
  winter foraging 50
duck, diving 151–2, 154
duck, heterospecific attraction 162, 164
duck, tufted (*Aythya fuligula*) 153, 154
duck, wood (*Aix sponsa*) 37
dunnock (*Prunella modularis*) 136

eagle, Bonelli's (*Aquila fasciata*) 93
eagle, booted (*Aquila pennata*) 93
eagle, Spanish imperial (*Aquila adalberti*) 93
ecological character release 203–5
  criteria 209
  criteria testing
    in coal tits 206–9
    in plethodonthid salamanders 209–11
  parallel *vs* individual release 204
ecological isolation 5, 6
ecological niche modelling 113–14
ecological release 113
eggs
  laying date 95–6
  mass 97
  second broods 97
  see also clutch size/quality
evolutionary effects of IC 203–23
  rate of change
    experimental data 213–16
      blue tits 214–16
      sticklebacks 213
    observational data 211–12
exploitation (scramble) competition 13

farmland birds, supplementary feeding in 47, 49
fecundity fitness surfaces 220
feeding *see* food/feeding
field experiments
  on cavity availability
    cavity *vs* open nesters 134, 135–9
    community effects 127–34, 135
    and foraging success 119–25
    single species effects 125–7
  evidence/effects of IC 117–70

quality of 169
shortcomings 170
finch
   cactus ground (*Geospiza scandens*) 36
   Darwin's 8, 114–5, 211–12
      evolution 211–12
      foraging niches 115
   house (*Carpodacus mexicanus*)
      competition with house sparrow 15–16, 140, 144–5
   large ground (*Geospiza magnirostris*) 36, 212
   medium ground (*Geospiza fortis*) 211, 212
fish 151–5
fitness 109–11, 123
flicker, northern (*Colaptes auratus*) 78, 130, 133
flocking, costs/benefits of 49–50
flycatcher, ash-throated (*Myiarchus cinerascens*) 129
flycatcher, collared (*Ficedula albicollis*)
   fitness, intra- and interspecific competition 123
   heterospecific attraction 162, 166–7
   nest site competition 126, 127
flycatcher, least (*Empidonax minimus*)
   heterospecific attraction 162, 164
   nest site limitation 121, 122–4
   removal experiments 140, 144
flycatcher, pied (*Ficedula hypoleuca*)
   competition with ants 146
   competition with open nesters 136
   heterospecific attraction 162, 165, 168
   nest site competition 126, 127, 135
foliage structure in nest-site safety 69
food/feeding
   and body fat management
      feeding site 50
      foraging success 52–4
   food availability
      and breeding 39
      effects on nestlings/parents 39
      and predation 63–5
   in mixed species flock 45
   and predation 44–5
      winter feeding effects 53–4
   supplementation *see* supplemental feeding
   and winter survival 42–9
   see also foraging; hoarding
food reduction experiments 59–60, 67
foraging
   density dependence 95–6
   foraging success 52–4
      field experiments 119–25
   niches 103–17
      altitudinal 113–14

and beak size 114
early observations of IC 105–8
ecological niche modelling 113–14
experimental data 108–9
interspecific differences 104–5
migrants *vs* residents 115–16
and proof of IC 109–11
replacement studies 113–14
seasonal variation in overlap 114–15

Gause's views on coexistence 5, 8
geographic races, interaction of 8
Ghent tit studies
   detailed analyses 234–44
   long-term IC experiments 171–201
goldcrest (*Regulus regulus*) 109
goldeneye (*Bucephala clangula*) 151–2, 154
goshawk (*Accipiter gentilis*) 61, 93
Gotland studies
   coal tit foraging 110, 111–13, 205, 207–9
   heterospecific attraction 165
   nest site competition 73–4, 122–4, 126
grebe, great crested (*Podiceps cristatus*) 153, 154
guillemot, common (*Uria aalge*) 93
gull, Audouin's (*Larus audouini*) 90

Habitat Heterogeneity Hypothesis 93
habitat heterogeneity, importance of 92–4
hermit, long-tailed (*Phaethornis superciliosus*) 150
heterospecific aggression 158–60, 170
heterospecific attraction 139, 160–8
Heterospecific Attraction Hypothesis 161
hoarding 39, 105
   food manipulations during breeding season 57–8
   and winter temperature 44
honeyeater evolution 211
hummingbird, ruby-throated (*Archilochus colubris*) 150–1

IC *see* interspecific competition
immigration rates 86
immune competence 62–3
indigobird, Jambandu (*Vidua raricola*) 140, 142
indigobird, variable (*Vidua funerea*) 140, 142
Individual Adjustment Hypothesis 93
insects
   IC with birds 137, 148–51
interference (contest) competition 13
intersexual competition 39
interspecific competition (IC)
   birds *vs* other classes 145–56
   birds *vs* ants 137, 145–8
   birds *vs* fish 151–5

interspecific competition (*cont.*)
  birds *vs* insects 137, 148–51
  birds *vs* mammals 155–6
  and climate change 199–200
  and community composition 216–17
  conditions for consideration/proof of IC 22–4
  continuous nature of 5, 11
  debate over importance 2–4
    in North America 11
  definition 13, 14
  evolutionary effects 203–23
    rate of changes 211–16
  experimental data 14–15, 225–6
  field experiments 117–70
    summary 168
  foraging niches 114
  importance 169
  and life-history changes 216–21
  non-experimental data 15
  in open populations 216–17
  as selection force 23–4
interspecific territoriality 158–60
intraspecific competition 1–2
  definition 13
  and density dependence 83–4
  ecological character release 208–9
  importance 83–4, 102
  and predation 86
isoclines 21–2

jackdaw (*Corvus monedula*) 118, 120, 121
jaeger, parasitic (*Stercocarius parasiticus*) 60
jay (*Garrulus glandarius*) 43
junco (*Junco hiemalis*) 140, 142–3
juvenile survival 89–90

kestrel, American (*Falco parverius*) 77, 94, 133
Kluijver *see* Kluyver's long-term studies
Kluyver's long-term studies
  Buffer Hypothesis 26
  cavities as roosting sites 198–9
  density-dependent reproduction 10
  food limitation 11
  great tit dispersal, long-term experiments 99, 198
  IC between great and blue tit 7, 10
  long-term studies 9
kittiwake (*Rissa tridactyla*) 62

Lack's views on the importance of interspecific competition 8, 9
landscape fragmentation 47
life-history changes with IC 216–21
  clutch size 218–21

linear models 16–19
logistic model 16–19
longspur, chestnut-collared (*Calcarius ornatus*) 59, 60
long-term experiments 171–201
Lotka–Volterra models 5, 16–9

macaw, scarlet (*Ara macao*) 75
magpie (*Pica pica*)
  clutch size 219–20
  effect of IC with jackdaw 120–1
  food limitation effects 118, 120, 121
  supplemental feeding effects 62
mallard (*Anas platyrhynchos*) 152, 154, 162, 164
mammals
  IC with birds 133, 155–6
  nest availability 131, 132
Mass-dependent Predation Hypothesis 52
mathematical models 16–22
  generating non-linear isoclines 21–2
  linear 16–19
  non-linear 19–22
mixed species flocks
  foraging niches 106
  winter foraging 45, 50
models, mathematical *see* logistic model; mathematical models; theta logistic model
'molesting' behaviour 99
mongoose (*Herpestes javanicus*) evolution 211
moorhen (*Gallinula chloropus*) 63
mutualism 1, 148
Myzomela
  ebony (*Myzomela pammelaena*) 211
  scarlet-bibbed (*Myzomela sclateri*) 211

nest boxes
  experimental configuration 178–81, 182–3, 185–9, 190–1, 192
  as roost sites 77, 80, 198–9
  suitability for cavity-nesters 76–7
nest sites
  and foliage structure 69
  as limiting resource 69–81
    cavity nesters 72–7
    cup-nesting species 69–71
    nest web communities 77–80
  variation during season 70
nest success 98
nest web communities 77–80, 226
  site availability experiments 78–80
Newton's definition of competition 14–15
niches
  ecological niche modelling 113–14

overlap and seasonal variation 114–15
    resource use by coexisting species 5, 9
niche shifts 23, 117
Niche Variation Hypothesis 203–5
non-linear models 19–22
nuthatch 226
    brownheaded (*Sitta pusilla*) 105
    Eurasian (*Sitta europaea*) 43, 93, 99, 130, 134
    pygmy (*Sitta pygmaea*) 135
    red-breasted (*Sitta canadensis*) 78, 79, 132, 133
    white-breasted (*Sitta carolinensis*) 44, 45, 140, 143

open populations, IC in 216–17
owl, eastern screech (*Otus asio*) 77, 133
owl, ferruginous pygmy (*Glaucidium brasilianum*) 76
owl, pygmy (*Glaucidium passerinum*)
    food limitation effects 121, 125
    tit predation 44, 51, 112–13
owl, Tengmalm's (*Aegolius funereus*) 125

parakeet, ring-necked (*Psittacula krameri*) 130, 133–4
parasitism 1
partridge, grey (*Perdix perdix*) 63
per-capita growth rate
    changes with competition 218
    intra-/interspecific competition
        blue tits 184–7
        great tits 182, 191–2
Perrins' Constraints Hypothesis 55, 56
petrel, band-rumped (Madeiran) storm (*Oceanodroma castro*) 157
petrel, Bermuda (cahow) (*Pterodroma cahow*) 156–7
pheasant (*Phasianus colchicus*) 91, 92
pigeon, wood (*Columba palumbus*) 43
polychaete predation 64
population density, and nest site quality 27
population growth
    logistic model 16–18
        expanded 19–21
        generating non-linear isoclines 21–2
        with interspecific competition 20–1
        non-linear 19
population processes, effects of IC on 83–102
population size/density
    and beech masting 40–1, 42, 43
    and nest site quality 26
Potential Nest Site Hypothesis 69
predation 1
    and body fat reserves 50
    vs competition 2
    and density dependence 85
    and food availability 44–5, 63–5, 120
        supplemental feeding 61, 67
    at nest sites
        cavity nesters 72–3
        experimental data 70, 71
        reduction 69
    predator–prey studies 2
    risk
        behavioural responses to 49–50
        and body fat management 51
ptilochronology experiments 110–11

quelea, redheaded (*Quelea erythrops*) 114

raptors 76
recruitment
    density dependence 89–90, 194–6
    IC in tits 188–9, 192
redstart, American (*Setophaga ruticilla*) 86–8, 94, 140, 144, 162, 164
redwing (*Turdus iliacus*) 164
removal experiments 139–45
    fish removal 152–3
    food as limiting resource 46–7
    space as limiting resource 28, 29–34
        black-capped chickadee 29, 31, 32–3
        Carolina chickadee 31, 33
        coal tit 31
        crested tit 30–1
        great tit 29–30
        tufted titmouse 31
        willow tit 30–2
        woodpecker 124–5, 140, 143
    testing for IC 139–45
        removal of all birds 140, 142–5
        removal of single territory holders 140, 141–2
reproduction
    density dependence 92–3, 98
    effects of supplemental feeding 55–8
    field experiments 119–25
    food reduction experiments 59–60
    and site quality 26
resource limitation
    field experiments 117–70
    food 39–67
    nest sites 69–81
    space 25–38
resource use by coexisting species 5, 8–9
roach (*Rutilus rutilus*) 153
robin, European (*Erithacus rubecula*) 39
robin, New Zealand (*Petroica australis*)
    conservation studies 65
    density dependence 91–2

saddleback (*Philesturnus carunculatus*) 65, 91–2
salamander evolution 209–11
scientific names
  birds 229–32
  other species 233
scramble (exploitation) competition 13
seedcracker, black-bellied (*Pyrenestes ostrinus*) 114
sexual dimorphism 205, 207
shag (*Phalacrocorax aristotelis*) 8, 103
shearwater, Cory's (*Calonectris diomedea*) 157
shearwater, little (*Puffinus assimilis*) 157
sheathbill, lesser (*Chionis minor*) 93
shrike, great grey (*Lanius excubitor*) 163, 164
shrike, red-backed (*Lanius collurio*) 163, 164
shrike, San Clemente loggerhead (*Lanius ludovicianus mearnsi*) 66
site-dependent regulation *see* Habitat Heterogeneity Hypothesis
Sitka spruce cone crop 43
skua, South Polar (*Stercocarius maccormicki*) 60
snake predation 69–70
song playback experiments
  and IC 158–60
  and heterospecific attraction 164
space
  Buffer Hypothesis 25–8
  effects of limitation 37
  IC in titmice 98–101
  as limiting resource 25–38
  and winter social organization 28–34
sparrow, American tree (*Spizella arborea*) 36
sparrow, field (*Spizella pusilla*) 70
sparrow, golden-crowned (*Zonotrichia atricapilla*) 140, 142–3
sparrow, house (*Passer domesticus*)
  competition with house finches 15–6, 140, 144–5
  supplemental feeding effects 62
sparrow, rufous-collared (*Zonotrichia capensis*) 34
sparrow, song (*Melospiza melodia*)
  density dependence 85–6
  supplemental feeding 48, 67
  and predation 61, 90
sparrowhawk (*Accipiter nisus*)
  density dependence 89–90, 93
  tit predation and mass 51, 53, 54, 67
squirrel, red (*Tamasciurus hudsonicus*) 132, 133
squirrel, Northern flying (*Glaucomys sabrinus*) 79, 132, 133
squirrel, Southern flying (*Glaucomys volans*) 124, 155–6

starling (*Sturnus vulgaris*)
  competition with great tits 128–9, 130
  food limitation effects 57
  nest site competition 130, 131
  supplemental feeding effects 61
starvation risk *see* body fat reserves
stickleback evolution 213
stitchbird (hihi) (*Notiomystis cincta*) 65–6
stonechat (*Saxicola torquata axillaris*) 56–7
storage effect 217
summer limitation 29–30
supertramp strategy 112–13
supplemental feeding
  as conservation tool 65–6
  effects
    on breeding 55–8, 67
    large-scale 47, 49
    on nestlings/parents 60
    surprising 61–2
  experimental data summary 48
  and flock size 50
  and habitat quality 61
  and immune competence 62–3
  vs natural food 43–4
  pre-breeding 55–7, 66
  and predation 61
  and winter temperature 45–6
  see also calcium; carotenes
surplus birds (floaters) 31–2
swallow, barn (*Hirundo rustica*) 62
swallow, tree (*Tachycineta bicolor*) 78, 131
swan, black-necked (*Cygnus melanocoryphus*) 153
swift (*Apus apus*) 219
sympatry *vs* allopatry
  evolutionary effects of IC 209–10, 211, 213, 227
  food as limiting resource 41
  foraging niche use 108
  heterospecific aggression 158
  nest site competition 121, 122

tarsus length variation
  blue tit 214–15
  coal tit 208
Tasmanian devil (*Sarcophilus harrisii*) 217–18
T-cells 62
teal (*Anas crecca*) 162, 164
territoriality
  interspecific 34–6
  non-breeding 28–9
  winter-territorial species 30–4
  year-round 34
theta-logistic model 19–22
thrush, song (*Turdus philomelos*) 38, 70

tit
  acoustic competition 35–6
  clutch size and density 96–7
  competition with flycatchers 121, 122–4
  dispersal 99, 101
  egg mass 97
  food limitation effects 57, 120–2
  foraging niches 104–5
  heterospecific attraction 162, 164
  hoarding 105
  interspecific competition 5, 7, 171–201
    avoidance 6, 7
    and dispersal 194–6
    evolutionary effects of 206–9, 214–16
    experiments *vs* correlational studies 190–1, 192–3
    for food 10–11
    long-term experiments 171–201
    origin of idea 172–6
  interspecific territoriality
    amongst tits 34–6
    with non-tit species 36–7
  natural *vs* supplemental food 42–3
  nest site limitation 73–4
  population regulation 4–8
  predation
    and body mass 51–2
    and food 44, 45
    and foraging niche 112
    at nest site 72–73
  recruitment 194–6
  supplemental feeding effects 43–4, 48, 57, 62, 98
  territoriality 25, 28–9
  winter-flocking species 29–30
  winter-territorial species 28–9
  winter temperature effects 43, 44
tit, blue (*Cyanistes caeruleus*)
  interspecific competition, with great tits
    and breeding density 199
    during breeding season 197
    and climate change 199–200
    detailed analyses 234–44
    for food 174, 176–8
    and nest-box configuration 178–81, 182–3, 185–9, 190–1, 192
    nest site competition 125, 127
    outside breeding season 176–8, 197–9
    and per-capita growth rate 184–7
    for roosting sites 176–8
    for space 196–7
    summary 196–201
    for territory 173–6
    varying intensity 179–82, 199–200
  predation and fledging mass 54
  roosting in nest boxes 77

surplus birds (floaters) 30
tarsus length 214–5
territoriality 28–9
tit, coal (*Periparus ater*)
  ecological character release 205–9
  flock foraging 49–50
  food as limiting resource 105
  foraging niches 109, 111–13
  foraging niches, and predation 112
  foraging niches, on Gotland 110, 111–13
  tarsus length 208
  territoriality 31
  winter body mass 51
tit, crested (*Lophophanes cristatus*)
  competition with coal tits 111–3, 206–7
  dominance and mass 51
  flock foraging 49
  foraging niches 106, 107–9, 110
  foraging niches, and predation 112
  overwinter survival 33, 44, 45, 48
  predation risk and mass 51–2
  space as limiting resource 33–4
  territoriality 30–1
tit, great (*Parus major*)
  and blue tit dispersal 195–6
  dispersal 99–101, 198
  fecundity and density 10
  heterospecific attraction 162, 163, 165
  interspecific competition, experiments varying intensity 179–82
  interspecific competition, with ants 146
  interspecific competition, with blue tits
    and breeding density 182, 199
    during breeding season 197
    climate change 199–200
    detailed analyses 234–44
    for food 174, 176–8
    long-term experiments 171–201
    and nest-box configuration 178–81, 190–1, 192
    nest site competition 125–6, 127
    outside breeding season 176–8, 197–9
    and per-capita growth rate 182, 191–2
    summary 196–201
  interspecific competition, with chaffinch 36–7
  interspecific competition, with starling 128–9, 130
  population and beech masting 40–1, 42, 43–4
  surplus birds (floaters) 29–30
  survival/dispersal 98–101
  territoriality, non-breeding 28–9
  territoriality, removal experiments 29–30
  winter temperature effects 43, 44

tit, Hume's ground (*Pseudopodoces humilis*) 72
tit, marsh (*Poecile palustris*) 6, 35, 43, 57, 62, 99, 111, 206, 207
tit, southern black (*Melaniparus niger*) 25
tit, stripe-breasted (*Melaniparus fasciiventer*) 25
tit, varied (*Parus varius*) 50, 57
tit,willow (*Poecile montanus*)
 body mass and predation 51–2
 competition with coal tits 206–7
 density dependence 86, 87, 96, 98, 101
 flock foraging 49–50
 foraging niches 35, 106, 107–8, 109, 110, 112
 habitat competition 35
 overwinter survival 30–3
titmouse, bridled (*Baeolophus wollweberi*) 25, 29, 34, 99
titmouse, tufted (*Baeolophus bicolor*)
 dispersal 99
 food supplementation effects 44–5, 48
 foraging niches 110–11
 in mixed species foraging 45
 territoriality 31
Total Foliage Hypothesis 69
treecreeper, Eurasian (*Certhia familiaris*)
 competition with insects 146–8
tropicbird, white-tailed (*Phaethon lepturus*) 156–7

vireo, blackcapped (*Vireo atricapilla*) 69–70
vireo, red-eyed (*Vireo olivaceus*)
 food reduction experiments 59–60
 heterospecific attraction 164
vulture, bearded (*Gypaetus barbatus*) 65
vulture, griffon (*Gyps fulvus*) 65, 93

warbler, *Acrocephalus*, heterospecific aggression 158–9
warbler, black-throated blue (*Dendroica caerulescens*) 59, 60, 86–9, 93
warbler, garden (*Sylvia borin*) 140, 143–4
warbler, golden-cheeked (*Dendroica chrysoparia*) 69–70, 140
warbler, Lucy's (*Oreothlypis luciae*) 129
warbler, orange-crowned (*Vermivora celata*) 70, 71, 140, 144
warbler, pine (*Dendroica pinus*) 105
warbler, Virginia (*Vermivora virginiae*) 70, 71, 140, 144

warbler, willow (*Phylloscopus trochilus*)
 competition with cavity nesters 136
 heterospecific attraction 164
 removal experiments 140, 141–2
Wiens' definition of competition 14
winter-flocking species 99–101, 102
Winter Food Limitation Hypothesis 45, 54–5
winter limitation 30–4
winter roosts, competition for 121, 125, 133
winter social organization
 costs/benefits 49–50
 and space 28–34
 winter-flocking species 29–30, 101–2
 winter-territorial species 30–4, 101–2
 and effects of winter feeding 44–5
winter survival
 and breeding population size 45–7, 48
 and food 42–9
winter temperature 51
 and flock foraging 49–50
 and food supplementation 45–6
 and species 44
wood ant (*Formica aquilonia*) 145–6, 148
woodcreeper, Planalto (*Dendrocolaptes platyrostris*) 76
woodpecker, downy (*Picoides pubescens*) 44, 45
woodpecker, Gila (*Melanerpes uropygialis*) 72, 129
woodpecker, great spotted (*Dendrocopos major*) 43
woodpecker, red-bellied (*Melanerpes carolinus*) 121, 124–5, 156
woodpecker, red-cockaded (*Picoides borealis*)
 nest site competition 77, 121, 124–5, 128–9, 130, 133, 140, 155–6
woodpecker, red-headed (*Melanerpes erythrocephalus*) 140, 143
wren, house (*Troglodytes aedon*) 135
Wytham Woods tit studies
 coal tit ecological character release 205, 207
 food as limiting resource 41, 53, 120
 species coexistence 6, 9
 nest site competition 126
 population processes, effects of IC on 98
 predation and great tit mass
 species coexistence 6, 9
 territoriality 3, 6, 95–6